ROWAN UNIVERSITY
LIBRARY
201 MULLICA HILL RD.
GLASSBORO, NJ 08028-1701

D1595546

MULTIFRACTALS *and*
1/*f* NOISE

Springer
New York
Berlin
Heidelberg
Barcelona
Hong Kong
London
Milan
Paris
Singapore
Tokyo

SELECTED WORKS OF BENOIT B. MANDELBROT

REPRINTED, TRANSLATED, OR NEW
WITH ANNOTATIONS AND GUEST CONTRIBUTIONS
COMPANION TO *THE FRACTAL GEOMETRY OF NATURE*

Benoit B. Mandelbrot

MULTIFRACTALS *and* 1/*f* NOISE

Wild Self-Affinity in Physics (1963–1976)

SELECTA VOLUME N

*Includes contributions by
J.M. Berger, J.-P. Kahane,
J. Peyrière, and others*

Benoit B. Mandelbrot
Mathematics Department
Yale University
New Haven, CT 06520-8283, USA
http://www.math.yale.edu/mandelbrot
and
IBM T.J. Watson Research Center
Yorktown Heights, NY 10598-0218, USA

QA
614.86
M27
1999

Library of Congress Cataloging-in-Publication Data
Mandelbrot, Benoit B.
   Multifractals and 1/f noise / Benoit B. Mandelbrot.
      p.   cm.
   Includes bibliographical references and index.
   ISBN 0-387-98539-5 (hardcover : alk. paper)
   1. Multifractals. 2. Electronic noise. I. Title.
QA614.86.M27   1998
514′.742–dc21   98-3970

Printed on acid-free paper.

© 1999 Benoit B. Mandelbrot
All rights reserved. This work may not be translated or copied in whole or in part without the written permission of the publisher (Springer-Verlag New York, Inc., 175 Fifth Avenue, New York, NY 10010, USA), except for brief excerpts in connection with reviews or scholarly analysis. Use in connection with any form of information storage and retrieval, electronic adaptation, computer software, or by similar or dissimilar methodology now known or hereafter developed is forbidden.
The use of general descriptive names, trade names, trademarks, etc., in this publication, even if the former are not especially identified, is not to be taken as a sign that such names, as understood by the Trade Marks and Merchandise Marks Act, may accordingly be used freely by anyone.

Camera-ready copy provided by the author.
Printed and bound by R.R. Donnelley and Sons, Harrisonburg, VA.
Printed in the United States of America.

9 8 7 6 5 4 3 2 1

ISBN 0-387-98539-5 Springer-Verlag New York Berlin Heidelberg SPIN 10679267

To my wife, Aliette

I dedicate this intellectual fruit of mine,
her adoptive child

List of Chapters

In this List of Chapters, the sources given after the titles include (in parentheses) the letter M followed by the year of publication and by the lower case letter that the Bibliography *uses to distinguish different texts published in the same year.*

In the Bibliography, *the items reproduced in this book and in Volumes N, H and L of these* Selecta *are marked by a star followed by a chapter number, which in some cases is incomplete or only tentative.*

NP	Preface (1998)	1
I	**INTRODUCTIONS AND SHORT PIECES**	**17**
N1	Panorama of grid-bound self-affine variability (1998)	18
N2	Sketches of prehistory and history (1998)	62
N3	Scaling, invariants and fixed points (1998)	100
N4	Filtering and specifications of self-affinity (1998)	111
N5	Short excerpts (1964–1986)	117
II	**UNIFRACTAL ERRORS AND LÉVY DUSTS**	**131**
N6	New model for error clustering on telephone circuits (Berger & M 1963). Additional tests on clustering (1998)	132
N7	Self-similarity and conditional stationarity (M 1965c)	166

III INTERMITTENT 1/f NOISES AND CONDITIONED RANDOM PROCESSES ... 207

N8 1/f noises and the infrared catastrophe (M 1965b) ... 208

N9 Co-indicator functions and related 1/f noises (M 1967i) ... 215

N10 Sporadic random functions and conditional spectra self-similar examples and limits (M 1967b) ... 247

N11 Random sets of multiplicity for trigonometric series (Kahane & M 1965) ... 284

IV TURBULENCE AND MULTIFRACTAL MEASURES ... 289

N12 Sporadic turbulence (M 1967k) ... 290

N13 Intermittent free turbulence (M 1969b) ... 292

N14 Lognormal hypothesis and distribution of energy dissipation in intermittent turbulence (M 1972j) ... 294

N15 Intermittent turbulence in self-similar cascades: divergence of high moments and dimension of the carrier (M 1974f) ... 317

N16 Iterated random multiplications and invariance under randomly weighted averaging (M 1974c) ... 358

N17 "On certain martingales of Benoit Mandelbrot" Guest contribution (Kahane & Peyrière 1976) ... 372

N18 Intermittent turbulence and fractal dimension: the kurtosis and the spectral exponent $5/3 + B$ (M 1976o) ... 389

N19 Fractal dimension, dispersion, and singularities of fluid motion (M 1976c) ... 416

Cumulative Bibliography ... 419

Index ... 432

NP

Preface

CERTAIN NOISES, MANY ASPECTS OF TURBULENCE, and nearly every aspect of finance, exhibit a level of temporal and spatial variability whose "wildness" vividly impressed itself upon me in the early nineteen-sixties. I soon concluded that those phenomena cannot be described – and much less handled, understood, or explained – by simply adapting or extending slightly the statistical techniques that had sufficed in earlier physics.

At an International Congress of Philosophy of Science (Jerusalem, 1964), I argued that successful existing statistical models represented a first stage of indeterminism in science. As I saw it, the study of finance and turbulence could not move forward without the recognition that those phenomena must involve a new second stage of indeterminism. Altogether new mathematical tools were needed.

A related need was to acknowledge that all possible questions one might raise must be sorted out in two categories. Some questions are "well-posed," and can expect an answer. Other questions are "ill posed," and – at least for the time being – are not even worth raising. Some questions that are well-posed in first-stage indeterminism should be expected to be ill-posed in the second stage.

This sharp distinction echoes one that Hadamard raised in mechanics. In my work, it first arose in finance but soon became essential elsewhere. In the context of climatology, this underlying idea is becoming widely known to everyone therefore deserves mention, even though the topic is pursued in M1998H rather than in this book. The question is that of anthropogenic global warming. If climate follows the (implicit) assumption that its fluctuations belong to first-stage indeterminism, it is legitimate to view those fluctuations as varying around a durable equilibrium level. If so, to evaluate Man's influence on climate may be a well-posed problem. If second-stage indeterminism prevails, the search for equilibrium might, to the contrary, be an ill-posed problem and may have to be rephrased.

The early technical papers in which I reported on all those issues are being reprinted in several books, of which this is the second to appear. Those books are largely independent of one another. They also contain substantial additional material, but will be referred to by the old-fashioned Latin word for "Selected Papers," which is *Selecta*.

Having met Edward Lorenz around the time of that 1964 Congress in Jerusalem, I was privileged to follow the development of the theory of deterministic chaos from the beginning of its flowering. In due time, I contributed to it by discovering the Mandelbrot set. However, this work does *not* concern the theory of chaos, but – to use a good old-fashioned term – the theory of chances. Incidentally, an unintended but amusing loose parallelism exists between what I called in 1964 a second stage of indeterminism and chaos viewed as a second stage of determinism. Both push back the "edge of messiness," beyond which simple descriptions are not yet available.

My talk at the above-mentioned 1964 Jerusalem congress was not actually printed until M1987r. This first reference is an opportunity to draw attention to the style adopted throughout this book. The principle is explained on the first page of the bibliography and can be illustrated by a few examples. The *Selecta* books include M1997E, M1998H and M1998L, as well as M1997FE. The present book is denoted in other writings as M1998N. My book-length Essays on fractals are denoted as M1975O, M1977F and M1982F, the third one being also denoted as *FGN*.

Another idiosyncratic usage is this: instead of writing "the model presented in M1972j{N14}," this book will often write "the M1972 model."

To continue in more specific terms, this book's subtitle lists two broad themes. Both are essential to the study of the substantive topic and tool listed in the title, and to many of my other contributions to science.

The theme of *wild* versus *mild* or *slow* variability amplifies and deepens my 1964 theme of successive "stages of indeterminism." It has become easiest to express by a metaphor that begins in physics. As is well-known, mechanics relies on a unique set of laws of great generality. However, there are several quite distinct states of matter. Best known are gases, liquids, and solids. It is equally well-known that probability theory relies on a unique set of axioms of great generality. However, this and many other works of mine imply or claim that it is not only useful but necessary to recognize the existence of several quite distinct "states of randomness" as well as of random or non-random variability.

In surprisingly clear-cut parallelism to the three basic states of matter, both randomness and variability can be *mild, slow or wild*. The mild state is characterized by the validity and direct relevance of certain classic limit theorems. As to the wild state, it brings together new observations with behaviors that were glimpsed in the past, but remained neglected, mishandled, or at best viewed as "anomalies" to be tackled independently of one another. It will be argued that many so-called "anomalies" coalesce in one major phenomenon that deserves to be investigated on its own.

Stating them aggressively for the sake of effect, this and other parts of my work argue strongly for a theme that runs through fractal geometry. The theme is that much in nature is ruled by "pathology." That is, diverse forms of behavior that were originally viewed as "anomalous" or "unnatural" could bloom after fractal geometry revealed their true nature, not by rooting them in mathematics but in science.

A secondary metaphor is less compelling yet useful. The definition of a nondecreasing function is of great generality, but Lebesgue recognized three very distinct "states of non-decrease:" differentiable, discrete and singular. This book largely deals with *singular* variation.

The importance of the contrast between mildness and wildness is in part due to its links with a contrast between *locality* and *globality*. The study of mild phenomena is the easiest, because their properties can be handled locally. The study of wild phenomena is harder, because the interactions extend, in one way or another, to infinity.

When the key factor is the integral of the correlation function, locality is expressed by a positive and finite integral, while globality is expressed by an integral that vanishes or is infinite. Alternative and synonymous statements associate globality with dependance that is infinite, or of infinite span, range or run.

In some of the works in finance collected in M1997E, the contrast between locality and globality is expressed differently, in terms of addition of random quantities. Locality expresses that relative contributions to a sum are negligible, even the largest. It follows that each part can be modified considerably without significantly affecting the whole. Globality is expressed by the fact that the largest contribution is non-negligible, therefore cannot be modified much without affecting the whole.

To end by a quick allusion to slow randomness, it is the difficult case characterized by locality in asymptotically large systems contrasting with globality within small systems.

This book's second theme is as follows. Since wildness and globality bring many novelties and severe complications, the necessity of accepting and studying them on their own would be bad news. A counteracting piece of good news is provided by a surprisingly ubiquitous empirical finding: in those natural phenomena that we shall examine, the wildness is *not* as general as it might have been; the "pathology" of those particular aspects of nature is *not* unmanageable. This is so because it obeys a form of invariance or symmetry called *scale invariance* or *scaling*. More precisely, this invariance is called *self-affinity*. This and the related but distinct and more familiar theme of *self-similarity* are central to my life work. Cases that combine wildness with failure of self-affinity bring truly bad news and I choose to avoid them.

Mathematical definitions of the themes of wildness and self-affinity will be provided later in this preface.

We now move to this book's title. It lists two separate topics, and seems to imply that this book is addressed to several non-overlapping communities. Indeed, noises are of primary concern in signal processing and applied physics, while multifractals began in the study of turbulence and dynamical systems – and also in finance as a result of M1997E. After a few words about each topic considered separately, it will be argued that they are not separate at all.

The term "power law" will denote a relationship that expresses one quantity (mathematical or natural) as being a power of another. Thus, a *1/f noise* (stated more carefully, a f^{-B} *noise*, where B is a numerical constant not too far from 1) is a physical phenomenon with a Fourier spectral density given in terms of the frequency f by the power law f^{-B}. Interest in those noises primarily arose from experiments in very applied sciences. As to the term *multifractal*, this book applies it to a measure having scaling properties, therefore ruled by diverse other power laws. Interest in those measures arose in pure mathematics but did not develop until it moved into an applied context, the key paper being M1974f{N15}.

This and related books of *Selecta* defend the view that, notwithstanding many real or apparent differences, $1/f$ noises and multifractals are intimately interrelated through the two fractal themes listed in the subtitle. They are best viewed as being distinct aspects of a single phenomenon. A subsidiary point (which was already mentioned) is that this link involves a powerful collection of technical tools, which are of use to diverse scientific communities, including finance.

This view of the nature of multifractals emerged from my original papers on this topic. Indeed one reason for reprinting those papers in this book is to make this view become widely understood and perceived as *natural*. This is especially true of Chapter N1, titled "Panorama of grid-bound self-affine variability." Until this unity becomes generally recognized, however, a redundant title remains appropriate.

In summary, combining the bad and the good news, a) $1/f$ noises and multifractals exemplify wild rather than mild random or non-random variability, and b) they belong to the broad and unified mathematical notion of *self-affine fractal variation*, which ordinarily implies *uniformly global statistical dependence*. *Self-affinity* overlaps nature and mathematics and takes three distinct forms: *unifractality*, *mesofractality* and *multifractality*. The step from uniscaling to multiscaling self-affinity was taken when I introduced the multifractals to model an aspect of the gustiness of the wind that eluded a primitive $1/f$ noise I had initially favored.

A form of mesofractal self-affinity that does not matter here is the topic of the papers from the 1960s reprinted in M1997E. Additional aspects of self-affinity are scheduled for M1998H. The strong links between those books and the present one are underlined by the inclusion of evolving forms of a chapter whose title includes the word *Panorama*.

The notion of wildness deserves amplification within the context of probability and statistics. It reinterprets diverse forms of "anomalous behavior" in unified fashion: as evidence that common research techniques were pushed beyond their domain of validity. Those common techniques rely on the validity of several separate but interlinked theorems : 1) the law of large numbers, 2) the limit theorem that asserts that properly renormalized sums are Gaussian, 3) the property that diffusion is asymptotically Fickian, that is, spreads proportionally to $\sqrt{T}$, and 4) the property that, in a sum of many random components, the largest is of relatively negligible size.

Those familiar tools were spectacularly successful. Forgetting that all theorems involve specific assumptions, they came to be treated as "folk theorems" valid under all circumstances to be encountered in science. More precisely, it was practically taken for granted that incremental extensions would in due time make them applicable to phenomena previously left aside such as price variation, anomalous noises, hydrology, motions of the atmosphere and the ocean, turbulence, etc., ...

However, once again, I came to believe very early that incremental extensions will never suffice. The need to recognize "anomalies" implies the recognition that the classical limit theorems are flatly contradicted by at least some uncontournable facts.

The proper response is to rethink the concept of "generalization." In mathematics, this is near-always a highly respected activity, but an important distinction must be kept in mind. The lower level of generalization changes the premises but not the conclusions. It buttresses already familiar facts and continue to hold unchanged in broader and perhaps unfamiliar contexts beyond the original environment. But many limit theorems of probability partake of the next higher level of generalization, which leaves an overall problem unchanged, but alters the premises and reaches *very different* conclusions.

At that point, to use a philosopher's words, "quantity changes into quality." In the context of matter, change is quantitative as long as a system's temperature is modified, but the system remains a gas or a liquid. Change becomes qualitative when a gas liquefies or a liquid freezes. It is to describe the corresponding step in probability that I suggest one should say that randomness changes to a different "state." To anticipate a possible criticism, it is good to point out that, under certain conditions (passage through special points called "critical"), states of matter can change from one another continuously. It is reassuring to know that the same is conceivable for states of randomness.

When the three folk theorems in the above list are valid and other conditions hold, I describe randomness as being in the "mild state". The point is that the common probabilistic thinking in the sciences does not give even a passing thought to the possibility that randomness could be anything but mild. This is why the numerous phenomena that do not fit tended to be described as "anomalous" and treated as deserving less careful attention than "normal" phenomena. Unfortunately, difficulties that are not faced do not simply go away. As applied to the anomalies, the classical tools brought confusion, not the clarification that is expected when the appropriate language is used.

My view is that anomalies are symptoms of the "wild"-"global" state of randomness. Early on, the task of creating a whole new toolbox consisted in identifying existing pure mathematical constructions that could be transformed into tools of scientific investigation. As the texts reprinted in this book were being published, and increasingly more often in recent years, the task consists in designing altogether new tools.

Let us stop for two examples of this view of generalization. In the context of the crucial notion of stationarity, M1982F {FGN} (p. 383-386) observes that "Ordinary words used in scientific discourse combine (a) diverse intuitive meanings, dependent on the user, and (b) formal definitions, each of which singles out one special meaning and enshrines it mathematically. The terms *stationary* and *ergodic* are fortunate in that mathematicians agree on them. However, experience indicates that many engineers, physicists, and practical statisticians pay lip service to the mathematical definition, but hold narrower views." That is, many *mathematically stationary* processes are not *intuitively stationary*. By and large, those processes exemplify wild randomness, a circumstance that provides genuine justification for distinguishing a narrower and a wider view of stationarity. As a matter of fact, M1965c{N7} shows how investigations of wild variability led to an even broader and wilder generalization, namely, conditional stationarity.

Another highly relevant example of the process of generalization concerns spectral analysis. That method arose through the search for hidden periodicities. In due time, the algorithm became well-understood. While it continues to be delicate, it no longer requires extraordinary care, as long as there is little energy in high and low frequencies. Khinchin and Wiener generalized spectral analysis from periodic functions to random processes called "stationary of the second order," such as light, which used to be spectral analyzed without rigorous justification. The Wiener-Khinchin theory was generalized further in several directions. However, much is lost along this series of generalizations, even when the Wiener-Khinchin theory is applied to Gaussian processes. To echo the end of the preceding paragraph, M1967b{N10} shows that my investigation of wild variability led to an even broader generalization, "conditional spectral analysis."

The beginning of this preface used too fleetingly some terms that demand elaborations. The term *self-affinity*, first used in M1977F and M1982F{FGN}, expresses invariance under some linear reductions and dilations. The far better known notion of *self-similarity* is the special case corresponding to isotropic reductions. This book deals with cases that differ greatly from the self-similar cases. A perspicuous illustration is provided by the widely-known fractal computer landscapes pioneered in M1975O and M1982F{FGN} Chapter 28. There, the lake and island coastlines are self-similar, while the relief itself is "self-affine".

A technical study of self-affinity introduces a formalism made of scaling laws, renormalizations and fixed points. This apparatus was developed in 1960-3 in my work in finance collected in M1997E, and immediately applied to physics, in Berger & M 1963{N6}. Starting about 1965, the same apparatus with considerable extensions allowed by physics was developed quite independently in the theory of critical phenomena of physics; it succeeded brilliantly, as will be recalled in Chapter N3.

Faced with phenomena I view as self-affine, other students take an extremely different tack. Most economists, scientists and engineers from diverse fields begin by subdividing time into alternating periods of quiescence and activity. Examples are provided by the following contrasts: between turbulent flow and its laminar inserts, between error-prone periods in communication and error-free periods, and between periods of orderly and agitated ("quiet" and "turbulent") Stock Market activity. Such subdivisions must be natural to human thinking, since they are widely accepted with no obvious mutual consultation. René Descartes endorsed them by recommending that every difficulty be decomposed into parts to be handled separately. Such subdivisions were very successful in the past, but this does not guarantee their continuing success. Past investigations only tackled variability and randomness that are mild, hence, local. In every field where variability/randomness is wild, my view is that such subdivisions are powerless. They can only hide the important facts, and cannot provide understanding. My alternative is to move to the above-mentioned apparatus centered on scaling.

Fractal is another term that demands comment here. It was coined in M1975O and used only in the last few papers reprinted in this book. Informally, fractal geometry is the systematic study of certain very irregular shapes, in either mathematics or nature, wherein each small part is very much like a reduced size image of the whole.

Is fractal geometry a new subject, or an old one? This is a surprisingly subtle issue that becomes clarified by reference to a well-known story from the past. Group theory dates to the early 1830s, but addition, multiplication and rotation had been known for millennia. They were not properly understood and mutually related until the advent of an organized group theory made them into groups. A loose analogy to that story helps understand the past and present status of several "monster" constructs known since around 1900. It is fractal geometry that made them into fractals, by incorporating them, putting them to work and bringing forth their true nature. For the sake of convenience and accuracy, they deserve, in my opinion, to be called *protofractals* and *protomultifractals*.

Every analogy is at best partial. The development of group theory moved from the top down. To the contrary, the goals and scope of fractal geometry were not developed according to a master plan; instead, they emerged very gradually as needs developed. The question of "what are fractals and multifractals today," is further taken up at the end of this preface.

Pure mathematics prides in being able to strip complicated notions to essentials that provide the foundation of rich and inspiring developments. This is an achievement I greatly admire; it inspired my work as well, insofar as scaling and other invariances are used prominently as organizing and motivating principles. However, reduction to essentials must never be rushed prematurely. In many important contexts to be encountered in this book, features that had tended to be labeled as inessential end up by playing a vital role.

Particularly illuminating are the many cases when a formula can be "visualized" by a picture, therefore both contain the same information. The formula is praised because it alone can be handled analytically. Moreover, it is concise and not burdened by anything irrelevant. But consider the chapter of early chaos theory that is concerned with the iteration of rational functions. In 1917, P. Fatou and G. Julia did marvels with formulas in the absence of pictures, but the theory they created stalled for sixty years due to the absence of new questions to analyze. It did not start again until, resisting the universal condemnation of pictures, I discovered what came to be known as the Mandelbrot set. Its baroque and anything-but-concise complication became the source of new mathematical questions no one could have imagine otherwise.

Closer to this book and equally illuminating are the many problems triggered by a sound or a picture. Only afterwards is a formula devised, and then proclaimed or implied to be a full representation of reality. The examples I am thinking of are noises that become stripped of everything deemed inessential and labeled as having $1/f$ spectrum. The concision of this description seemed admirable, but turned out to be a drawback.

Whatever contribution I managed to bring to this study came from forsaking concision and paying respect to the reputedly inessential geometry. For example, arrange for $1/f$ noises from different sources to be traced on an oscilloscope. Even the untrained observer will notice deep geometric differences that turn out to be symptoms of deep physical differences. Therefore, I think that a unique physical explanation for all $1/f$ noises is hopeless. Chapters N1, N2, and N3, and Part III will argue that the identification of a $1/f$ spectrum is *never enough*.

Let me continue in praise of explicit and visual geometry. A theme that runs increasingly strongly in my work is that its usefulness in science and mathematics has revived and should again be recognized.

Blind analytic manipulation *is never enough*.

Formalism, however effective in the short run, *is never enough*.

Mathematics and science are, of course, filled with quantities that originated in geometry but eventually came to be used *only* in analytic relationships. In many cases, those analytic relationships *are not enough*.

For example, fractal codimension is often the exponent of a correlation. But correlation *is not enough*. To identify D with an analytic exponent destroys much of its meaning; *it is not enough*.

Yesterday's mathematicians claimed to have completely reduced geometry to analysis. But in fact they have not: the same geometric shape, when examined with increasing care, often goes on to reveal fresh geometric features that are obvious to the eye and moreover affect physics. Many of those geometric features have not been tamed by the existing analytic tools and therefore demand entirely new analytic tools. This is how I was led to introduce the notion of lacunarity.

Scaling geometry is not merely a reflection but a genuine counterpart of scaling analysis. It is rich in features of its own and celebrates the power of simple rules to create geometric shapes of extraordinary and seemingly chaotic complexity, which analysis is then called upon to explain.

The allegation that geometry is dead deserves scorn.

One last topic is *explanation*. I fully agree that it is the ultimate goal of science, and regret that wild variability in nature (or finance) is fully explained in relatively few cases. That is, many brilliant investigators did their best but did not go far. A generic explanation of fractality, as created by partial differential equations, is advanced in Chapter 11 of M1982F{*FGN*}; it has not yet been implemented in technical detail. The reader knows that – in my opinion – the passage from mild-local to wild-global variability faces the scientist with a steeply increased level of complexity, therefore of difficulty.

In any event, the engineers (including financial engineers) need not feel unduly apologetic when they disregard the scientists' goals and keep to their own. On many occasions, excessive concern with ultimate tasks of explanation harms the performance of other tasks one cannot postpone.

When explanation is unavailable, I keep away from mere curve-fitting "phenomenologies." A "middle way" often found successful is described in Chapter N3. *Ecclesiastes* told us that "for every time there is a reason,... a time to be silent and a time to speak." It is also true that on the frontiers of science there is "a time to explain and a time to explore."

Once again, this is one of several books described together as *Selecta*. Soon after my first scientific paper came out on April 30, 1951, the impression arose that each of my investigations aimed in a different direction. In due time, the accumulation of diverse works created two needs. My book-length *Essays*, M1975O, 1977F and 1982F{*FGN*} elucidated the strong and long-term unity of purpose that always underlied my work but developed slowly. However, the more technical material was necessarily left out. The rushed *Mathematical Backup* found in Chapter 39 of M1982F was not enough. Placing these belated technical *Selecta* next to the book-length *Essays* establishes a balance between overall thrust and technical detail.

As work moved on slowly, each *Selecta* volume evolved in its own special way and to avoid imposing an artificial sequence, each is denoted by a letter related to a key word; this is Volume N, because of *noise*. The material most clearly related to finance and economics, as well as a detailed technical discussion of the concept of states of randomness, are already available as M1997E.

The question of how this book relates to M1997E and M1998H is complex. There is a sketch in the *Panorama* in Chapter N1 and a fuller discussion is withheld for M1998H.

Like the clothing worn by a traditional English bride, each *Selecta* volume incorporates something old, something new, something borrowed and even something blue.

In this volume, the *old* predominates: it consists in reprints of effectively every paper that I published on this topic between the dates included in the sub-title: 1962 and 1976. A fresh reading after 20 to 35 years showed that overlaps were minimal and wholesale repeats non existent, even when the originals appeared in hard-to-find journals or Proceedings. The referees and editors of major journals rarely welcomed my work – but it was a blessing never to be told that the themes that eventually led to fractal geometry were already familiar! Unfortunately, what may be called "underpublishing" contributed to the fact that my work on multifractals had a wide *indirect* impact, but a *direct* impact limited to few readers. This book also includes texts that never received attention, that is, ideas that have not yet been rediscovered and investigated seriously. Sporadic processes are but one example.

In the historical Chapter N2, fully documented and dated claims of priority are stated more bluntly than in the original papers.

Continuing with the components of a bride's clothing, the *borrowed* refers to a *Guest contribution*, Chapter N17, an important text that deserved attention, two co-authored papers that gave Chapters N6 and N11, and extensive quotes from diverse sources, especially in Chapters N2 and N6.

I worked for IBM for half of my life, continued at Yale, and, long ago, started at the Paris Laboratory of Philips of Eindhoven. Those institutions' official colors are three shades of *blue*.

Finally, the *new* consists in new chapters and in Forewords and/or *Annotations* that precede and/or follow some reprints. Each reprint concerns one or the other of this book's substantive topics, and the new chapters adhere (in a minor key) to the goals Dirac assigned to theoretical science, namely, to remove inconsistencies and unite theories that were previously disjointed.

In the preparation of the original papers, help was provided in 1961-2 by the assistants of J. M. Berger (co-author of Chapter N6) and later by my assistants, H. Lewitan in 1970-2 and S. W. Handelman in 1974-6. Otherwise, I worked largely alone, but several chapters were discussed in varying detail with U. Frisch. Original individual acknowledgements are found after each paper.

Warm thanks go to J.M. Berger for allowing our joint work to be reprinted and to the authors of the *guest contributions*, especially of the important paper by J.P. Kahane and J. Peyrière.

In the long-drawn actual preparation of this book, invaluable help was provided by several long-term secretaries, L. Vasta, followed by P. Kumar and K. Tetrault, and their successors beginning with J. Lemarié. Several short-term assistants were of great help. Clumsy English made the old papers collected in this book disappointingly difficult to read. To help on this account, the reprints were copy-edited by diverse hands (not mine), in particular, by H. Muller-Landau. Of course, extreme care was taken not to modify the original thoughts. This can be verified by comparisons with the originals, which are available in good libraries and may be provided, if need arises, by the author. In addition, the notation was systematically unified and updated. Occasional corrections and after-thoughts that seemed necessary are set out in curly brackets {} and preceded by the mention "P.S. 1998."

For over thirty-five years, the Thomas J. Watson Research Center of the International Business Machines Corporation provided a unique haven for a variety of investigations that science and society demanded but Academia neither welcomed nor rewarded. I worked at that haven through most of that period and finished this book at Yorktown while staying on as IBM Fellow Emeritus. I am deeply indebted to IBM for its continuing support of mavericks like myself, Richard Voss, with whom I worked closely from 1975 to 1993, and Rolf Landauer. Many other names rush to mind, and among long-term friends and colleagues, the Yorktown of its scientific heyday will remain most closely associated with Martin Gutzwiller and Phillip Seiden.

As to management, at a time when the old papers in this book were being written and my work was widely perceived as a wild gamble, it received whole-hearted support from Ralph E. Gomory, to whom I reported in his successive capacities as Group Manager, Department Director, and IBM Director of Research and Senior Vice-President. Gomory reminisced on the old times in the *Foreword* he kindly wrote for M1997E.

Last but not least, this book is dedicated to my wife, Aliette. It would not have been written without her constant and extremely active participation and unfailingly enthusiastic support.

New Haven, CT & Yorktown Heights, NY

The old papers collected in Parts II, III and IV follow the chronological order, which was also the order of natural development. Chapter N5 collects short excerpts. The bulk of Part I is made of introductions especially written for this book. The reader may wish at this point to scan the page-long texts in italics that begin each part. Taken together, they add up to a summary of this book.

Appendix: What is it that we call fractal geometry?

The terms *fractal* and *multifractal* remain without an agreed mathematical definition. Let me argue that this situation ought not create concern and steal time from useful work. Entire fields of mathematics thrive for centuries with a clear but evolving self-image, and nothing resembling a definition. Here are several examples.

First, consider geometry. Chern 1979 confides that "I am glad that we do not know what it is and, unlike many other mathematical disciplines, I hope it will not be axiomatized. With its contact with other domains in and outside of mathematics and with its spirit of relating the local and the global, it will remain a fertile area for years to come." Chern 1990 writes that "A property is geometric, if it does not deal directly with numbers or ... the coordinates themselves have no meaning." O. Veblen and J.H.C. Whitehead wrote that: "A branch of mathematics is called geometry, because the name seems good on emotional and traditional grounds to a sufficiently large number of competent people." A. Weil said: "The psychological aspects of true geometric intuition will perhaps never be cleared up. At one time it implied primarily the power of visualization in three dimensional space. ...Some degree of tactile imagination seems also to be involved."

Next, "What is it that we call complex analysis?" In the preface of a splendid textbook, Boas 1987 notes that " Complex analysis was originally developed for its applications; however, the subject now has an independent and active life of its own, with many elegant and even surprising results." A review, Piranian 1989, quotes from the book approvingly: "Boas avoids the folly of an impossible definition by making a modest declaration. It does not characterize complex analysis; but complex analysts know that no reasonable description of their territory could ever have remained satisfactory for more than a quarter century ... . Today, complex analysis is primarily the study, by analysis and synthesis, and with geometric, topological, algebraic, number-theoretic, or other cultural orientations, of complex-valued functions in spaces of one or more complex variables."

The third example, chosen for the sake of contrast, concerns the question "what is it that we call a curve?" Camille Jordan (1838-1922) and Giuseppe Peano (1858-1932) transformed a seemingly straightforward notion into one that is obscure and full of controversies. P. S. Urysohn (1898-1924) and Karl Menger (1902-1985) attempted in the 1920s to create a proper "theory of curves," but the attempt petered off after Menger 1932.

Attempts to answer the question of "what is it that we call probability theory" fare even worse, with a quirky complication. After the probabilist Paul Lévy failed to be accepted in his lifetime as a full-fledged mathematician, some of his heirs gained acceptance by willingly narrowing the scope of probability theory to fit a clear definition. Yet, uncertainty remains about who is, and who is not, a *real* probabilist.

Finally, let us move sideways to a field of physics, namely, quantum theory. Griffiths 1995 observes that "there is no general consensus as to what its fundamental principles are, how it should be taught, or what it really 'means'. Every competent physicist can 'do' quantum mechanics, but the stories we tell ourselves about what we are doing are as various as the tales of Scheherazade, and almost as implausible. Richard Feynman (one of its greatest practitioners) remarked: 'I think I can safely say that nobody understands quantum mechanics'."

This background is uncontournable, and in particular, it is clear that the "modest declaration" by Boas remains true when one replaces *complex analysis* with *fractal geometry*, a field that began within the memory of many who practice it today. In any event, the book that first used the term *fractal*, M1975O, took a very informal approach to its meaning; its Chapter 1X discussed *multifractals* even more informally (even abstaining from coining a word for them). To my later regret, M1977F and M1982F{*FGN*} gave in and wrote down a "tactical definition" that linked the very general notion of fractal with the specialized notion of Hausdorff-Besicovitch dimension. I have been trying ever since to loosen this link. One unintended consequence was to classify the devil staircases as nonfractal. I prefer to keep the devil staircases and jettison the 1977-1982 definition.

To summarize, the question of "what are fractals and multifractals?" may continue with no mathematical answer. However, whether or not fractal geometry proves sufficiently useful to survive, the question will cease to be asked.

PART I: INTRODUCTIONS AND SHORT PIECES

Chapters 1, 2 and 3 provide newly written alternative introductions to the book. There is enough overlap to allow them to be read in any sequence. Chapters 4 and 5 are inserted in this part because there is no better place.

Chapter N1 describes a unified conceptual and technical context that relates diverse models and themes of this book to one another. It also takes a broader look at wild self-affinity as encountered in other volumes of these Selecta *and looks forward to diverse challenging new topics.*

On casual reading, Chapter N1 may seem to set a grand plan for future research and publications. However, fractal geometry did not develop from top down, according to plan, but from bottom up. No Panorama *could be put together until a good part had been implemented.*

Chapter N2 surveys the prehistory and history of the topics that join in contributing to this book. The prehistory ranges back to isolated and largely forgotten early mathematical esoterica, and the history is meticulously documented in the hope of straightening out some credits.

Fixed points and renormalization arguments were pushed farthest and became most widely known through the modern statistical physics that L.P. Kadanoff in 1966 and K.G. Wilson and M.E. Fisher in 1972 built around the theory of critical phenomena. Shortly afterwards, however, G. Jona-Lasinio established a close connection between that theory and the stable distributions of probability theory.

This connection is essential from the viewpoint of this book, because my work has relied heavily on the stable distributions since M1956c, especially so in the papers reprinted in M1997E and Part III of this book. In fact, I have been using fixed points and renormalization arguments for a decade before 1966. My work had no direct influence whatsoever on the birth of modern statistical physics, but this deep intellectual kinship deserves a special Chapter N3.

Chapter N4 starts tackling the technical task of distinguishing quantitatively between functions with $1/f$ spectrum that greatly differ from one another.

Chapter N5 collects diverse short excerpts that are of high historical interest from this book's viewpoint. They first appeared between 1964 and 1984 as portions of books and papers that do not otherwise relate to the present volume.

N1

Panorama of grid-bound self-affine variability

✦ **Abstract.** Multifractals and $1/f$ noises arose independently, as will be seen in Chapter N2. Therefore, it is not surprising that they continue near-unanimously to be viewed as separate and unrelated concepts.

This chapter argues the opposite case: that both concepts are specific aspects of a far broader underlying reality denoted as "wild self-affine variability." The term "self-affine" may be unfamiliar to some readers, but everyone soon perceives it as being self-explanatory. It was first used in M1977F, in an entry reproduced with little change in M1982F{*FGN*}.

To achieve its goal, this chapter describes straightforward constructions that are recursive, therefore automatically yield self-affine outputs. They are "grid-bound," that is, constrained to a recursively refined grid, in the style of the devil staircase, the binomial measure and other familiar proto-fractal constructions. Hence, they are sharply simplified and will be referred to as "cartoons" of reality.

Nevertheless, their outputs include the following structures:

A) Cartoons of the (ordinary) Wiener Brownian motion and of its fractional Brownian generalizations investigated in M1998H.

B) Cartoons of varied forms of $1/f$ noise, some of them already familar and others less so or not at all.

C) The simplest multifractals, namely, C') the binomial and the multinomial measures and the "microcanonical" random measures investigated in M1974f{N15} as well as C") the cartoons of oscillatory variation investigated in M1997E.

Much of this book deals with multifractals and $1/f$ noises that are immensely more realistic than any cartoon, because they are gridless. One example is the original limit lognormal multifractal introduced in

M1972j{N14}. However, those gridless constructions are not simply related to one another. The cartoons include the essentials of the topics treated in this book and in its companions (M1997E, M1998H, and M1998L). The word *Panorama* in the title is justified more fully (perhaps to excess!) because this chapter glances towards further forms of self-affine variation that have not been "visited" yet. The cartoons underline both the similarities and the differences between those various cases.

For self-similar shapes, the key parameter is fractal dimension, a concept that keeps splitting into ever-new variants. For self-affine shapes, the same role is played by an exponent H. It has modest double origins in the work of Hölder and Hurst. But it has taken off on its own and branched out, and now keeps appearing under an increasingly wide range of distinct roles. Section 1 describes many examples as introductions to a general definition of the diagonal-axial self-similar cartoons. For most of those examples, H has a single implementation.

Section 2 takes up the general idea behind H from a very different angle and introduces important new variants and the corresponding terminology. The most general diagonal-axial cartoons will be called multifractal. Those in an important subclass will be called unifractal. A different important subclass will be called mesofractal.

Section 3 is perhaps the most novel in this chapter. It throws light on the nature of cartoons, therefore of multifractals and $1/f$ noise, by transforming clock time in diverse ways. Section 3.2 introduces a trail exponent H_T, in effect, by projecting on a space orthogonal to the time axis: this amounts to adopting as time any monotone increasing function of clock time. Section 3.3 finds it convenient to play with "isochrone time." Section 3.4 introduces the notion of "intrinsic time." Section 3.5 describes a simple but fundamental result dubbed "baby theorem." It permits every cartoon to be represented as a compound function: a unifractal cartoon of a multifractal intrinsic time.

Section 4 describes graph and spectrum implementations of H.

Section 5 comments briefly on several distinct topics. ✦

IN THE LANDSCAPE OF MATHEMATICS AND PHYSICS, multifractals are often perceived as an isolated island ruled by an esoteric formalism. I believe, to the contrary, that it is useful to think of them as being special examples of a far broader natural and mathematical phenomenon that occurs in diverse fields. Different instances having been identified independently, they go under a variety of distinct names. Physicists and engineers call this phenomenon $1/f$ or f^{-B} *noise*, the exponent B being a constant. Narrower groups of scientists speak of *non-Fickian diffusion*. Students of turbulence associate some instances of the phenomenon with *intermittency*. And the unifying effect undertaken in this book and especially in this chapter identifies the phenomenon with the mathematical notions of *uniformly global statistical dependence* and *wild self-affine fractal variation*. This identification has already been characterized in the Preface as being (in a very minor key) in the spirit of the goals Dirac set to formal science, to remove inconsistencies and to unite theories that were previously disjointed.

The word "phenomenon" in the preceding paragraph should not be misunderstood. By definition, all $1/f$ noises share the same Fourier power spectral density, but from other viewpoints my studies show them to be far from unified, either physically or mathematically. They involve considerations of bewildering diversity that are or will be at the center of at least three volumes of these *Selecta*.

This chapter investigates a versatile family of recursive constructions that proceed by self-affine interpolation and converge in the limit to fractal curves. Those constructions serve as a kind of "guidebook" to the old publications reprinted starting with Chapter 5. With no special effort at either inclusion or avoidance, this chapter also includes a nonnegligible amount of new research material.

0. INTRODUCTION: THE CONTRAST BETWEEN SELF-SIMILARITY AND SELF-AFFINITY, FIRST EXAMPLES AND COMMENTS

M1982F{*FGN*} mostly tackled constructions that yield self-similar fractals. In the simplest non-random case, those sets decompose naturally into "parts," S_m, such that each part is obtained from the "whole", S, by a linear transformation that changes circles into smaller circles. Self-similar recursive constructions became very popular. As everyone knows by now, they are very versatile, that is, leave room for many varied implementations. They also have innumerable uses in science.

A major generalization of self-similarity is provided by self-affinity. This term expresses that the transformations from a whole to its parts are affinities: they are linear but at least some of them change circles into ellipses. It is useful to also allow degenerate affinities that collapse circles into intervals. Self-affinity creates a baroque wealth of mathematical structure that is so versatile as to seem impossible to organize usefully. It also seems insufficiently rich for any concrete purpose in science.

Most fortunately, both fears are completely unfounded. Quite to the opposite, this book hopes to show that self-affinity helps apprehend and organize the baroque wealth of structure found in nature. More precisely, this chapter describes a tightly organized special collection of recursive self-affine constructions that allow for a great diversity of behavior. This collection includes simplified "cartoons" of Wiener Brownian motion, fractional Brownian motion, $1/f$ noises of several distinct kinds and diverse other behaviors, including binomial and multinomial multifractals.

The virtue of the cartoons resides in their being invariant under *diagonal* self-affinities. That is, each reducing affinity is defined by a matrix and we shall only consider affinities whose matrices reduce to diagonal terms. This collection is tightly organized but diverse: sufficiently narrow to follow general rules that can be studied easily and sufficiently broad to be useful.

Non-diagonal self-affinity will be mentioned in Section 5.1.2 but not otherwise considered in this book, because it is less enlightening and fails to provide a unifying framework.

Diagonal cartoons do not include all the contents of this book. In particular, cartoons are "grid-bound," while Parts I and III, as well as M1972j{N14}, describe "grid-free" construction.

The word *Panorama* is also found in an introductory chapter of M1997E (Chapter E6). The present version combines a significant expansion of Section E6.4 with a summary of Section E6.3. Many obscurities, typographical errors and inaccuracies found in M1997E were corrected (or at least made less disruptive). A futher version planned for in M1998H will be more extensive than either predecessor.

0.1 Self-similarity and on to self-affinity

Self-similarity. Many geometric shapes are approximately isotropic. For example, no single direction plays a special role when coastlines are viewed as curves in the plane. In first-approximation fractal models of a coastline, small pieces are obtained from large pieces by an isotropic reduction (homothety) followed by a rotation and a translation. This property defines the fractal notion of self-similarity. Self-similar constructions make free and essential use of angles, and distances can be taken along arbitrary directions in the plane.

Beyond self-similarity. This book deals mostly with geometric shapes of a different kind, namely, graphs of functions and the like, for which the abscissa may be time and the ordinate, price, pressure, or perhaps another quantity whose scale is totally unrelated to the scale of time. Each coordinate scale can be changed freely with no regard to the other. A distance continues to be defined along the coordinate axes. But one cannot say that "time increment = pressure increment," until arbitrary units have been selected. That is, would-be Pythagorean expressions exemplified by

$$(\text{distance})^2 = (\text{time increment})^2 + (\text{pressure increment})^2$$

make no sense. It follows that circles are not defined. For an angle and rectangle to be defined, the sides must be parallel to the axes. Squares are not defined, even squares with sides parallel to the axes are not intrinsic, even though they will be useful in Section 2.1.

The notions of diagonal affinity and self-affinity. Leonhard Euler introduced the unfortunate term *affinity* to denote a linear operation that generalizes similarity, maps the plane onto itself and can be inverted. M1997F proposed *self-affine* to denote an object that is left invariant by a collection of maps that include affinities, but also allows linear transformations that fail to be invertible, because (for example) they collapse a set having interior points into an interval.

Once again, when the matrix of each contraction has only diagonal terms, the matrix and the corresponding affinity are both called *diagonal* Diagonal affinities are the simplest of all because in that case the two coordinates are contracted independently of each other.

0.2 Prototypical self-affine cartoons

The gist of this chapter's content is illustrated in this section, using examples to be amplified later.

Surrogates of Brownian notion. Figure N1-1L implements self-affinity by recursive interpolation from time 0 to time 1. The construction begins with an initiator, that is, a square crossed diagonally by a straight trend line. To avoid clutter and make this illustration easier to read, the square is stretched horizontally into a rectangle and the trend is not drawn. Next, this trend is interpolated by a generator, which is simply a broken line formed of three intervals, whose horizontal and vertical lengths, from left to right, are (4/9,2/3), (1/9,1/3), (4/9,2/3). (At least three intervals are needed to achieve a cartoon that can oscillate up and down). Then each interval is interpolated by three shorter ones that reproduce the generator after it has been squeezed to fit the same endpoints. The white lines in the second figure from the top emphasize the need for different horizontal and vertical reduction ratios. For the first interval, they are 4/9 and 2/3 = 6/9, respectively.

The same process continues without end. Each interval of the first approximation ends up with a shape just like the whole, except for a diagonal affinity. Therefore, scale invariance is present in the limit, simply because it was built in. This demonstrates a wholly unexpected possibility: a great wealth of structure can be generated by an extremely simple formula. This possibility is crucial to fractal geometry and the theory of chaos.

Figure N1-1R concerns (in part) the "randomization" of Figure N1-1L; it consists in changing the generator before each interpolation by scrambling at random the sequence of its three intervals.

The vertical extent of each generator interval is precisely the square root of its horizontal extent. The same is true of every interval of the "prefractal" broken line constructed at a later stage of recursion and is approximately the case for arbitrary increments or in the fractal limit.

This "square root" property, called "Fickian," also characterizes Brownian "diffusion" and expresses that increments are not correlated. Cartoons with the Fickian property converge to limits that are "surrogates" of Brownian motion.

Figure N1-2 is an example using a different generator with the Fickian property. In many figures in this chapter, the plot of successive increments merges into a strip. This artefact, a very useful one, is intended and achieved by choosing a suitable width for the computer's "pen."

More general "unifractal" cartoons. One can arrange for each interval's vertical extent to equal its horizontal extent raised to a power H such that $0 < H < 1$ but H is not 1/2. The resulting constructions are cartoons of processes called "fractional Brownian motions." Not only are their increments definitely correlated, but their essential property is "global dependence." Stated in terms of prices, global dependence expresses that the correlation is the same between price changes over successive days, months, or years. Global dependence is essential to the task of modeling nature, and it is also found in less obvious forms in the multifractal processes to which we now proceed.

Multifractal intrinsic time. Public opinion and the financial profession know that big fortunes are made and lost when the market seems to speed up and the volatility of prices soars. For example, it may be that prices follow a Brownian motion, but instead of clock time, proceed in a trading time that flows slowly on some days and fast on others. Altogether, one has to face a multiplicity of variable speeds – one of several explanations of the prefix in "multi-fractal!" The technical challenge was to give technical meaning to multifractal trading time.

FIGURE N1-1. Figure N1-1L (the left-side half of this figure) shows the generator and three stages in the construction of a non-random grid-bound cartoon of the Wiener ("ordinary") Brownian motion. Details are found in Section 0.2.

In constructing Figure N1-1R (the right-side half of this figure), each stage of interpolation is preceded by shuffling the sequence of the generator's intervals at random.

N1 ◊ ◊ PANORAMA OF GRID-BOUND SELF-AFFINE VARIABILITY

A straightforward implementation that preserves self-affinity is illustrated in Figure N1-3 and consists in modifying the generator as follows: the intervals' vertical extents are not affected, but their horizontal extents are either lengthened or shortened, while preserving a sum equal to 1.

The gradual effect of increasingly strong slowing and speeding up is illustrated in Figure N1-4. Going down the stack, one begins with near-Brownian motion and continues with cartoons that diverge increasingly from Brownian motion. To increase the strip's width and spikiness, it suffices to move the generator's first break to the right or the left!

Mesofractal trading time. After the first break is moved as far as it can, the first or second intervals becomes vertical. If so, the limit process becomes discontinuous in every interval, however small.

Furthermore, one can hold on to the first and third interval and split the middle one into two, one of which is horizontal. In this case, the limit

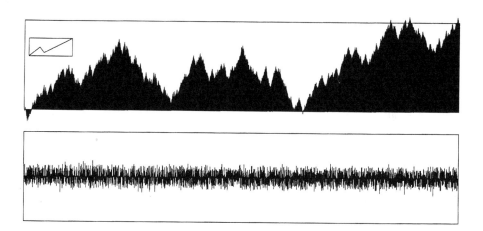

FIGURE N1-2. The top portion illustrates a cartoon of Wiener Brownian motion carried to many recursion steps. Each step begins by shuffling at random the three intervals of the generator. The generator, shown in a small window, is not the same as in Figure N1-1L but identical to the generator A2 of Figure N1-6. Randomization insures that, after a few stages, no trace of a grid remains visible to the naked eye.

The bottom portion represents the corresponding increments over successive small intervals of time. This strip with no large spike is for all practical purposes a diagram of Gaussian "white noise."

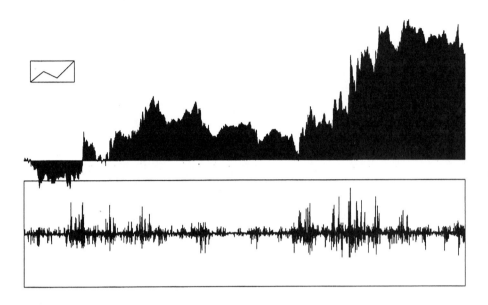

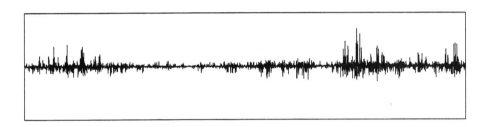

FIGURE N1-3. The top portion illustrates a cartoon of Wiener Brownian motion following a multifractal intrinsic time. Starting with the three-box generator used in Figure N1-2, the box heights are preserved, so that the Fickian $H = 1/2$ is left unchanged, but the box widths are modified.

The middle portion shows the corresponding increments. Spectrally, this sequence remains a "white noise," but it becomes very far from being Gaussian. In fact, this spiky strip exhibits a conspicuously high level of serial dependence and a high level of resemblence to records from physics (and also finance.)

The bottom portion repeats the second, but with a different "pseudo-random" seed. The goal is to demonstrate the very high sample variability that characterizes wildly varying functions in general and the present cartoon in particular.

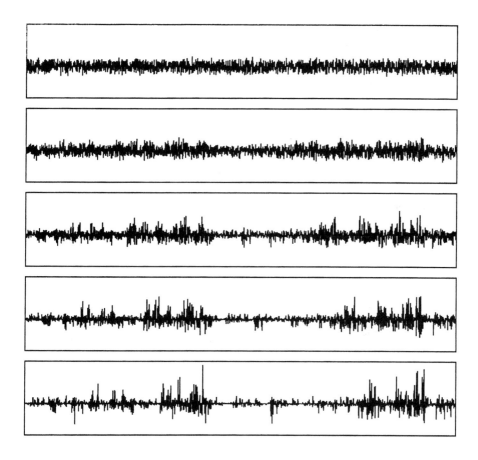

FIGURE N1-4. Simulations of a sequence of white noises.

Each line in this stack is obtained by performing many iterations of a recursive cartoon, and then evaluating the increments. The top line uses the same randomized generator as Figure N1-1R. To obtain the following 4 lines, the first break in the first-line generator is displaced to the left by .06, then .12, .18 and .24. The "spikiness" of the main strip keeps increasing.

Spectral whiteness is preserved because the intervals' vertical extents are unchanged. As a result, terms of the theory to be elaborated in Section 3, the trail dimension is unchanged at the "Fickian" value $D_T = 2$. As to the intrinsic time, it is clock time on the top, then becomes multifractal. Therefore, this figure shows the increments for a sequence of cartoons of Wiener Brownian motion, as followed in increasingly variable multifractal times.

process becomes constant except for instants of time that belong to a Cantor dust. It is called "intermittent."

Playing with cartoons of this sort, a multiplicity of levels and kinds of volatility can be obtained at will. Many are multifractal measures, others are $1/f$ noises of all kinds, either fractal or multifractal.

0.3 Fractal techniques provide quantitative measurements of texture, irregularity or roughness

The very irregular and rough shapes often encountered in Nature excite the imagination, but no serious attempt was made to define and measure numerically the irregularity of a coastline or a price record.

Topology does not provide the answer its name promises. For example, a time record, when filled-in to be continuous, can be obtained from the line without a tear, using a one-to-one continuous transformation. This property defines *all* records as being topological straight lines!

Nor does statistics provide a useful answer. For the roughness of physical surfaces, statistics suggests the following steps: fit a trend-like plane (or perhaps a surface of second or third degree), then evaluate the root-mean-square (r.m.s.) of the deviation from this trend. What is unfortunate is that the r.m.s., when evaluated in different portions of a seemingly homogeneous surface, tends to yield conflicting values.

Does the inappropriateness of topology and statistics imply that irregularity and roughness must remain intuitive notions, inaccessible to mathematical description and quantitive measurement? Fractal geometry shows that in many cases roughness can be measured by scaling exponents.

The fractal dimension. Coastlines are nearly self-similar, and the most obvious aspect of their roughness is measured by a called *fractal dimension*, which is described in M1982F{*FGN*}. Many alternative definitions look different but near always end up with a unique value. (Fractal dimension does *not* fully specify a set. In particular, I have investigated additional texture features collectively referred to as "lacunarity".)

The exponent H and kin. For physical surfaces, roughness can also be measured reliably. In the simplest (unifractal) case, one needs two numbers. One is an exponent H; engineers have now adopted widely and call it simply the "roughness exponent," but I have introduced it as being a Hurst-Hölder exponent and a fractal co-dimension. The second characteristic number is a scale factor similar to a root-mean-square, but more appropriately defined. (This topic is treated in M1998H). The study of roughness in terms of self-affinity has become a significant topic in physics; see Family & Vicsek 1991.

Hurst, Hölder, and a way to conciliate mathematical and concrete needs. H has independent roots in two very different fields. The hydrologist H. E. Hurst (1880-1978) was concerned with observable scales. The mathematician L. O. Hölder (1859-1937) was solely concerned with infinitesimally small scales. The considerable effort required to make these concrete and mathematical contexts fit comfortably together proved fruitful in may ways; in particular, the original "H" split into many separate notions, as documented later in this chapter.

0.4 The discriminating power of the eye

The purpose of this section is to refer to Figure N1-5. The eye discriminates better between records of *changes* of a function than between records of the function *itself*. Therefore, Figure N1-5 plots the changes of a miscellany that combines simulations with actual data.

The top "line" repeats the bottom line of Figure N1-2, with a different seed. It could be confused with a record of physical thermal noise.

Lines 2 and 3 and at least one of the lower 5 lines illustrate my successive models of price variation, as discussed in Chapter N2 and Chapter N5 and in M1997E. Those models date to 1963, 1965 and 1972, therefore predate modern statistical physics (as emphasized in Chapter N3.)

Of lines 4 to 8, at least one is a simulation of either a multifractal cartoon or the increments of Brownian motion in multifractal time. At least one is a real record drawn (as it happens) from finance, but certain records drawn from physics would not have seemed out of place.

No one would want to be restricted to the eye and foresake numerical measurement, but the message of Figures N1-2 to N1-5 is that the eye can be a very worthy associate.

0.5 Distinctions among recursive construction: grid-bound versus grid-free, fractal versus random, and additive versus multiplicative

The reader may skim this section now, and come back before proceeding to Section 2.

0.5.1 The role of grids in providing simplified surrogates to grid-free models of nature. Nature very rarely involves boxes subdivided into smaller boxes and so on recursively to infinity. Therefore, the best models of nature are "grid-free," as exemplified in Parts II and III and

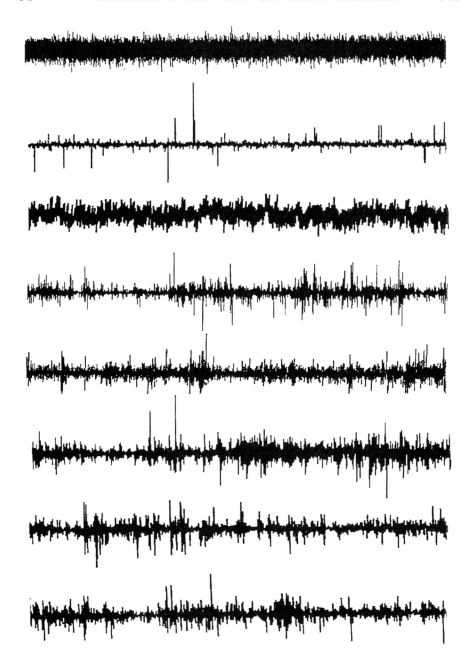

FIGURE N1-5. Mystery stack of increments meant to prove the power of the eye. The top line could well concern yet another Wiener Brownian cartoon or be a sequence of independent pseudo-Gaussian variables; it could also be a record of thermal noise. Other lines are surely non-Gaussian.

M1972j{N14}. Unfortunately, those best models are, in many ways more difficult than the fractal constructions that proceed within a grid as in Section 0.2 and can be called "grid-bound." For example, the grid-free model in M1972j{N14} is more complicated than the grid-bound model in M1974f{N15}. Granted that all grid-based constructions must, emphatically, be viewed as "cartoons," many investigations are easier when carried out on cartoons. Besides, the grid-bound construction in M1974f{N15} is more versatile than the grid-free construction in M1972j{N14}. This example is not unique.

Being grid-bound brings pluses and minuses. The latter are in the process of being investigated and will be reported elsewhere. The major plus is that suitable grid-bound constructions can be used as "surrogates" to grid-free random processes that proved or promise to model nature.

Finally, the cartoons in this chapter have another great asset: there exists an overall "master structure" that relates them to one another as special cases. That master structure is enlightening and "creative" in that it suggests a stream of additional variants. I came to rely on it increasingly in modeling new or old problems. Striking parallelisms were mysterious when first observed in the grid-free context, but became natural and obvious in this master structure.

Grid-bound constructions risk looking "creased" or "artificial." A conspicuous "creasing" effect is present when recursion subdivides each cell in the grid into equal cells. This was the case of some early commercial computer-generalized fractal landscapes. Around 1984, creasing became a serious issue in the back-offices of Hollywood involved in computer graphics. A limited randomization called "shuffling" creates a contrast between Figures N1-1L and N1-1R. To reduce creasing at small cost in added complication, the grid is subdivided into unequal cells, and/or the grid is preserved but the generator is selected among several variants.

0.5.2 Fractality versus randomness. *Respective roles from the viewpoint of describing variability in nature.* While the bulk of useful models in science are random, the cartoons in this chapter may be completely non-random.

However, an important lesson emerges from near every study of random fractals. At the stage when intuition is being trained, and often beyond that stage, a construction's being random is for many purposes less significant than its being fractal. That is, the non-random counterparts of random fractals exhibit analogous features, and also avoid, postpone, or otherwise mitigate some of the notorious difficulties inherent to randomness. In the case of recursive grid-bound fractals, it helps if the rules of construction are not overly conspicuous. For example, the grid is best obtained by a subdivision into unequal pieces.

This chapter also includes non-random fractals for which acceptable randomizations are absent or limited. They too are of high educational value.

Fractality versus randomness in the historical distinction between the contributions of Wiener and Khinchin to spectral analysis. In the spirit of the preceding remarks, it is enlightening to recall that the mathematical theory of covariance and spectrum had two simultaneous sources. Wiener's work (and also Bochner's) aimed at an understanding of non-random solutions of differential equations, with the hope of understanding turbulence. This hope was unsuccessful but contributed a mathematical tool: the theory of non-random functions satisfying certain conditions that make them harmonizable. Khinchin studied second-order stationary random functions. From the viewpoint of modeling of reality, it is significant that the two approaches yield identical formalisms.

0.5.3 Uses of addition versus multiplication in recursive interpolation. As known to specialists and discussed in Section 6, the best known fractal functions are introduced through additive operations. To the contrary, the best known multifractal constructions use multiplicative operations. Hence, between those two structures, a lack of unity and parallelism that is an artefact and highly misleading. In this light, a great virtue of the diagonal fractal cartoons in this chapter is that some could be viewed as additive, but all are best viewed as multiplicative, special consideration being given to discontinuities. Thus, reliance on those cartoons makes it unnecessary to change gears in the middle of an argument.

1. DIAGONAL-AXIAL SELF-AFFINE CARTOONS

This section adds to the substance of Section 0.3 and repeats some points in far greater detail. Leaving generality to the final subsection, it examines all the special examples in Figures N1-6 and N1-7. They are identified by a letter (A, B or C) denoting a column and an integer (1 to 4) denoting a row. When a process is denoted by XYZ, its cartoon surrogate will be denoted by C(XYZ). A fractional Brownian motion (FBM) carries an aura of inevitability, as the only unifractal self-affine Gaussian process. By contrast, every cartoon carries an inevitable aura of artificiality that is confirmed by the existence of many alternative cartoons. Therefore, in cases of ambiguity, the symbol C is followed by a numerical index.

N1 ◊ ◊ PANORAMA OF GRID-BOUND SELF-AFFINE VARIABILITY

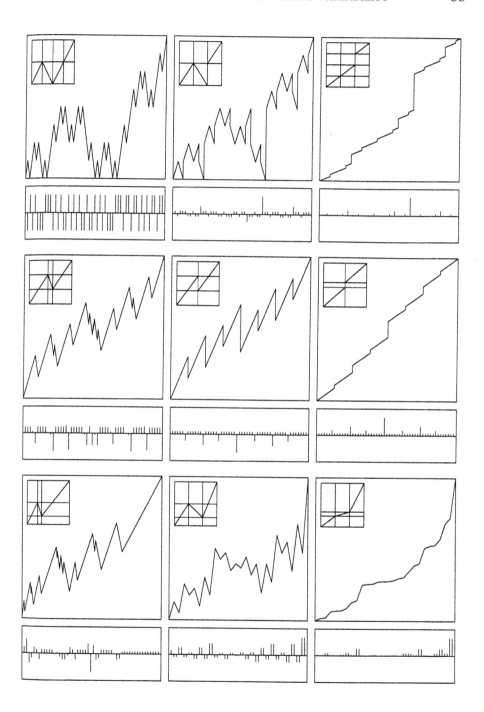

FIGURE N1-6. Cartoon constructions explained in the next caption.

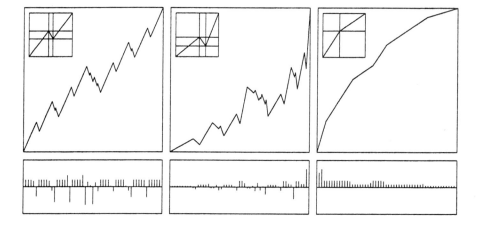

FIGURE N1-7. Additional row of cartoons. The following explanation applies to this figure and the preceding one. The columns are denoted, from left to right, by A, B and C. The rows are numbered from top down on Figure N1-6, then continuing on Figure N1-7. Each construction is illustrated by a square and a rectangle. The generator is shown in a small window within the square: it is either diagonal or diagonal-and-axial. The square shows the level-2 approximation. The rectangle graphs show the increments of the level-2 approximation, taken over equal time increments. Those graphs are "underdeveloped" forms of the spiky or non-spiky strips illustrated in other figures.

The A, B and C diagrams juxtaposed on each horizontal row are intimately related by the fundamental "baby theorem" described in Section 3.5. Had we saved the computer program that drew Figures N1-6 and N1-7 (but unfortunately we did not), two rows would have been added. They would have illustrated two dust-born $1/f$ noises relative to M1967i{N9}. An increasing generator that includes a horizontal interval like the inverses of C1 and C2 creates a positive intermittent $1/f$ noise. An oscillating generator that includes a horizontal interval creates an oscillating intermittent $1/f$ noise.

1.1 Examples of grid-bound recursive constructions with diagonal generators, related to the Wiener and fractional Brownian motions, hence to Gaussian $1/f$ noises

The Wiener and fractional Brownian motions $B_H(t)$. They are random functions with Gaussian increments such that, for all t and T,

$$E[B_H(t+T) - B_H(t)] = 0, \text{ while } E[B_H(t+T) - B_H(t)]^2 = t^{2H}.$$

The exponent H must satisfy $0 < H < 1$. The special value $H = 1/2$ yields the ordinary Wiener Brownian motion.

The Fickian rule of diffusion and spectral properties. $B_H(t)$ diffuses as $\Delta x \sim (\Delta t)^H$. The special case $H = 1/2$ yields ordinary Fickian diffusion. Other values of H yield non-Fickian diffusion.

The "continuing," or "humming" fractional Gaussian noises $B'_H(t)$. The "generalized derivatives" $B'_H(t)$ of $B_H(t)$ are noises having the spectral density $\propto f^{-2H+1} = f^{-B}$, with $B = 2H - 1$ ranging between -1 and 1. Those noises are the main topic of M1998H. They were introduced in M. & Van Ness 1968 as models of diverse phenomena that exhibit cyclic non periodic variability at all time scales. The oldest recorded example concerned the annual discharge of the Nile River and is associated with the biblical story of Joseph the son of Jacob. Therefore, I refer to non-periodic cyclicity as the Joseph Effect.

• C_1 **(WBM).** Diagram A1 of Figure N1-6 yields a simple cartoon of WBM that is detailed and carried further in Figure N1-8. Variants are described in M1986t{H} and other articles collected in M1997H.

The construction begins with a box to be described as being of "level-0" and called "initiator". The definition of H to be given in Section 2.1 makes it convenient to choose the two box sides as units of the two coordinates. Therefore, it is useful to think of the initiator as being a square. It is subdivided into "level-1" rectangular boxes that are congruent because they all are of width b_t^{-1} and height b_x^{-1}. Then, it is subdivided recursively into rectangles of width b_t^{-k} and height b_x^{-k}. In the present C_1 (WBM) case, $b_t = 1/4$ and $b_x = 1/2$. The level-1 boxes are crossed by diagonal intervals that add to a continuous string, namely, a broken line along the sequence UDUU (up, down, up, up). In the second stage, each box crossed by a diagonal string is interpolated by being replaced by an image of the generator, reduced in two *distinct* ratios: 1/4 horizontally and 1/2 vertically.

After k construction stages, the box widths are $\Delta t = b_t^{-k}$ and their heights are $\Delta x = \pm b_x^{-k}$. Each stage of construction reduces the area of each box in the ratio $1/8$. The prefractal collection of boxes of level k is contained in the prefractal of level $k-1$, resulting in a "nested" sequence of prefractals. For $k \to \infty$, this sequence converges to a connected planar set of vanishing area (Lebesgue measure.) (A connected domain of area tending to zero provides one of the possible definitions of the surprisingly elusive notion of curve.)

Relation with the Wiener Brownian motion $B(t)$. Diffusion and spectral properties. When the interval Δt belongs to the generating grid, $\Delta X = \pm \sqrt{\Delta t}$ with equal probabilities. Over other intervals $X(t)$ satisfies the approximate relation $\Delta X \sim (\Delta t)^H$, where $H = \log b_x / \log b_t = 1/2$. This coincides with the Fickian diffusion of WBM. Therefore, the spectrum of the increments ΔX is flat (except for a periodic disturbance). In effect, ΔX is a white

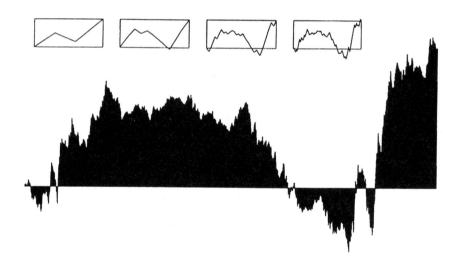

FIGURE N1-8. Yet another random cartoon of Wiener Brownian motion. This construction uses the generator A1 of Figure N1-6. It is made of 4 equal length intervals; that is, what seems like a longer interval is really two intervals following each other. The top of the figure shows the first few construction stages and the bottom shows the outcome of many stages.

noise. This cartoon has become widely used in physics (M1985l, M1986t, Family & Vicsek 1991).

However, the distribution of ΔX is binomial (on the grid), not Gaussian. The move from the cartoon to $B(t)$ preserves $H = 1/2$ and consists in adopting a flat spectrum and a Gaussian distribution for the ΔX.

• **C_1(FBM)**. This cartoon of FBM is too simple to require an illustration. The boxes continue to be congruent to one another and obtained by a reduction in which horizontal and vertical coordinates are linearly reduced in the respective ratios b_t^{-1} and b_x^{-1}. The novelty is that the bases are no longer linked by the Fickian property $b_t = b_x^2$, only by the weaker constraint that the generator must join the bottom left to the top right. The two bases b_x and b_t define a fundamental scaling exponent $H = \log b_x / \log b_t$ that rules all the fractal properties.

To allow a continuous limit, one must be able to draw a "hidden string" through the boxes of the generator; hence $b_t - b_x$ is constrained to be positive. (In the case $b_t = b_x$ the string through the boxes remains straight and invariant as recursion proceeds.) A continuous self-affine limit demands that $b_t - b_x$ be even. It follows that the present exponent H satisfies $0 < H = \log b_t / \log b_x < 1$, as does the exponent $B_H(t)$.

A rational $H = p/q$ in [0, 1] can be implemented exactly: it suffices to take $b_t = b^p$ and $b_x = b^q$, with b an arbitrary integer. An irrational H in [0, 1] can be approximated arbitrarily closely.

Asymptotic local isotropy of C_1(WBM) and C_1(FBM). After k stages, denote by A_k and B_k the respective numbers of diagonal intervals with $\Delta x > 0$ and $\Delta x < 0$, respectively. The identities $A_k + B_k = b_t^k$ and $A_k - B_k = b_t^k$ yield $A_k / (A_k + B_k) = 1/2 + (b_x/b_t)^k / 2$. In the limit $k \to \infty$, one finds $A/B \to 1$. That is, the "drift" over the unit time interval is increasingly subdivided among the boxes; the local behavior of the asymptotic function $x(t)$ involves no up or down drift.

• **C_2(WBM)**. (Generator A2). The generator of this cartoon was described in Section 0.2. The basis $b = 3$ is the smallest that allows oscillations. Denoting the side intervals' heights by x, the middle interval is of height $2x - 1$. Since $H = 1/2$ for WBM, the generating identity becomes $2x^2 + (2x - 1)^2 = 1$, yielding $x = 2/3$.

• **C_3(WBM)**. (Generator A3). The novelty is that all three intervals were made unequal, to add realism to the construction. A form of it is carried over many stages in Figure N1-6.

- **C_2(FBM).** (Generator A4). Cartoon C_2 (WBM) is readily generalized to $H \neq 1/2$. It suffices to take for x the positive root x_0 of the equation $2x/H + (2x-1)/H = 1$. For $H > 1/2$, $x_0 < 2/3$; for $H < 1/2$, $x_0 > 2/3$.

1.2 The multiplicity of distinct (but functionally related) scaling exponents for $B_H(t)$ and its self-affine cartoons

The definition of FBM by $E[B_H(t+T) - B_H(t)]^2 = t^{2H}$ is an analytic scaling relation of a kind that is very common in condensed matter physics (see Chapter N3), fluid mechanics and non-linear dynamics. Two independently defined geometric or physical quantities x and y are said to scale analytically when they are linked by a power-law relation of the form

$$y \propto x^e.$$

The symbol $\propto$ will mean that the plot of $\log y$ versus $\log x$ is near linear and of slope $e(x; y)$.

Geometrically achieved scaling is the source of many exponents. For self-similar fractals, the most important is the measure of roughness or irregularity called fractal dimension.

Self-affine fractals are intrinsically far more complicated in many ways. This will be particularly apparent when the general case is discussed in Sections 3 and 4.

There are several Gaussian fractals defined as graphs relative to $B_H(t)$; some are self-affine with the fractal dimension $2 - H$, others are self-similar with the dimension $1/H$, which is larger. This multiplicity of values is confusing but intrinsic and unavoidable. In addition, it will be seen in M1998H that the notion of fractal dimension further splits into local and global forms, whose values differ from one another.

The two values $2 - H$ and $1/H$ refer to different geometric objects. The graph of $B_H(t)$ has a (local box) dimension equal to $2 - H$. The value $1/H$ corresponds to a "trail," which is a geometric object that is different from, but related to $X(t)$. Its definition injects a second function $Y(t)$ that is independent of $X(t)$ but follows the same process. The trail is the set of values of the plane motion $\{ X(t), Y(t) \}$, when considered irrespective of time.

There is a simple counterpart of the trail for a slight variant of the cartoon C_1(WBM) of Wiener Brownian motion. Take the top right half of Plate 65 of {FGN}, and imagine that the triangular white "rivers" have been thinned down to vanishing. The limit obtained in this fashion is an example of Peano motion often improperly called a "curve") that I called "Cesaro triangle sweep." It fills a triangle, moving from the lower left to the upper right corner, through the upper left corner, and is of course of dimension 2. Now imbed this curve into a three dimensional one by adding a vertical "time" coordinate from 0 at top left to 1 at bottom right. Successive approximations of the Cesaro triangle sweep are represented by broken lines $\{X(t), Y(t)\}$.

We are done now with explaining the peaceful coexistence of two values of D: the dimension $D = 1/H = 2$ applies to that three-dimensional curve, as well as to the trail obtained by projecting on the plane (X, Y). However, the projections of the three dimensional curve on the planes (t,X) and (t,Y) are of dimension $D = 2 - H = 1.5$.

1.3 Examples of grid-bound recursive constructions related to L-stable motion or other purely discontinuous processes

A random process that plays a central role in M1997E is the "Lévy flight" or L-stable motion (LSM). In a first approximation, it reduces to the running sum of independent increments whose probability distribution is $\Pr\{U > u\} = u^{-D}$.

- C_1(LSM). (Generator B2). Begin with C_2(WBM) and modify the generator's boxes as follows: keep the heights constant; expand the first and third box to be of width 1/2, hence $H_1 = H_3$ and reduce the second box width to 0, hence $H_2 = 0$. All the jumps are negative.

- C_2(LSM). (Generator B1). A more versatile surrogate of LSM is one with both positive and negative jumps. To achieve this goal, it is necessary that $b \geq 4$. Begin with C_1(WBM), and modify the generator's boxes as follows: keep the heights constant, expand the first, second and fourth box to be of width 1/3, and reduce the third box to be of width 0. Now, $H_1 = H_2 = \log_3 2$, $H_3 = 0$, and $H_4 = \log_3 2$.

Observe that some of the linear transformations used to define C_1(LSM) and C_2(LSM) cannot be inverted, therefore are not affinities. But it is necessary to allow them when defining self-affinity.

1.4 Examples of grid-bound recursive constructions with continuous generators, related to "dustborne" intermittent 1/f noises

The standard example of such a generator is the triadic Cantor devil staircase, which is so well-known by now that it need not be illustrated. The inverses of two variants are shown in Figure N1-6 as generators C1 and C2.

To increase versatility, take a 5-interval generator in which the fourth interval is horizontal. The variation of the limit curve is restricted to a Cantor dust of dimension log4/log5. Those numbers 4 and 5 were selected to allow this set of variation to be compatible with the two behaviors that follow.

The diagonal intervals may all satisfy $\Delta_i t = 1/5$ and $\Delta_i x = 1/4$. Section 3 will interpret that case as uniform motion in a cartoon of fractal time. It will appear in M1967i{N9} as the "$V(t)$- process" of a dust.

Alternatively, the diagonal intervals may satisfy $\Delta_i t = 1/5$ and $\Delta_1 x = \Delta_2 x = \Delta_5 x = 1/2$ and $\Delta_3 x = -1/2$. Section 3 will interpret that case as a cartoon of WBM varying in a cartoon of fractal time. It will reappear in M1967i{N9} as the "white process" of a fractal dust.

1.5 Examples of grid-bound recursive constructions that yield strictly increasing limits to be used to define an intrinsic time

- C_1(FIT) and C_2(FIT). (Generators C1 and C2). These are inverse functions of variants of the classical devil staircase. The letters FIT refer to "fractal intrinsic time," as defined and discussed in Section 3.

- C_1(MIT). (Generator C4). A point was chosen at random in the unit square, and joined to the lower left and upper right corners by a nondecreasing broken line. Almost surely, this generator's intervals are diagonal with $\Delta_1 t \neq \Delta_2 t$ and $\Delta_1 x \neq \Delta_2 x$. In the limit, this generator yields the integral of a multifractal measure called "binomal". The letters MIT refer to "multifractal intrinsic time," as defined and discussed in Section 3.

- C_2(MIT). (Generator C3). This is a three-interval generator yielding a multifractal measure, hence a multifractal intrinsic time in the specific sense to be introduced in Section 3.

1.6 Other examples

The two remaining generators in Figures N1-6 and N1-7 are listed with scant explanation.

- C_1(MFM). (Generator B3). This example of oscillatory multifractal motion is a much simplified version of one due to Bernard Bolzano (1781-1848). Begin with C_3(WBM thru FBM), and modify the generator's boxes by keeping the heights constant and changing all the widths to 1/3.

- C_2(MFM). (Generator B4). A three interval oscillating generator was chosen haphazardly.

1.7 General definitions: grid-based recursive constructions with prescribed initiator and diagonal or axial generator

The preceding examples are easily generalized by allowing the generator to take diverse forms.

To resume the problem from scratch, each of our cartoon diagrams is drawn within a *level-0 box*, whose sides are parallel to the coordinate axes of t and x. Its width and height are chosen as units of t and x; consequences of this choice are discussed in Section 2.1. The "initiator" is an ordered interval (an "arrow") acting as a "hidden string" that crosses the level-0 box from bottom left to top right.

In addition, each diagram contains a *string generator* that joins the bottom left of the initiator to its top right. Alternative descriptions for it are "string of arrows," "broken line," and "continuous piecewise linear curve." The number of intervals in the generator, b, is called *"generator base."* When the generator is increasing, there is a lower bound $b \geq 2$; when the generator is oscillating, the lower bound becomes $b \geq 3$. The larger b the greater the arbitrariness of the construction. Hence, the illustrations in this chapter use the smallest acceptable values of b.

Axial and diagonal generator intervals. To insure that the recursive construction generates the graph of a function of time, the string generator must be the "filled-in graph" of a function $x = G(t)$, to be called *generator function*. To each t, the ordinary graph attaches a single value of x. To each t where $G(t)$ is discontinuous, the filled-in graph attaches a vertical oriented interval that spans the discontinuity. The generator may also include horizontal intervals.

Vertical or horizontal intervals are called *axial*, all others are called *diagonal*.

Recursive construction of a self-affine curve joining bottom left to top right, using successive refinements within a prescribed self-affine grid. Step 0 is to draw the diagonal of the initiator. Step 1 is to replace the diagonal of the initiator by the filled-in graph of the generator function $G(t)$. Step 2 is to create a line broken twice, as follows. Each diagonal interval within the generator is broken by being replaced by a downsized form of the whole generator. To "downsize" means to reduce linearly in the horizontal and vertical directions. In the self-affine case, the two ratios are distinct. It is often allowable or even necessary to begin by transforming two generators by symmetry with respect to a coordinate axis.

As to the generator's axial intervals, the recursion leaves them alone. One may also say that they are downsized in the sense that one linear reduction ratio is 0, collapsing the generator into an interval.

The resulting "prefractal" approximations of self-affine graphs consist of increasingly broken lines. Figures N1-6 and N1-7 take up important generators and draw corresponding approximations.

Alternative interpolation of the construction, via self-affine "necklaces." Each diagonal interval may be replaced by a *level-1 box* with sides parallel to the axes, of which it is the diagonal. If so, the prefractal approximation consists in "necklaces" made of boxes linked by flat boxes identical to axial pieces of string. These prefractals are "nested," that is, the k-th is contained within the $(k-1)$-th.

Degenerate affinities. A definition given earlier in this chapter deserves amplification. The collapse of the generator upon an axial interval is a linear contraction whose matrix reduces to one non-zero term: therefore it has no well-defined inverse, such a transformation is *not* an affinity. Nevertheless, the term, *self-affine* had no previous meaning, and I saw no harm and great convenience in defining it to allow degenerate affinities.

Trivial generators that reduce to axial intervals. In that case, each generator interval remains unchanged by recursion.

At this point, the reader may want to visit Section 0.5 again.

2. THE H_i EXPONENTS AND A CLASSIFICATION OF CARTOONS: THE NOTIONS OF UNI-, MESO-, AND MULTIFRACTALITY

Examples having established that cartoons are very versatile in their behavior, the next task is to classify the cartoons; their versatility is reflected by a classification that is rich and somewhat complicated.

Except for location and scale, a fractional Brownian motion $B_H(t)$ is fully specified by the single scaling exponent H. This H determines the spectral exponents and two forms of fractal dimension discussed in Section 1.3. In other words, those diverse specialized exponents are functionally related to one another through H.

This and the next sections show that a general cartoon exemplifies an altogether different situation. There, the multiple distinct roles that H plays in the FBM case become split between a continually growing multitude of specialized exponents that are independent functions of the inputs and *not* functionally related to one another.

2.1 The H_i exponents of the intervals in a generator

Diagonal boxes and their finite and positive H exponents. Given a diagonal box β_i of sides $\Delta_i t$ and $\Delta_i x$, an essential characteristic is the expression

$$H_i = \frac{\log \Delta_i x}{\log \Delta_i t} = \frac{\text{log of the absolute height of the box } \beta_i}{\text{log of the width of the box } \beta_i}.$$

In other words,

$$\Delta_i x = (\Delta_i t)^{H_i}.$$

This H_i concerns a non-infinitesimal box, hence is a *coarse* coefficient.

Axial intervals and the values H=0 and H = ∞. One may say that horizontal intervals yield $H = \infty$, and vertical discontinuities yield $H = 0$.

Reasons for selecting the units of t and x so that the initiator is a square. It suffices to make the point in the case when all boxes of the same level are congruent with sides $\Delta t = b^{-1}$ and $\Delta x = b'^{-1}$ for level-1, therefore $\Delta t = b^{-k}$ and $\Delta x = b'^{-k}$ for level-k. In this case, $H_i = \log b' / \log b = H$ for all boxes at all levels; level 0 yields $\log 1 / \log 1 = 0/0$, which can be interpreted as equal to H. Now examine what would happen if the level-0 box had sides 1 and B. If so, all the level-k boxes would yield

$$H = \frac{\log b' + \log B / k}{\log b}.$$

The fact that this ratio depends on k does not matter in the pure mathematical context originating with Hölder; there, H is a local concept that concerns the limit $k \to \infty$, which is not affected by B. By contrast, the concrete interpretation of H that I pioneered does not concern local asymptotics but concrete facts, therefore applies uniformly to all sizes. To serve a concrete purpose, the "coarse" H must be independent of all units of length; this is achieved by setting $B = 1$.

2.2 Unifractal cartoons and relation with Wiener and fractional Brownian motions

Definition. If all the H_i have a common value H satisfying $0 < H < \infty$, the cartoon construction will be called *unifractal*. The conditions $H > 0$ and

$H < \infty$ exclude both kinds of axial generator intervals. This case is very special, but includes cartoons of WBM and FBM.

Unibox versus multibox constructions. It may be possible to superpose all boxes by translation (preceded, if needed,) by a symmetry with respect to an axis); if so, the generator will be called *unibox*. Unibox constructions are necessarily unifractal. Many of their properties depend only on H, but other properties depend on the boxes' two sides, and some properties also depend on the details of the arrangement of the boxes.

When a generator is not unibox, it will be called *multibox*. Multibox constructions depend on a larger number of parameters; they are less regular, hence less "artificial-looking," making their fractality a better surrogate for randomness.

2.3 Beyond unifractality, taking two steps instead of the usual one

The concept of unifractal curve has extremely important implementations, which is surprising because its definition makes extreme demands on the generator. For example, choose the filled-in generator function $G(t)$ at random, only the base b being prescribed. In the corresponding space of possible generators, the identity "H_i = a constant independent of i" holds on a very small subspace. Under a typical probability distribution for the generator, it is almost sure (the probability is equal to 1) that the H_i will not be equal to one another and that one will instead obtain a generalized form of multifractality, a notion to be described in Section 2.5.

However, specific behaviors needed to tackle pratical problem are not bound by those probability considerations. Some cases that chance yields with probability zero deserve special attention. Unifractality is one. Measure-generating multifractality, which will correspond to monotone non-decreasing generators, is another. And a third important special case of fundamental importance has grown to warrant a special name. Since "in between" is denoted by the Greek root *meso*, this case will be called *mesofractal*, a word used for the first time in M1997E.

2.4 Mesofractal cartoons and discontinuous and/or intermittent mesofractals

Definition. The term *mesofractal* will denote cartoons whose generator intervals fall into two non-empty classes: a) diagonal intervals sharing a unique H satisfying $0 < H < \infty$, b) vertical intervals with $H = 0$ and/or horizontal intervals with $H = \infty$.

A finer subdivision is needed.

Discontinuous mesofractals correspond to generators with a mixture of diagonal and vertical boxes. A simple basic example is C(LSM), a cartoon of the L-stable motion LSM.

Intermittent mesofractals correspond to generators with a mixture of diagonal and horizontal boxes.

Cases when discontinuity and intermittence co-exist may be called *mixed mesofractals*.

2.5 Multifractal cartoons

Definition. The term *multifractal* denotes the most general category of multibox cartoons. It allows the generator to combine axial boxes and diagonal boxes with non-identical values of H_i from $H_{min} > 0$ to $H_{max} < \infty$.

This definition includes cartoons that a) can but need not oscillate, and b) combine three behaviors: continuous variation over any time interval, existence of intervals of no variation, and jumps.

The definition of multifractality used in this book and almost everywhere else in the literature is *far narrower*. It is limited to singular non-negative measures constructed using continuous non-decreasing generators. The only allowable axial intervals are horizontal. Therefore, the general definition of multifractality demands several finer distinctions.

Oscillating multifractals correspond to oscillating generators, while *multifractal measures* correspond to non-decreasing generators.

Multifractal measures cannot be differentiable. Those which satisfy the narrow definition are singular, that is, continuous but not differentiable. Discontinuities are allowed, for example, in M & Riedi 1997 and Riedi & M 1998.

Among oscillating multifractals, a different finer subdivision is needed.

Pure multifractals correspond to generators with diagonal boxes, while *mixed multifractals* correspond to generators with both diagonal and axial boxes. (The term *mixed* is unattractive, but *multimesofractals* would not fly.)

Mixed multifractals must be further refined.

Discontinuous multifractals correspond to generators whose boxes are either diagonal or vertical.

Intermittent multifractals correspond to generators whose boxes are either diagonal or horizontal.

2.6 Non-random or random shuffling of the generators

Digression concerning lacunarity. In the C_1(WBM) cartoon described in Section 1.1, the sequence UDUU of the generator intervals may be replaced by the sequence UUDU (up, up, down, up)

More generally, one can perform an arbitrary permutation in the time sequence of a generator's contributing arrows, then restore continuity by displacing the arrows vertically, as follows. The beginning of the first moves to the point {0, 0}, and the beginning of the remaining arrows move to the endpoints of the preceding ones. The resulting broken line again ends in the point {1, 1}, hence is an acceptable generator.

This generalization allows the replacement of the sequence UDUU by either UUUD or DUUU. Furthermore, the construction can be redone with any continuous generator satisfying $b_t = 4^k$ and $b_x = 2^k$. Increasing k preserves $H = 1/2$ but increases the number and variety of alternative generator shapes. When k is large, one extreme shape is symbolized by a string of U followed by a string of D; the opposite extreme is symbolized by 4^k letters U and D that mostly alternate, except that U is replaced by UU in 2^k positions that are uniformly spread from 0 to 1.

The resulting limit curves are said to differ in their "lacunarity." Lacunarity is a delicate topic, touched upon in M1982F{FGN}, Chapters 34 and 35, later discussed in M1995f and M1998e, and scheduled for a detailed discussion in M1998L.

Iterated generator systems (IGS). The procedure described when introducing lacunarity can be used to construct a finitely indexed "generator system." Each branch of the recursion then begins by selecting a generator index, then applies the corresponding generator. The selection can be systematic (for example follow a finite "dynamical system") or random.

The term "iterated generator system" is patterned on "iterated function systems." The latter is the accepted term, due to M. Barnsley, for an old notion that M1982f{FGN} (p. 196) had used and called "decomposable dynamical systems." See Section N5-7.

Randomized generators. The IGS start with prescribed elements and continue with a permutation, but randomization can be subjected to fewer constraints. An example is a generator of base $b = 3$ including the two end intervals {(0, 0) to (r, d)} and {$(1 - r, 1 - d)$ to (1, 1)}. In this case, randomness beyond the IGS is achieved by choosing the point of coordinates (r, d) at random, with some prescribed distribution.

3. TRAIL DIMENSION AND EMBEDDING; INTRINSIC TIME AND "COMPOUND CARTOON" REPRESENTATIONS

This section is the most novel in the whole chapter, that is, moves farthest from the strict interpretation of the concept of *Selected Papers*.

3.1 Description of goals

This section continues the exploration of different exponents related to the Holder-Hurst H. The cartoon exponents H_i defined and examined in Section 2 depend equally on the quantities $\Delta_i t$ and $\Delta_i x$. In this section, to the contrary, the two coordinates play very different roles: the $\Delta_i x$ are preserved, while the $\Delta_i t$ are either disregarded or adjusted in one of two ways.

Section 3.2 disregards them and defines a cartoon's dimension trail D_T, the inverse of one of several generalizations of the H exponent of FBM.

Section 3.3 replaces clock time by an alternative time, described as "isochrone," in which all the $\Delta_i t$ are made identical.

Section 3.4 replaces clock time by an alternative time θ, described as "intrinsic," for which the definition $\Delta_i \theta$ is based on the $\Delta_i x$.

Section 3.5 proves a very simple yet fundamental result, to be called "baby theorem" for easy identification, that builds on intrinsic time to bring order to an otherwise messy situation. It also connects the cartoons to a process introduced in M1997E and called fractional Brownian motion of multifractal time.

To lead to the baby theorem, let us step back. The major examples in Section 1 include cartoons of WBM, FBM and LSM and Section 2 adds a multitude of other grid-bound self-affine functions that combine long tails and long memory. In many ways, the generating functions are of great diversity, and innumerable additional examples immediately come to mind. Their very multiplicity might have been a source of disorder and confusion. But it is not, for the following strong reason.

"Every oscillating cartoon construction can be rephrased as a compound function, namely, as an *intrinsic* function of an *intrinsic* time, more precisely, a *unifractal* function of a *multifractal* time."

A full description is this: "an *intrinsic* function that is a unifractal cartoon C(FBM) – of which C(WBM) is a special case – and is not followed in clock time, but in an *intrinsic* time related to clock time by a monotone non-decreasing auxiliary function."

For a unifractal cartoon, the intrinsic time trivially reduces to clock time. The most general non-degenerate intrinsic times will be called "multifractal." Mesofractal time will be shortened to "fractal time," for the sake of brevity, to follow existing practice, and to make it easier, when needed, to add the qualifiers "discontinuous" or "intermittent."

Oscillating cartoons beyond the simplest may often be conveniently described by injecting discontinuity, intermittence and pure multifractality separately. This amounts to repeated steps of compounding.

The notion of intrinsic time is compelling and inevitable.

3.2 Definitions of a "trail dimension" D_T and "trail diffusion exponent" $H_T = 1/D_T$; a spectral density of the form f^{-B}, where $B = 1 - 2/D_T$

We begin by defining the quantities D_T and $H_T = 1/D_T$.

A simple identity that characterizes unifractal generators. In all cartoons, the box widths satisfy $\sum \Delta_i t = 1$. In the unifractal case where $H_i = H$ for all i, let us *define* D_T as $1/H$. It follows that the box heights satisfy

$$\sum_i |\Delta_i x|^{D_T} = 1, \text{ with } D_T = 1/H.$$

A simple identity that characterizes the monotone non-decreasing generators of multifractal measures. When $\Delta_i x > 0$ for all i, the equality $\sum \Delta_i x = 1$ trivially implies

$$\sum_i |\Delta_i x|^{D_T} = 1, \text{ with } D_T = 1.$$

The dimension-generating equation of a cartoon construction. This term will denote the following equation for the unknown σ:

$$Q(\sigma) = \sum_i |\Delta_i x|^\sigma = 1.$$

The multifractal formalism (see Chapter N2, Section 3) defines $\tau(\sigma)$ as proportional to $\log Q(\sigma)$, hence the dimension-generating equation reads $\tau(\sigma) = 0$.

In the unifractal case the only positive root is $\sigma = D_T = 1/H$, and in the monotone case the only positive root is $\sigma = 1$. We now proceed to the remaining possibility.

N1 ◇ ◇ PANORAMA OF GRID-BOUND SELF-AFFINE VARIABILITY   49

A new and highly significant concept: generalized D_T and $H_T = 1/D_T$, as defined for oscillating multifractal cartoons. In the oscillating multifractal case, the quantities H_i cease to be identical. Therefore, the generating equation becomes highly significant. Its only positive root D_T satisfies $D_T > 1$ and turns one to be a *fundamentally* important characteristic of the construction. It only depends on the positive $\Delta_i x$ and is unchanged if the sequence of generating intervals is shuffled.

The dimension-generating equation taken after the k-th interpolation stage. The output of k interpolation stages is a prefractal broken line made of b^k intervals is easy to see (by induction) that their ordinate $\Delta_{k,i} x$ satisfy

$$\sum |\Delta_{k,i} x|^{D_T} = 1 \text{ for all } k.$$

Instead of stopping after k stages, one may select a threshold ε, and stop as soon as the interval height becomes $< \varepsilon$. If so, the interval lengths $\Delta_{\varepsilon, i}$ satisfy

$$\sum |\Delta_{\varepsilon, i} x|^{D_T} = 1 \text{ for all } \varepsilon.$$

The exponent value D_T is "critical" in the sense that, as $\varepsilon \to 0$, the same sum written with an exponent $q \neq D_T$ would tend to zero if $q > D_T$ and to infinity if $q < D_T$.

Digression: The dimension-generating equation extends to functions that are not obtained by self-affine interpolation. That is, given any function $X(t)$, a critical D_T can be defined for the integral $\int (dX)^q dt$.

3.3 Embedding a scalar cartoon in an intrinsic vector cartoon trail and the geometric interpretation of D_T as trail dimension

For a self-similar fractal, for which all r_i reduce to r, one has $D_T = D = \log N / \log(1/r)$. When the reduction ratios r_i are not identical, the dimension of a self-similar set is given by a more complicated classical equation due to Moran, which is obtained by replacing $|\Delta_i x|$ by r_i in the above-written generating equation for σ. This formal identity is endowed with deep significance that is revealed by a procedure to be called "cartoon embedding."

Embedding in the Brownian context. A model for embedding is provided by some well-known results on the Wiener and fractional Brownian motions, WBM and FBM. It is well-known that the scalar WBM $X(t)$ becomes more natural when embedded in an E-dimensional process, whose $E > 2$ coordinates are statistically independent WBM's. The Hölder exponent is $H = 1/2$ for both the scalar and vector process, and $D = 1/H = 2$ is (almost surely) the dimension of the vector process in space and also of its projection on the plane or space $t = 0$. (Observe that this projection, therefore its dimension, remains unchanged if time is transformed by an arbitrary monotone non-decreasing map. This invariance will play an essential role in Section 3.5.

Furthermore, when $E > 2/H = 4$, the projection of the vectorial WBM (almost surely) does not self-intersect. More generally, a scalar FBM $X(t)$ can be interpreted as a projection of a vectorial FBM $X(t)$ of $E > 1/H$ coordinates. $D = 1/H$ is (almost surely) the dimension of the samples of that vectorial FBM, and also of its projection on the plane or space $t = 0$. When $E > 2/H$, the projection does not self-intersect.

Embedding a scalar cartoon $X(t)$ into the smallest vector cartoon $P(t)$ whose projection on $t = 0$ is self-avoiding. Here, "smallest" is meant to express that E is as small as is compatible with self-avoidance. This topic would take too far afield and the simplest is to give meaning to D_T when $E = \infty$.

Embedding a scalar cartoon $X(t)$ into a vector cartoon $P(t)$ in an infinite-dimensional space. The key step is to construct a vectorial cartoon $P(t)$, to project it into the space $t = 0$, and show that this projection is of dimension D_T. The vectorial cartoon can be parametrized by any monotone non-decreasing function of clock time. The easiest is to proceed as follows. Disregard the generator's horizontal intervals, because they do not affect $P(t)$. Set the number of other intervals to b and choose the "isochrone" time defined by the condition that the time increment is $\Delta_i T = 1/b$ for all i. The argument is simplest when $H = 1/2$.

Introduce a first Euclidian "phase space" of dimension b and consider in that space the point $P(1)$ of coordinates $\Delta_i x$. The distance from the origin O to $P(1)$ is 1. This follows from the assumption that $\sum |\Delta_i x|^2 = 1$. Next, the vector of length $\Delta_i x$ along the i-th coordinate axis can be interpreted as $P(i/b) - P((i-1)/b)$. This defines the function $P(t)$ for t of the form i/b. The last step in relating $P(t)$ to $X(t)$ is to project on the main diagonal of our b-dimensional phase space. The increment $P(1/b) - P((i-1)/b)$ projects on an interval of length $\Delta_i x/\sqrt{b}$, and the vector $OP(1)$ projects on an interval of length $1/\sqrt{b}$. Choose the unit of length along the diagonal as being $\sqrt{b}$ times smaller than elsewhere in the phase space. Using the isochrone time T, we have constructed for integer i a map from the values of $P(i/b)$ on the values of $X(i/b)$.

The second stage of interpolation demands a Euclidean "second stage phase space" of dimension b^2. This is taken to be the product of b subspaces orthogonal to one another. The i-th subspace is a replica of the first-stage space, reduced in the ratio $|\Delta_i x|$. Furthermore, the i-th subspace is positioned so that the vector $OP(1)$ is mapped on the i-th coordinate of the first-stage phase space. (An arbitrary rotation around this vector is allowed.) The second phase space suffices to interpolate $P(t)$ to all values of time of the form $t = ib^{-2}$. And so on.

Continuing interpolation constructs a self-similar curve in an infinite-dimensional space. This interpolated curve is covered by Moran's theorem. This is the reason why the fractal dimension of the representative curve is the root D_T of the dimension-generating equation.

The preceding construction relies on the Pythagoras theorem, which is why the case $D_T = 2$ is the simplest possible, but the result holds for all $D_T > 1$.

3.4 A cartoon's intrinsic time: it is linear for unifractal cartoons, fractal for mesofractal cartoons, and multifractal for multifractal cartoons

Examine the three sets of cartoons that are shown in Rows 1, 2 and 3 of Figures N1-6 and N1-7, and label the coordinates as follows: θ and X in Column A, t and X in Column B and t and θ in Column C. This reinterprets each cartoon in Column B as a compound cartoon obtained from its neighbor in Column A, by replacing the clock time by the fractal or multifractal time defined by its neighbor in Column C. Let us now show that such a representation can be achieved for every cartoon.

The notion of "intrinsic duration of an interval in the generator." Start with a recursive construction as defined in Section 1.6. To make it into a cartoon of FBM with the exponent $H_T = 1/D_T$, the recipe takes two steps:

- keep the height $\Delta_i x$ constant,
- define $\Delta_i \theta = |\Delta_i x|^{D_T}$ and change the i-th generation interval's width from $\Delta_i t$ to $\Delta_i \theta$. Vertical generator intervals automatically become diagonal, and horizontal intervals vanish.

Definition of a cartoon's intrinsic time. Intrinsic time is itself generated by the cartoon, whose generator is defined by the quantities $\Delta_i t$ and $\Delta_i \theta$. Each recursion stage ends with an "approximate intrinsic time" that becomes increasingly "wrinkled" as the interpolation proceeds. The limit intrinsic time may, but need not, involve discontinuity. Each stage of interpolation makes the intrinsic time become increasingly concentrated and in the limit manifests a high degree of concentration in very short periods of time. How multifractals acquire this property is a very delicate topic that constitutes a core topic of this book.

3.5 The "baby theorem;" existence of a unique intrinsic representation of a general cartoon as a one- or multi-step "compound cartoon"

We are now ready to show that an oscillating but otherwise arbitrary cartoon can be represented as a unifractal oscillating cartoon of exponent $H_T = 1/D_T$ of a multifractal (possibly fractal) intrinsic time. This possibility is fundamental, yet obvious by inspection of Figures N1-9 and N1-10. It will be called a "baby theorem.", an easily remembered term. It will be stated first for the purely multifractal case, then extended.

Special one-step baby theorem. Cartoons whose generator intervals are diagonal can be represented as intrinsic functions of an intrinsic time. That is, they are unifractal cartoon functions of a purely multifractal cartoon time. Degenerate examples are uniform velocity functions of clock time, unifractal cartoon functions of clock time, and uniform velocity functions of purely multifractal time.

General one-step baby theorem. In the presence of axial intervals in the generator, the multifractal time is not pure.

Time-compounding baby theorem. A non-pure multifractal time can be represented as a mesofractal function of a pure multifractal time, or as a discontinuous mesofractal function of an intermittent mesofractal function of pure mathematical times.

Multi-step baby theorem. A general multifractal cartoon function can be represented as a unifractal function of a discontinuous mesofractal time, itself a function of an intermittent fractal time, itself in turns, a function of a purely multifractal time.

The general case. Generalized multifractal formalism. M1997E, Section 3.9 of chapter E6, generalizes in two ways the function $\tau(q)$ that the body of this book applies to multifractal measures. The resulting functions are denoted as $\tau_C(q)$ and $\tau_D(q)$. The same definitions apply to the cartoon constructions.

4. THE SPECTRAL EXPONENTS H_G AND B

Once again, that in the case of FBM the single exponent H connects several conceptually separate specific exponents. Indeed, $D_T = 1/H$ is the trail

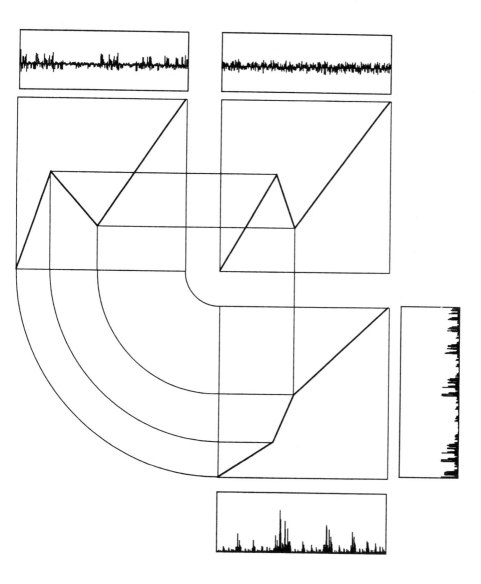

FIGURE N1-9. The fundamental "baby theorem" of Section 3.5 is proved by simple inspection of this figure. The square in the top right quarter represents the generator of a unifractal cartoon that is a surrogate for fractional Brownian motion. Its increments, represented above the square merge into a strip of more or less constant variability or "volatility." The top left quarter represents in the same way the generator of a multifractal cartoon with its increments forming a very spiky strip. The bottom right quarter represents the monotone increasing relation between a) the clock time that applies in the upper left cartoon and b) the intrinsic clock time that applies in the upper right cartoon.

dimension, $D_G = 2 - H$ is the graph dimension, and the spectral density of the increments of FBM is proportional to f^{-B}, with $B = -1 + 2H$.

In the cartoons, those various exponents cease to be functionally related. The generalization of $H_T = 1/D_T$ was examined in Section 3. Now we proceed to $H_G = 2 - D_G$ and the generalization of B.

4.1 The graph dimension D_G of a cartoon; it is not functionally related to the trail dimension D_T

The next goal is to generalize D_G beyond the value $2 - H$ which it takes in the FGN case. Non-linear changes in the scale of time do not modify D_T but modify the graph dimension.

Graph dimension in intrinsic time. Its value is $D_G = 2 - H_T < 1/H_T = D_T$.

Graph dimension in the isochronal time satisfying $\Delta_i t = 1/b$ for all i. As in the derivation of D_G for FBM, the idea is to cover the graph with stacks of square boxes of side $\Delta t = b^{-k}$. Their number is obtained by dividing by b^{-k} the expression

$$\chi(1, b^{-k}) = \sum |\Delta X| = \left\{\sum |\Delta_i x|\right\}^k = \left\{\sum b^{-H_i}\right\}^k.$$

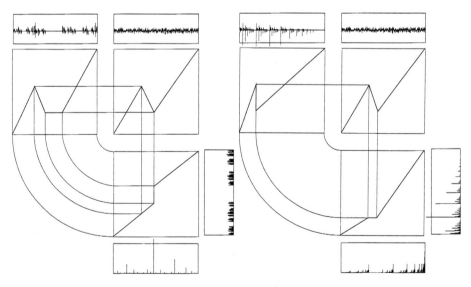

FIGURE N1-10. Baby theorem for an intermittent and a discontinuous self-affine cartoon generator.

Using the multifractal function $\tau(q)$ (see Chapter N2, Section 3),

$$D_G = \log[X(1, b^{-k})b^k]/\log(b^k) = 1 - \tau(1).$$

If one defines $H_G = 2 - D_T$ so as to generalize the relation $D = 2 - H$ valid for FBM, one finds the odd average

$$b^{-H_G} = \frac{1}{b}\sum b^{-H_i}.$$

The special case of the measures obtained when $\Delta_i x > 0$ yields $D_G = 1$. This degenerate result was expected, because $D = 1$ is known to hold for the graphs of all non-decreasing continuous functions (either differentiable or singular), and also for the filled-in graphs of non-decreasing discontinuous functions.

The general case of oscillating functions implies $b \geq 3$, therefore requires $b - 1 \geq 2$ independent parameters H_i. In general, the two exponents H_G and H_T are distinct functions of those parameters. A brief argument shows that $1 - \tau(1) < 2 - H_T$, a quantity smaller than $1/H_T = D_T$.

4.2 The 1/f noise character of cartoon increments; the exponent B

This section sketches, briefly and belatedly, a central point of the introductory chapters written specially for this book of *Selecta*, namely, the link between 1/f spectrum and self-affinity. A more thorough treatment would bring this chapter beyond its definition as a Panorama.

Discontinuous generators lead to limit functions with an infinite variance, for which a power spectrum is not defined. Therefore, we can assume that the generator is continuous. Working in isochronal time, namely $\Delta_i t = 1/b$ for all i will avoid extraneous complication that would slow the story without benefit.

Properties of the increment ΔX of $X(t)$ taken over a randomly selected time interval $\Delta t = b^{-k}$. This increment is the product of k the factors of the form $\Delta_i x$. Therefore,

$$E\{|\text{ increment } \Delta X \text{ over a random in-grid } \Delta t = b^{-k}|^q\} = \{E|\Delta_i x|^q\}^k.$$

Write

$$-\log{}_b E|\Delta_i x|^q = -\log{}_b[(1/b)\Sigma|\Delta_i x|^q] = \tau(q) + 1,$$

with the understanding that both negative and positive powers of 0 are set to 0. Further, the above moment becomes

$$E\{|\text{ increment } \Delta X \text{ over a random in-grid } \Delta t = b^{-k}|^q\} = (\Delta t)^{1+\tau(q)}.$$

Write $\sigma(q) = [\tau(q) + 1]/q$. Our moment's q-th root defines the increment's q-th scale factor. It takes the form

$$\left\{E\left[|\text{ increment of } X \text{ over a random in-grid } \Delta t = b^{-k}|^q\right]\right\}^{1/q} = (\Delta t)^{\sigma(q)}.$$

The quantity $\sigma(q)$ was apparently first used in M1997E. It is more important than the quantity $D(q) = \tau(q)/(q-1)$ to be mentioned in Chapter N2, Section 5.6.

The correlation exponent. The correlation is the second difference of $(\Delta t)^{1+\tau(2)}$, therefore the large lag correlation exponent takes the form $\tau(2) - 1$. This form is familiar in the study of multifractal measures, but the present generalization introduces an important new distinction.

The case where $D_T < 2$, therefore $\tau(2) > 0$, is the rule for multifractal measures. A positive correlation expresses that the varia tion is *persistent*. Also, $\tau(2) - 1$ can be shown to let between 0 and -1, hence the correlation is non-integrable.

The case $D_T > 2$, therefore $\tau(2) < 0$, is unfamiliar and very different. The large lag correlation is negative, hence the variation is *anti-persistent*. Also, the correlation is integrable (its sum is 0).

The notions of persistence and antipersistence were defined in the context of FBM (M1982E{FGN}, starting with p. 250). In the present context the "critical" role made familiar by the Fickian $H = 1/2$ attaches to $H_T = 1/D_T = 1/2$.

The spectral density of the increments of the cartoon $X(t)$, taken in isochrone time. Being the Fourier transform of the correlation, it is proportional to f^{-B}, with $B = \tau(2)$.

4.3 Fractional Brownian motion in multifractal time

The cartoons' numerous drawbacks are described early in this chapter, as well as their assets. The baby theorem did not come out of the blue, and its spirit comes from a certain process $\tilde{B}_H(\theta)$. It was referred to obliquely in the last remark in M1972j{N14}, and investigated in M1997E, Section E6-3. That process $\tilde{B}_H(\theta)$ is obtained when the fractional Brownian motion $\tilde{B}_H(t)$ is compounded with a multifractal time $\theta(t)$, where $B_H(t)$ and $\theta(t)$ are statistically independent random functions. Thus, $\tilde{B}_H(\theta) = B[t(\theta)]$.

The most important special case $H = 1/2$ yields Wiener Brownian motion of multifractal time.

The compound process $\tilde{B}(\theta)$ was first considered in M1972j{N14}, for reasons of elegance and self-affinity. Besides, it generalized the Brownian motion in fractal time, which was investigated in M & Taylor 1967. Only later was it tested as a possible model of financial price variation; see M1997E, especially Chapters E6 and M, Calvet & Fisher 1997. In the old papers reprinted in this book, $\tilde{B}_H(t)$ plays no role, therefore it only deserves a brief mention in this chapter.

Motivation for the Moran-like equation that defines D_T. The most natural setting is one in which $\tilde{B}_H(t)$ is not scalar but vectorial of dimension $E \geq 1/H$. The trail of $\tilde{B}_H(t)$ is then defined by projecting the spatial graph along the time axis. When t is replaced by θ, the time axis and the trail remain unchanged. The different aspects of $\tilde{B}_H(\theta)$ become separated. First, the value of H is obtained by projection along the time axis, and after that the multifractal time is treated on its own.

Consequences for statistics. To develop a full theory, the statistical independence between the functions $\tilde{B}_H(t)$ and $\theta(t)$ is essential. This is a strong restriction on the compound function $\tilde{B}_H(\theta)$, but leads to enormous simplification in statistics, as discussed in M1997E, Section E6-3.14.

Of specific importance for statistics is a major but unrecognized "blind spot" of spectral analysis. $D_T = 2$ characterizes spectral whiteness. This property is insensitive to change of time; in the absence of additional tests, it may be misleading. Successive increments ΔB of Wiener Brownian motion are independent, therefore uncorrelated ("orthogonal"), and their spectral density is a constant, defining a white spectrum. Let us now follow Brownian motion in an arbitrary alternative time. The increments of the compound motion become dependent but remain uncorrelated, therefore *spectrally white*. In other words, spectra as applied to a compound process solely reflect the whiteness of the directed function, and are completely blind to the properties of the directing function.

Indeed, given two non-overlapping time increments $d't$ and $d''t$, the corresponding increments $d'B(t)$ and $d''B(t)$ are, by definition, independent. It is obvious that this property continues to hold when B is followed in a trading time θ that is non-linear non-decreasing function of t, and $B(t)$ is replaced by $\tilde{B}(\theta) = B[t(\theta)]$. The increment of the compound function exhibit very strong dependence, yet they are white, that is uncorrelated.

Remark concerning statistical method. This serious "blind spot" was noted, but not developed in M1967i{N9}. In the non-Gaussian and non-Brownian context, it constitutes a fundamental limitation of spectral analysis that statistics must face.

Disgression concerning the spectral whiteness of financial data. During the 1960s spectral analysis was introduced into economics with fanfare, but never lived to its promise. The "blind spot" of spectra suffices to account for many puzzling observations reported in the literature. Indeed, Voss 1992 and the contributors to Olsen 1996 are neither the first nor the only authors to report on such whiteness. Both parties also examined the "rectified" records of absolute price change, or of price change squared. They found that the spectrum is no longer white but instead takes the "$1/f$" form characteristic of FBM. Now we see that this apparent contradiction is characteristic of Brownian motion in multifractal time. See also Chapter E3.

The functions $\tau(q)$ and $f(\alpha)$ for the directing and the compound function. Chapter N2 sketches the formalism of multifractal measures, as developed in M1974f{N15}, M1974c{N16}, Frish & Parisi 1985 and Halsey et al 1986. This formalism applies to multifractal time, and Section E6-3.9 of M1997E and M, Calvet & Fisher 1997 describe the elegant generalization of this formalism to the compound function.

5. MISCELLANY

5.1 Alternative cartoon constructions; a step back from multiplicativity to additivity

The signal advantage of the cartoon scheme described in this chapter is that self-affinity is achieved by *multiplicative* operations. As exemplified in M1974c{N16}, this is precisely how I introduced grid-free multifractals and their grid-bound cartoons. Other cartoons in this chapter prove that the same procedure can also yields unifractals or mesofractals. In the example of continuous cartoons with $\Delta_i t = 1/b$, a weight can be defined as $W_i = b\Delta_i x$. In those terms, cartoons are interpreted as follows. The linear initiator of the construction is $X_0(t) = t$, and its derivative is 1 for all t. The first prefractal $W_1(t)$ has at every point a derivative or at least a left and right derivative. In the interval i, this derivative is obtained from the initial value 1 by applying the multiplicative weight W_i. And so on.

To provide a broader background it is useful to mention a few cases of self-affinity obtained by an additive procedure.

5.2 The additive character of the graphs of standard unifractal functions

The first unifractal function ("proto-unifractal" because it was defined long before fractal geometry) was written down by K. Weierstrass in 1872. It is discussed in many books and surveyed in M1982F{FGN}, pages 388-390. Its being unifractal restates a theorem Hardy proved in 1916.

The Weierstrass function is defined by a trigonometric series. That is, it is the limit of "prefractal" approximations, each of them the sum of trigonometric expressions. The passage from one approximation to the next involves an *additive* perturbation.

Many early 20th century authors simplified the Weierstrass series by replacing the sine by a more general periodic function. Takagi and then Landsberg used "sawtooth" functions. The functions they defined are unifractal and self-affine, but their affinities are *not* diagonal. Crafting additive approximations became a fine art in Knopp 1918.

In the sciences, strict diagonal self-affinity is merely a matter of convenience carried to a very high degree.

5.3 Novel $1/f$ noises obtained as fractal sums of pulses ("FSP")

The fractal sums of pulses are an alternative recursive construction that is additive and yields random self-affine functions. I introduced this concept in the 1970s and it was first used in modeling rain fields (Lovejoy & M1985{L}). The theory was sketched in M1995n and is scheduled to be developed in detail in M1998L. (See also Cioczek-Georges & M1995, 1996, Cioczek-Georges et al 1995). The FSP bring in a very new point that is important for the present story. Therefore, it is good to define the simplest case, which is partly random.

One begins by choosing a real exponent $\delta > 0$ and a *pulse template*, which is a function $\Omega(t)$ that vanishes except in an interval. The length of the smallest such interval is taken as unit of width, and the template's maximum (actually, the least upper bound) is taken as a unit of height. Within the smallest width interval a center is chosen arbitrarily and denoted by $t = 0$. Given positive quantities σ, ν and ε, a pulse is obtained from the template by moving its center from 0 to t_m, where the position of the pulse center t_m follows a Poisson distribution of density ν. For each pulse center t_m one changes the pulse width from 1 to W_m, a random number with the distribution $\Pr\{W > w\} = (w/\varepsilon)^{-1}$. The height is changed from 1 to $W^{1/f}$. Finally, a fractal sum of pulses is defined as

$$X(t) = \Sigma (W_m/\sigma)^{1/\delta} \Omega(\frac{t - t_m}{W_m}).$$

For a wide range of combinations of a template and δ, this expression converges. Moreover, letting $\varepsilon \to 0$ and $\nu \to \infty$ suitably yields a non-degenerate limit that is referred to as a "fractal sum of pulses," or FSP.

A basic property is that, by construction, the FSP are self-affine, with $H = 1/\delta$ independent of the template and the remaining input σ. In other words, all FSP are unifractal, but different choices of the template and σ allow a great diversity of behavior. A second basic property is that as $\sigma \to \infty$ the limits are as follows: fractional Brownian motion when $\delta > 2$, L-stable motion when $\delta < 1$, and either FBM or LSM (depending on the template) when $1 < \delta < 2$.

From many viewpoints, FBM and LSM are profoundly different. We already know that the diagonal cartoons provide one environment connecting them. We now find that the preasymptotic FSP provide another such environment.

For some combinations of a template and δ, the increments of an FSP define a $1/f$ noise. For other combinations, however, the sample increments come out as the sum of a white noise and a $1/f$ noise. This last possibility cannot be developed in this text, but is conceptually novel and may account for some concrete examples where a $1/f$ noise seems to be added to an underlying white noise. This appearance may reflect a genuine mixture of disparate phenomena, but it may also reflect two indissolubly linked aspects of a single phenomenon.

5.4 The co-indicator function introduced in M1967i{N9} and the random functions introduced in M1969e

Of the several $1/f$ noises described in M1967i{N9}, some have cartoon surrogates. But the most important, called co-indicator function, does not.

A little known article, M1969e examined the sum of many independent co-indicator functions. It was purposely contrived to appear to be Gaussian in one form of analysis and extremely non-Gaussian in another. Its properties are sketched in a Post-Publication appendix to Chapter N9.

N2

Sketches of prehistory and history

✦ **Abstract.** This book's variety of purpose was played down when Chapter N1 put forward a modern synthesis centered around the concept of self-affine variability. This chapter, to the contrary, describes how the reprints classified into Parts II, III and IV fit together historically. It also examines conceptual connections and relevant historical events; all assertions are documented.

A book of *Selecta* being unavoidably subjective, it is fitting Section 1 should involve a bit of autobiography.

Section 2 concerns Part III of this book, namely, the fluctuations called $1/f$ noises and defined as having a spectral density proportional to f^{-B}. The focus is on empirical observations that began in electrical engineering and very applied physics. The concept of $1/f$ noise is sometimes restricted narrowly, but this book interprets it broadly, as having additional sources in a multitude of empirical observations of diverse kinds that can be interpreted as symptoms of self-affinity in nature.

Section 3 concerns Part II of this book, namely the self-similar unifractal model in nature, as introduced in Berger & M 1963{N6}. A "cartoon" form, bound to a regular cascade, was sketched in Novikov & Stewart 1964, and is also known as "beta-model."

Sections 4 and 5 deal with the topic of Part IV of this book, which is multifractal measures. Their development proceeded largely in the style of theoretical physics or of rigorous mathematics. Yet the notion of multifractal was introduced and defined in several distinct ways.

Without attempting to make this chapter self-contained, Section 4 sketches the basic formalism, together with a number of historical and other side remarks and new wrinkles. A few may prove significant.

Section 5 comments on successive stages of development that led to the discovery of the key functions $\tau(q)$ and $f(\alpha)$. "Proto-multifractals" began during the "prehistory" of fractal geometry, as esoteric curiosities in a subfield of pure mathematics that Bandt 1997 charmingly calls a "Cinderella." Section 5 also incorporates reprints of some brief but telling documents. ✦

SELF-AFFINITY IS THE POWERFUL ELEMENT OF UNITY that binds together the topics of this book and also of M1997E, M1998H, M1998L, and many other works. As introduced in Chapter N1, that term is more or less self-explanatory but apparently did not become needed until M1977F coined it and defined it (p. 276) in a paragraph reproduced in M1982F{FGN}, p. 349. Chapter N1 describes this element of unity and looks forward and sideways to include additional structures. This chapter looks back, and its organization is due to special circumstances.

Firstly, multifractals and $1/f$ noise developed as two separate subjects and their histories carry very distinct "flavors." Secondly, selected papers are usually edited by a person other than the author, and relate and contribute to a pre-existing field of knowledge. In such cases, few words suffice to sketch the historical relations of the reprints to one another and a known environment. This book is atypical. The editor and the author are the same person, and fractal geometry was being built as my old papers were being written (the word, *fractal*, was first used in one of those papers.) For both reasons, and because I am fascinated by the history of ideas, this chapter is longer than originally planned. In particular, many readers will not be interested in all the topics, therefore several are kept more or less separate. That is, for better or worse, many of the repeats in this chapter are intentional. The bulk of this chapter is based exclusively on published and clearly dated documents, but it begins with personal recollections.

Formal analysis versus constructive geometry in the context of multifractals and f^{-B} noise. These two contexts provide the most clearcut, the most often repeated, and the most unexpected vindications of the praise of explicit and visual geometry that is found in this book's Preface. A mechanical use of the analytic formalisms has proven to be *not enough*. The formalisms' geometric meaning *must not* be ignored or disregarded, and it *must not* be left incompletely understood.

My approach generalized the classical binomial measure and involved a recursive geometric construction, a multiplicative random cascade. It automatically generates shapes that are self-affine, i.e., *geometrically* scaling.

To the contrary, the approach followed in some later investigations, including Halsey et al. 1986, is formal and exclusively analytic.

A mismatch between geometry and analysis became critical after the publication of the papers on multifractals reproduced in Part IV. Nevertheless, those readers who are already familiar with the purely analytic multifractal formalism may be puzzled by the apparently pointless complications of my geometric approach. Let me elaborate.

Every random fractal is specified by a number of probability distributions, therefore by a number of functions. A random variable is often summarized by one or more "typical values." The fractal dimension D of a self-similar fractal is a particularly important typical value, but (being trained as a probabilist) I knew well that typical values are not enough. Instead, my approach is based upon probability distributions and their limits, hence, it can be called the *probabilistic method*, or the *distributions method*. It relies upon Harald Cramèr's theory of large deviations. By contrast, the approach in Halsey et al 1986 is entirely based upon moments of distributions, hence can be called the *moments method*.

The Cramèr theory relies on a thermodynamical formalism. But M1995k shows a) that, in due time, the need arose to move well beyond thermodynamics, and b) that I had to acknowledge belatedly that my early papers involved a new notion of dimension.

1. OUTLINE OF THE MOTIVATIONS BEHIND THIS BOOK, TOLD THROUGH THE STORY OF A PERSONAL INVOLVEMENT

The story told in this Section was *not* a purposeless chase from one field to another. Orderly it was not, but it pursued a single-minded line of thought, each step being taken in a substantive field that raised important issues. The opening lines of this chapter refer to fractal geometry as a field that did not exist but was continually being built-up. Two elaborations will be useful.

Sources in esoteric old mathematics. Innumerable mathematical papers are referenced in this chapter and throughout this book and especially Chapters N11 and N17. They prove that I benefited repeatedly from acquaintance with harmonic analysis, probability theory and other established streams of research in pure mathematics. Conversely, my work had an effect on those streams, raising a number of new problems, including very difficult ones. However, fractal geometry does not identify with any previously existing activity.

Relations with physics. Beginning in 1952 with my doctoral dissertation, statistical thermodynamics has been central to my work. This fact is witnessed by the titles of several papers, such as M1956c, 1956w and 1957p, which will be analyzed in Chapter N3. However, the papers reprinted in this book nowhere mention the modern statistical physics that arose around critical phenomena with Kadanoff 1966, Wilson & Fisher 1972 and Wilson 1972. The simple reason is that, by the time that this major subfield arose, my old papers were already written and most were in print. The topic is so important that Chapter N3 is devoted to documenting that the notions of scaling, fixed points and renormalization were routinely used in my work. Of course, I used different names, but close interaction with physicists since the 1980s made me adopt their terminology and "retrofit" it in the forewords to many of this book's chapters.

The fully documented facts mentioned in the preceding paragraph surprise many physicists. Some continue to take it for granted that developments like multifractals arose in modern statistical physics. An unwarranted disbelief also sometimes greets my work on finance (M1997E) and the fractality of the distribution of galaxies (M1998L).

1.1 Summary of the author-editor's principal contributions to the topics of this book

My old papers stressed that the innovations being introduced were "natural," and were otherwise subdued in claims of novelty hence priority. This chapter aims at a higher level of fairness.

Since not every reader will be interested in every topic in this long chapter, this subsection summarizes the principal contributions brought by old papers reprinted in this book.

• I recognized that f^{-B} noise is not a narrow and conceptually unified concept, but a widely encountered analytic property. That is, a f^{-B} spectrum occurs in many geometric constructions that must absolutely be distinguished from the viewpoints of both mathematics and physics. They must also be studied separately.

• To handle the f^{-B} noises, I showed the central importance of an exponent H that merged the intellectual inheritance of the mathematician Hölder and the hydrologist Hurst.

• Also, to handle the "infrared catastrophe" for certain f^{-B} noises, I developed several distinct analytic tools which include conditional stationarity (M1965c{N7}), conditional spectral analysis and sporadic processes (M1967b{N10}).

• The two papers reprinted in Part II of this book introduced the unifractal model, by using a randomized Cantor dust in *real* space as tool of physics. To my knowledge, this step had never been taken before. That model had no prehistory within the sciences. To the contrary, mathematically "pathological" sets in phase space had a prehistory in the "protochaos" discovered by H. Poincaré and J. Hadamard in dynamics. And the prehistory of the Cantor dust within mathematics is a well-known topic touched upon in the text and the references of M1982F{*FGN*}.

• The unifractal model described in Berger & M {N6} was the first to interpret the previously esoteric notion of fractal dimension as a fundamental physical quantity. The papers that introduced multifractals (to be mentioned momentarily) also included and investigated from the fractal viewpoint a construction later relabeled "beta-model." Experiments inspired by Berger & M 1963 started me towards multifractals.

• Variants of fractal dimension keep multiplying but I adopted the conservative viewpoint that it is best to restrict dimension to sets and not to generalize it to measures.

• Credit for explicitly introducing the notion of multifractal measure belongs to several papers reprinted in this book, M1972j{N14}, M1974f{N15}, M1974c{N16}, and M1976o{N18}. Indeed, one finds in those papers the first mathematical statement of the underlying mathematical problem. That statement was not quite written in a standard "definition-theorem-proof" style. Nevertheless, it was complete, correct and phrased in a conceptual context (measure-valued martingales) that incorporates the major themes. Moreover, those works include significant theorems and mathematical conjectures that generalize those theorems. The conjectures guided later developments, beginning with Kahane & Peyrière 1976{N17}.

• At a time when computer graphics were in infancy, M1972j{N14} provided the first illustrations of multifractals in the form of coarse-grained measures. These illustrations long remained the only ones to be available; one of them is reprinted on this book's jacket and cover. Their preparation was in the spirit of my attitude towards geometry. I had felt that it was not enough to provide an analytic derivation of the subtle conditions under which a multiplicative cascade becomes degenerate. That transition should also be understood visually.

Ten years later, Plates 198 and 199 of M1982F{*FGN*} seem to have been the first illustrations of multifractals in the form of a cloud of points in a discrete orbit. The generators of such constructions later came to be called "iterated function systems" (IFS), as described in Section 7 of Chapter E5. A recent rendering of Plate 199 is on this book's jacket and cover.

• Multifractals involve a formalism based on the Legendre transform. In M1974c{N16}, the quantity now denoted by $-1 + f(\alpha)$ arose for the first time, as documented photographically in Section 5.5. It entered via the Cramèr theory of large deviations, as the limit of a suitably rescaled sequence of probability distributions. Unfortunately, this correct derivation of $f(\alpha)$ was not written in a style that makes the results immediately usable. In fact, it occurred in a paragraph that also included a wrong *Ansatz*. More importantly, this first derivation was not stated forcefully and was not repeated and refined until the 1980s, for example in M1989g.

This last paper, and related ones that respond to alternative and later approaches to multifractals, are prepared for reprinting in M1998L. The long lag of publication between 1976 and 1989 has a straight explanation (if not a good reason), that will be described in Section 1.7.

The random multifractals in M1974f{N15} are "cartoons" created by cascades. They take two forms: microcanonical (or conservative) and canonical. The former is in "competition" with alternative approaches. The latter is not, remains much less well-known, and is expected to become important. Two of its features deserve mention.

• The distribution of a canonical multifractal measure often involves infinite population moments beyond a critical exponent q_{crit}, which was denoted by α in the originals of M1974f{N15} and M1974c{N16}. See Section 5.6.

• The canonical multifractals are exactly renormalizable measures (see Chapter N3.) This feature led to the important concept of fixed point under *randomly* weighted averaging and induced an important functional equation that generalizes the concept of Lévy stability (See Section 2.5 of Chapter N3.) In the 1980's, L. Pietronero, S. Redner and others independent thinkers noted that all multifractals are little more than exactly renormalizable multiplicative measures. They were surprised when shown the words "random multiplicative iterations" in the title of M1974c{N16}.

By contrast, the limit lognormal multifractals in M1972j{N14} are *not* created by cascades and are *not* exactly renormalizable. However, even in contexts where cascades cease to be acceptable as approximations, the canonical random multfractals continue to be significant because they identify behaviors that characterize multifractality, but were missed in formal analytic studies.

Background. Not contradicting the claim made earlier in this section, multifractals had a prehistory and an important early history that was dominated by Kolmogorov 1962.

Follow-up. This chapter's bibliography contains few references beyond the first stages of the theory, in which very few persons were involved. However, this chapter will refer to two important and very different later developments.

• Section 1.7 mentions, with references, several innovations, including *negative dimension.*

• There will be many references throughout to two very well-known and important later publications, namely, Frisch & Parisi 1985 and Halsey et al 1986. Both greatly contributed to the popularity of multifractals and there are two reasons for discussing them.

A practical reason is that many readers familiar with those two publications may otherwise feel lost in the very different conceptual environment presented in this volume.

A durable reason is that the alternative approaches to multifractals prove to differ fundamentally. Events have shown that my constructive approach to multifractals is the best pedagogically. It is also most naturally extended from measures to oscillating multifractal functions, as is done in Chapter N1 and more broadly in M1997E and M, Calvet & Fisher 1997. Finally, only my approach is of sufficient generality to accommodate randomness and diverse features present in the most important applications, such as negative dimension. In practice, it follows that the notion of multifractal can be examined on two levels: a narrow one and a broad one; broad multifractals are both natural and essential, but from the narrow viewpoint they appear as "atypical" or "anomalous."

The remainder of this section elaborates on the preceding brief comments, following the chronological flow of events.

1.2 Finance as prelude to a personal involvement in physics

My life-time study of self-affine variability began before 1960, with papers in finance that are reproduced in the second half of M1997E and might even suggest new thoughts to the physicist or the engineer.

In M1962c (photographically reproduced in M1997FE) and M1963b{E14}, the key claim is that the essential role of a *Bourse* is to manage the discontinuity that is natural to financial markets. M1963b{E14} showed that several persuasive-sounding get-rich-quick schemes deliver "buy" or "sell" orders at levels that the actual price crosses without stopping. Shooting down a financial "perpetual motion" is a "negative" result, but we do praise thermodynamics for having shot down a host of *physical* "perpetual motions."

The M1963 model reduces to a one-line formula, yet its sample time series invariably exhibit an extraordinary wealth of apparently meaningful structure. Professionals could not distinguish the samples from actual price series and forecasters had a heyday. But no forecasting is possible in that model, because its generating mechanism is simply a "L-stable random process with independent increments." M1975O coined for it the colorful term, "Lévy flight." In 1962, the power of simple formulas to generate such rich patterns was a novelty.

The M1963 model assumed a) that price increments are independent, and b) that, except for a change of scale, the "renormalized" price increments over days, months and years follow the same distribution. That distribution is a fixed point under averaging. Around 1960 no one else was active in exploring this line of thought. This issue is carefully discussed in M1997E, Chapter E3, and sketched in Chapter N3.

1.3 Collaboration with J. M. Berger on noises in telephone lines

In 1962, Harvard invited me to teach a course in Economics based on M1962i. A colleague who attended a farewell lecture at IBM, Jay M. Berger, asked for help in dealing with computers, telephone lines and the like. His field was as murky as economics and lesser in a grand view of things. But it greatly affected IBM technology. The empirical tests of renormalizability that I suggested yielded positive results that led to Berger & M 1963{N6} followed by M1965c{N7}. The direction of my research, hence my whole life, was changed by this collaboration.

1.4 The year 1963-4: the Nile, fractional Brownian motion and 1/*f* noises

After a year teaching economics, I stayed on at Harvard in the applied sciences division. Expanding on unchanging views and ambitions, I set out to explain the "Hurst puzzle" concerning river discharges (Section 2.3 of this chapter.) The M1965 model assumed that discharges are Gaussian but non-independent and that their correlation function is another fixed point under averaging. It characterizes a process I called fractional Gaussian noise, the sequence of increments of fractional Brownian motion (FBM). The M1965 model of persistence is introduced in M1965h{H} and described in other papers reprinted in M1997H, many of them co-authored by J.W. Van Ness or J.R. Wallis. Its applications in finance are discussed in M1997E.

In my models, the telephone errors and the Nile floods both involved spectra of the form f^{-B}, but this common property seemed incidental and inconsequential, since in all other regards those processes were of totally different character. That is, a common spectrum did not imply any deeper commonality. At that point, D. Tufts heard me talk and reported the rumor that an $1/f$ spectrum also characterizes certain electronic noises that hardly anyone knew at the time. This led to the papers in Part III.

1.5 The year 1963-4: intermittent turbulence

The same year at Harvard, R. W. Stewart described observations concerning turbulent intermittence. Having studied turbulence at Caltech in the mid 1940's, I recognized instantly that the techniques used in Berger & M 1963{N6} (*fractal* techniques, without the term *fractal*) also concerned the phenomenom of intermittence, and provided a starting point for a new assault on turbulence.

Specific models of turbulent intermittence were found in Obukhov 1962 and Kolmogorov 1962 (*O & K*), and in the draft of Novikov & Stewart 1964 (*N & S*). This last paper advanced a "toy model" identical to the randomized Cantor set that Berger and I had considered before moving on to the grid-free model described in Berger & M 1963{N6}. Of course, our work was not known to *N & S*. (Had it been known, *N & S* would have been a throwback.) It took me minutes to rephrase and amplify Novikov & Stewart 1964 into the model, now called unifractal, of grid-bound, fractally homogeneous turbulence.

The first statement was in M1964i, an IBM External Report, of which two fragments are included in this book, in Section N5-1 and the *Foreword* of Chapter N10. There are echoes of it in Kahane & M1965{N11}. But the earliest formal announcement was M1967k{N12}.

1.6 The years 1963-1976: multifractals

Berger & M{N6}, presented a model of errors in which no physical activity occurred between the errors. This model's success mystified everyone. Later, while writing M1967k{N12}, I was mystified by the conflict between, on the one hand, *O&K* and, on the other hand, Berger & M1963{N6} and *N&S*. The mystery was solved by 1968 and reported in M1969b{N13}. I had found the Kolmogorov model to be deeply flawed, and its flaws spurred me to develop the singular measures now called *multifractal*.

Once again, a constructive and rigorous approach to multifractals as physical models was developed to the physicist's standard of rigor in M1972j{N14}, M1974f{N15}, M1974c{N16} and M1976o{N18}. The 1974 papers also provided a mathematical setting in terms of measure-valued martingales. They raised a large number of new questions and included theorems with conjectured generalizations. Several of the short pieces reprinted in Chapter N5 are excerpts from early surveys of this work.

Few of my papers match M1974f{N15} in either complication or importance. The function $\tau(q)$, that it uses is a classical notion. But it contains most of the essential facts about two structures that are central to this book, and became basic to fractal geometry under later labels. The term *beta-model,* due to Frisch, Sulem & Nelkin 1978, is discussed in Section 3. Section 5.5 documents how Frisch & Parisi 1985 chose the excellent term *multifractal* to denote processes that are effectively equivalent to my fixed-point multifractals. M1974f{N15} was published with extraordinary slowness (as mentioned in the *Annotations* of Chapter 15), and its direct influence, as well as that of the short expository pieces in Chapter N5 was largely restricted to three persons.

The first was Jean-Pierre Kahane, who spread the word to Jacques Peyrière, leading to Kahane & Peyrière 1976{N17}. The mathematics of my approach to multifractals moved to a high level and continued to develop with Durrett & Liggett 1983, Ben Nasr 1986, 1987, Guivarc'h 1987, 1990, and many works since.

Uriel Frisch became familiar with fractals by commenting upon early versions of two works in very disparate styles. He thought correctly (as I did) that M1975O would be well-received, but M1974f would have few readers in its time. Frisch spread the word to Giorgio Parisi and beyond.

The third person (many years later) was my then-post-doctoral student S. Lovejoy, who spread the word to his circle of meteorologists.

1.7 A personal involvement interrupted for twelve years, then resumed

M1974f{N15} fitted in none of the established scientific disciplines. Subtle arguments concerning *singular measures* had little chance of being understood by physicists until after the simplest fractal *sets* (like those occurring in Section 1.3) had become thoroughly explored, made widely understood, and accepted as "natural."

Those goals made it necessary to construct a fractal geometry. The advocacy of its use in science relied on a side trip through the study of the neutral example of coastlines. Early on, most of my papers began by raising a question that was by no means considered natural, for example, "How long is the coast of Britain?" By 1982, however, fractal geometry had made this question appear quite obvious. I shall explain this side trip when M1967s is reprinted in a forthcoming volume of these *Selecta*. These tasks occupied all my attention; as a result, beyond the surveys sampled in Chapter N5, I wrote nothing on multifractals from 1976 to 1988.

After years of commenting from the sidelines, Frisch & Parisi and Halsey et al 1986 made me return to multifractals and bring to fuller fruition the project that had occupied me from 1968 to 1976. The resulting papers, scheduled for reprinting in M1998L, responded to several needs

A) *The need for a more developed and systematic presentation of the approach I pioneered in 1974.* As of now, the most complete statement is M1989g, which appeared in an overly specialized periodical found in few libraries. It will be revised and reprinted in M1998L.

B) *"Heuristic dimension" and its negative values.* In the 1970s I was not aware that my work was bringing yet another split among the distinct variants of the notions of "H" exponent and dimension. In M1972j{N14} and M1974f{N15}, diverse arguments that were informal but rigorous led to a quantity one might call today a "heuristic dimension," D_{heur}. I conjectured, and Kahane & Peyrière 1976{N17} proved that when $D_{heur} > 0$ its value is equal to the Hausdorff-Besicovitch dimension D_{HB}.

But some cascades, for example those using lognormal multiplying weights, are peculiar: they allow D_{heur}, and even a generalization of the Hölder exponent α, to be negative. Those cases were without interpretation in existing mathematics. It was important to fully understand those behaviors and illustrate them by examples that are simple, explicit, and require no reference to subtle probability theory.

Their long investigation was reported in M1984e, M1988c, M1989e, M1989g, M1990d, M1990r, M1990t, and M1991k; the latest stage is (over concisely) described in M1995k. To summarize all those works (scheduled for inclusion in M1998L), I concluded that the above-mentioned expression D_{heur} should be viewed in *all* cases as a different and independent kind of fractal dimension. M1995k denoted this expression by D_{elna}, the initials of "entropy-like, but non-averaged." It deserves to be better understood on its own. The key novelty is that D_{elna} is not attached to a set, but to a constructive algorithm that leads to a set.

1.8 Credit due for the multifractals' acceptance in the mainstream

When writing M1982F{*FGN*}, I felt that multifractals' time had not come and restricted their discussion to a short entry in Chapters 39, titled "non-lacunar fractals." Then, almost overnight, fractal measures blossomed in many hands and under many names. Intuitive ideas that underlie the mathematical work of Besicovitch and his school were rediscovered and presented in publications too numerous to cite. I shall mention Farmer, Ott & Yorke 1983 and Amitrano, Coniglio & Di Liberto 1986 and comment on two already mentioned very influential papers.

The excellent term *multifractal* was first used in an excellent heuristic work, Frisch & Parisi 1985 (an excerpt is reproduced in Section 5.5.) It follows directly from, and refers extensively to, my papers of the 1970s.

Furthermore, another excellent heuristic paper, Halsey, Jensen, Kadanoff, Procaccia & Shraiman 1986 received wide attention. We read in Kadanoff 1993, p. 388 that Halsey et al 1986 "is in some ways an embarrassment to me. It is an excellent expository work which in fact helped make a particular form of fractal behavior very fashionable. The focus of analysis... follows from the work of Mandelbrot in 1974."

Leo Kadanoff deserves praise for this prominent and strong statement. In turn, I hasten to acknowledge that widespread interest in multifractals was unquestionably triggered and nourished by Halsey & al 1986. For this reason, I adopted their notation and arranged for the reprints of old papers to be edited accordingly.

The written-out *algorithms* in Halsey et al 1986 were easy to program and could be used mechanically. They led to legitimate uses in various non-random environments, such as dynamical systems. But in all too many instances, multifractal manipulations yielded incomplete understanding, and/or led to error and confusion. By contrast, the theory of random multifractals becomes perfectly clear when studied against the background of my original approach (as I repeatedly urged it should be). While I prefer to see my work acknowledged when appropriate, the important issue was and remains substantive. Everything comes out best when one begins with fully specified geometric scaling.

2. THE PERVASIVE BUT SHADOWY 1/f NOISES; THE FRACTAL NOTION OF GEOMETRIC SELF-AFFINITY

This section recalls that a $1/f$ noise is defined by its spectrum and argues that an understanding of such a noise demands more than the spectrum.

Geometric facets are essential, those facets are fractal, and fractal geometry goes well beyond expressing the spectra and correlation functions in different language.

2.1. Definition of 1/f noises; absence of a general explanation

A sequence of measurements or numbers ordered in time can be transformed into a sequence of Fourier coefficients. Plot the squared Fourier moduli as function of the frequency f. In nature, this function is often the product of a numerical prefactor F and the function f^{-B}, where the exponent B is a constant that varies from case to case. In early references, B was usually close to 1, hence the term $1/f$ noises.

Formally integrating a f^{-B} noise transforms it into a f^{-B-2} noise and differenciating, into a f^{-B+2} noise. In the Gaussian case, the range between -1 and 3 includes both stationary stochastic processes ($1 < B < 3$) and non-stationary processes with stationary increments ($-1 < B < 1$).

Physics explained very rapidly the origin and properties of "shot noise," which is a Gauss-Markov (Uhlenbeck-Ornstein) process, with the spectral density $1/(1+f^2)$. This explanation created the hope that one could similarly explain all f^{-B} noises, at least for some values of B. But this effort failed. The link with self-affinity described in Sections 1.5 and 1.6, and the extraordinary diversity of self-affinity exemplified in Chapter N1 make me feel that a unique model was unrealistic all along.

2.2 Renormalizability

An informal "renormalization" argument is useful. In an f^{-B} noise $X(t)$ for which $B > 1$, the cumulative spectral energy in frequencies $>f$ is f^{1-B}. Observe $X(t)$ across the computer screen at a speed faster or slower in the ratio $p > 0$. This transformation will move energy around, changing the tail energy in frequencies $>f$ from f^{1-B} to $p^{1-B} f^{1-B}$. It suffices to turn down the display control knob to achieve a gain in amplitude in the ratio $p^{-(1-B)/2}$, that is, a gain in energy in the ratio $p^{-(1-B)}$. After both the speed and the gain have been changed, the tail spectral energy in frequencies $>f$ will return to its original form f^{1-B}. When the noise is Gaussian,

$B = 2H - 1$, where H is a basic parameter; hence the linear transformation ratios are p for time, p^{-2H} for energy, and p^{-H} for amplitude.

Letting the Wiener Brownian motion run faster or slower is a well-defined procedure. By contrast, how does one change the speed of white noise? An analog or digital plotting routine meant to observe a white noise incorporates a high-pass filter to eliminate high frequencies. Speeding up demands decreased filtering, and the heuristic argument, if tightened up, ends up by not applying to the white noise but its integral, which is the Wiener Brownian motion.

2.3 The multiplicity of direct or indirect analytic symptoms of 1/f

Warning against the "Spanish moss" phenomenon. The experimental observations of $1/f$ noise near-always begin by plots on doubly logarithmic coordinates. An effect is to collapse the high frequencies and stretch out the low spectral components. When one expects a linear doubly logarithmic plot and the spectral components are not averaged, the effect is dramatic. As f increases, the natural sampling variability becomes increasingly strong. The uninformed user may expect mild fluctuations around the expected straight line, but in fact will observe a broadening fan that a wit compared to the "Spanish moss" hanging from tree branches in wet tropical climates. Publications are filled with complaints that a straight line drawn through such a mess is unreliable and even meaningless.

In fact, the fault is not with nature but with the chosen method of plotting. Spectral components must never be plotted raw, only after suitable averaging. The best procedure is to divide the high frequency portion of the scale of log (frequency) into equal bins and average the squared Fourier components in each bin. Plotting those averages dissolves the Spanish moss and the evaluation of the slope B becomes easy.

Moreover, the evaluation of B becomes reliable and comparable to the best analytic algorithms. Once again, geometry must not be sold short, but must be handled carefully. The evidence to be examined now is blameless in that respect.

Direct evidence of Fourier power spectra proportional to f^{-B}, in electronics and elsewhere. In the 1920s and the 1930s, $1/f$ noises were encountered by several physicists and electrical engineers who worked more or less independently on seemingly distinct phenomena. The earliest examples were found in vacuum tubes (Johnson 1925), thin films (Bernamont 1934-1937) and carbon resistors (Meyer & Thiede 1935, Christensen & Pearson 1936).

Less ancient references in physics concern semiconducting devices (van der Ziel 1970), continuous or discontinuous metal films (Hooge & Hoppenbrouwers 1969, Voss & Clarke 1975, Williams & Burdett 1969), ionic solutions (Hooge 1970), films at the superconducting transition (Clarke & Hsiang 1976), and Josephson junctions (Clarke & Hawkins 1976). A recent reference with an extensive bibliography is Kogan 1996.

$1/f$ spectra in music and medicine. $1/f$ noises characterize music (Voss & Clarke, 1975, 1978). They were recognized in nerve membranes (Verveen & Derkson 1968, Caloyannides 1974). Chronic illness turns out to have a $1/f$ spectrum (Campbell & Jones 1972). This striking discovery is due to a physicist who had severe diabetes, monitored his own vital signs and evaluated their spectrum. His finding disagrees with the common notion that such illnesses can be described as alternating between remissions and acute periods.

Indirect evidence from the "power law" decay of perturbations. Such a decay is closely related to $1/f$ noise. Page 417 of M1982F{FGN} is devoted to *Scaling: old empirical evidence.* It reports on evidence of perturbations whose decay follows a power law, while the universal expectation (even in the mid 19th century, when those perturbations were first examined) was that the decay should be exponential. Page 418 of M1982F{FGN}, devoted to *Scaling: durable ancient panaceas*, reports an old, all-purpose "explanation" that attempts to reduce the power law to a sum of either a few, or many, exponentials. This all-purpose explanation is endlessly rediscovered, submitted to journals, and sometimes even published.

Indirect evidence in Berger & M 1963{N6} concerning inter-event intervals in communication errors. The work in Chapter N6 was performed with no awareness of the electric $1/f$ noises, but clearly involved a $1/f$ spectrum.

Indirect evidence from R/S plots proportional to δ^H. Harold Edwin Hurst (1880-1978), a civil engineer whose story is told in Chapter 20 of M1982F, introduced a quantity I later called R/S, more precisely $R(\delta)/S(\delta)$. He used R/S to analyze the variations of yearly discharges of the Nile River and other hydrological and related records and near-invariably obtained $R(\delta)/S(\delta) \sim \delta^H$, where the exponent H satisfies $1/2 < H < 1$. The details cannot be stated here, but – as Hurst knew, and probabilists led by W. Feller proved rigorously – uncorrelated Gaussian discharges lead to $H = 1/2$. Therefore, the odd but incontrovertible findings in Hurst 1951, 1955 became known as a "puzzle" or "paradox." M1965h{H} "tamed" the puzzle by showing how Hurst's observations can be accounted for. In current terminology, those observations provide indirect evidence that the

Nile's yearly discharges are self-affine, namely, a $1/f^B$ noise with $B = 2H - 1$.

M & Wallis 1969b{H} also showed that $1/f$ noises rule many geophysical and planetary phenomena not related to water. The "secular" low frequency component of sunspot numbers is a striking example confirmed in other publications.

Indirect evidence from "anomalous" (non-Fickian) diffusion. Adolf Eugen Fick (1829-1901) was a physician and physiologist in Würzburg, Germany who discovered the law of diffusion in liquids. His name survives in terminology: *Fickian* became the (awkward) term to denote phenomena that diffuse like $\Delta X \propto \sqrt{\Delta t}$, and *non-Fickian*, to denote phenomena that diffuse like $\Delta X \propto (\Delta t)^H$, where $H \neq 1/2$. This ΔX is often equivalent to R; if so, the evidence of diffusion is equivalent to the evidence of R/S, therefore, the exponent is written as H in both cases. Fickian diffusion is closely related to white noise, and non-Fickian diffusion, to $1/f$ noise. Many characteristics of turbulence and meteorology are non-Fickian.

Self-affinity and the scaling laws of turbulence; the correlation of intermittent dissipation. Kolmogorov's $-5/3$ spectrum is an example of spatial $1/f$ noise. It assumes homogeneity, that is, uniformily distributed dissipation. In fact, Part IV centers on the fact that turbulent dissipation is not homogeneous and its correlation is proportional to d^{-Q}, where Q is a constant. The corresponding spectrum gives another example of $1/f$ behavior. This correlation plays a fundamental role in the theory of multifractals, Q being sometimes called "correlation dimension" (see Section 5.6).

Minerals. There is a reference to deWijs in Section 5.4.

Proto $-1/f$ noises in mathematics, Section 5.3. The fundamental "counter-examples" of 1900 mathematical analysis include the continuous but non-differentiable function of Weierstrass and the Cantor devil staircase, which also originally related to trigonometric series. Their construction was motivated by the needs of mathematics rather than physical observation and Section 5.3 notes that the overall shape of their spectrum did not attract attention. To my mind, however, both can be fairly said to involve $1/f$ noises. We see yet another deep relation between physics (represented by an "unrefined" notion, a noise) and mathematics (represented by questions that were not only very "pure," but esoteric). As is often the case (and fractal geometry brought more than its share of examples), this mutual relation was beneficial to both sides. The benefit to mathematics started with Kahane & M1965{N11}, to be previewed in Section 5.3.

2.4 The infrared catastrophe shown to be a mirage

If extrapolated formally from $f=0$ to $f=\infty$, a spectral density $\sim f^{-B}$ leads to paradox: it implies that energy is infinite in high frequencies when $B \leq 1$, and in low frequencies when $B \geq 1$. Those divergences are, respectively, called *ultraviolet* (UV) and *infrared* (IR) *catastrophes*.

• *The case $B \leq 1$; high-frequency divergence and its classical resolution by coarse-graining.* The UV "catastrophe" is known to disappear under local smoothing or coarse-graining, which are inevitable in physics and well-understood in mathematics. In physics, "derivative" really means "finite differential over an arbitrary small time increment ε." In mathematics, "derivative" means "Schwartz distribution." Smoothing is standard; the Wiener-Brownian motion $B(t)$ has no derivative, but one can study its increments over uniform time increments $B(t + ε) - B(t)$. continuous values of t; the resulting spectral density is defined for all f and is The "generalized derivative" $B'(t)$ is "white Gaussian noise," and its spectral density is f^{-0} for all f.

The infrared catastrophe. Divergence near $f=0$ is a far more serious issue and challenge impossible conclusions from competent extrapolation from unquestionable data raise basic questions. The square moduli of the Fourier coefficients of a data set are expected to open "windows" on the process that generated the data, with the help of a theory due to Khinchin (in the case of random functions) and to Wiener (in the case of functions that are not random but behave as if they were.) But the Wiener-Khinchin theory demands an integrable spectral density near $f=0$ therefore fails for $1/f$ noises. Its failure can be met in diverse ways.

Experimentalists expect the f^{-B} behavior to fail below a positive lower cutoff frequency (as well as above the finite upper cutoff.) The troubling surprise is that many important cases exhibit no accessible cutoffs.

As presented in this book, my reaction was to seek a more general theory that explains how a sample spectral density could seem to take the form f^{-B} near $f=0$. The Fourier components have an intrinsic meaning for light, sound, and the like, and can be physically separated from one another. But in many other cases, they can be evaluated but not separated intrinsically. Thus, the question "what is the *real* meaning of the value yielded by a spectral estimator?" is always legitimate. In the cases when the Wiener-Khinchin theory leads to infrared catastophe paradox, this question beomes unavoidable hence plays a central role.

• *The case when $B \geq 1$ and a process is Gaussian; the low-frequency divergence is an unavoidable token of non-stationarity.* Recall the definition of $B(t)$:

$E[B(t+s) - B(t)] = 0$, and $E[B(t+s) - B(t)]^2 = s$ and $B(t+s) - B(s)$ is Gaussian. The Wiener-Khinchin notion of spectrum fails to apply to the Wiener Brownian motion $B(t)$, but a generalized spectrum does apply, and shows that $B(t)$ is a $1/f^B$ noise with $B = 2$, as well as an ordinary function. The cause of the infrared catastrophe is that $B(t)$ is nonstationary.

• *The case when $B \geq 1$ and the process is non-Gaussian; low frequency divergence may be a mirage that is resolved by introducing a new, conditional form of the Fourier spectrum.* In this case, the situation is altogether different. Two of the main points of several chapters in this book are as follows. A) Gaussianity is a very subtle notion (See Chapter N4). B) In an environment that is not exactly Gaussian, a numerical spectrum found by experimentalists to be proportional to $1/f^B$ need *not* be an estimate of an underlying Wiener-Khinchin spectrum. It may well be a more generally valid expression that is described in M1967b{N10} and called *conditional spectrum*. In addition to f, a conditional spectral density $S(f)$ depends on a conditioning length T and takes the form $TG(Tf)$, where $0 < G(0) < \infty$, but $G(f) \sim f^{-B}$ for $f \to \infty$.

One may think of $G(fT)$ as being roughly of the form $T[1 + (fT)^B]^{-1}$. This expression is coined in analogy with $\tau[1 + (f\tau)^2]^{-1}$, which is the Wiener-Khinchin spectral density of "shot noise" (Markov-Gauss or Ornstein-Uhlenbeck process). But there is a fundamental difference. In the case of shot noise, the time τ is an intrinsic physical time factor. In my generalization, the role usually played by τ is played by the non-intrinsic sample length T.

The infrared mirage. In a finite sample, the accessible frequencies do not range down $t0$, only down to $f = 1/T$. Therefore, an effect of letting $T \to \infty$, is to modify this the accessible range, and force an unchanging total energy to "flow along" toward increasingly low frequencies.

As a result, the threatened low frequency divergence or infrared catastrophe *never materializes* and the self-consistency of nature is preserved. However, the interpretation of spectra is deeply affected. The fact that the additional prefactor T^{1-B} is not numerical but a function of T expresses that the measured square Fourier moduli do *not* estimate a Wiener-Khinchin spectral density, but something *different*. Thus, the differences in geometry have obvious practical consequences that one could not deduce from the form of the spectrum alone.

2.5 Towards geometry: the evidence of the eye and the ear

The term "noise" has long become metaphorical, but every recorded fluctuation can be processed so that it is actually heard, and – even more often – actually seen on an oscilloscope or computer screen. In addition to quantitative criteria, it is essential to remind sober scientists not to despise the evidence of their senses.

The evidence of the ear. Early authors described some $1/f$ noises as *flickering, popping, or frying,* and other noises as *unsteady or variable.* These distinctions happen to be meaningful. A noise's large energy at very low frequencies is evaluated correctly by a spectral algorithm, but not by the ear.

Part III studies many noises that are "dustborne," that is, vary when time belongs to a Cantor or Lévy dust. In that case, the low frequency energy comes entirely from the "intermissions" during which a random function $X(t)$ is constant. As seen in M1967i{N9}, the probability distribution of the intermission lengths is reflected by the f^{-B} spectrum near $f = 0$. However, the ear does not hear what happens *during* the long intermissions, only what happens *between* them. That is, the ear only registers *high* frequency phenomena. This explains the terms *popping* and *frying* applied to some $1/f$ noises.

In other examples, the high power in low frequencies manifests itself as slow variability in a high frequency noise.

The evidence of the eye. Like the ear, the eye is very sensitive to features that the spectrum does not reflect. Seen side by side, different $1/f$ noises, Gaussian, dustborne and multifractal, obviously differ from one another. The analogies revealed by a common $1/f$ spectrum are far from obvious to the eye. Better still: watch a $1/f$ noise on an oscilloscope or a computer screen. Such output watching is commonplace outside the "hard science.", and the financial economists' notion of "market volatility" is ill-defined but useful. Volatility can be described as being near-constant for Gaussian noises but extraordinarily variable for multifractals and dustborne noises. It is also highly variable in turbulent *intermittence.* Very variable volatility was the reason why the already mentioned old-fashioned but careful engineers reported on *flickering, popping* and *frying noises.*

2.6 The renormalizability of 1/f noises is an analytic property; it very strongly suggests an underlying geometric property called "self-affinity"

By now, everyone knows that a set S is called self-similar if $S = \sum S_k$, where the part S_k is obtained from the whole S by a linear reduction having the same ratio in every direction, and the S_k do not overlap. Self-affinity is a more general concept and it is useful to repeat a definition given in Chapter N1.

The graph of a function $X(t)$ is a geometric object. When it is invariant under a combination of distinct linear transformations, each involving one coordinate, this graph is called *diagonally self-affine* with respect to the fixed points of the transformations. When the fixed point can be chosen freely, the object is simply called *diagonally self-affine*.

Using these terms, the heuristic argument in Section 1.2 yields the following conclusion, which is one of the main points of this book. In order for a noise to be f^{-B}, it is *sufficient* that a) the second moments implicit in the valuation of spectra behave nicely, and b) the noise possess the property of diagonal self-affinity. Less demanding forms of self-affinity also suffice, but cannot be described here.

Can a $1/f$ noise fail to be self-affine? Not having encountered any example, I use the terms $1/f$ *noise* and *self-affine* as synonymous. On occasion, this usage extends to apply the term $1/f$ noise to fluctuations whose variance is infinite, hence no spectrum can be defined.

2.7 The notion of self-affinity, hence that of " $1/f$ noise," is grossly under-specified from the viewpoints of both mathematics and physics

Two nearly synonymous words may cast totally different lights. The analytic term $1/f$ *noise* is misleadingly specific, while the geometric term *self-affinity* is realistically general. Today, everyone is aware of the immense variety of the self-similar fractals; being more general, self-affinity may be compatible with even more immense variety.

Two conclusions come to mind. Reducing the notion of "$1/f$ noise" to self-affinity does not make it any less interesting, but shows it to be very severely under-specified. One cannot understand a $1/f$ noise without searching beyond its spectrum and the search for a single explanation becomes hopeless. Secondly, the new tools that are needed are not to be found in the literature. Chapter N4 sketches a broad analytic method of discrimination.

A mantra The study of $1/f$ noise, like the study of multifractals, provides clearcut, repeated, and unexpected vindication of the view presented in the Preface in a section *In Praise of Explicit and Visual Geometry*. An analytic relationship is *never enough*.

3. ABSOLUTE CURDLING AND THE UNIFRACTAL MODEL

In this Section, "unifractal" will denote something like a randomized Cantor dust.

3.1 From random Cantor dusts to Lévy dusts: Berger & M 1963{N6} and Novikov & Stewart 1964

Berger & M 1963{N6} faced data that exhibit very strong hierarchical clustering at a large number of superposed levels. The published model is a discretized Lévy dust, but the authors first considered a model based on a Cantor dust.

As is well-known, a Cantor dust is strongly hierarchical by construction. Its generating process consists in a regular cascade that operates in the recursive grid obtained by dividing the interval [0, 1] in b pieces and preserving N organized in a fixed pattern. The next thought was of a Cantor dust constructed in the same lattice, but with a randomized pattern of pieces. The inputs N and b are not present in more realistic models, nor in reality. They are not expected to be measurable separately, only through the similarity dimension

$$D = \log N / \log b.$$

The Lévy dust involves no counter-physical grid and cascade, hence N and b are not needed. There, the sole parameter D is an easily measured concrete quantity. It is not a similarity dimension but a mass dimension. That is, an interval of length R centered on the dust contains a mass of the order of R^D.

Compared to the Lévy dusts, the Cantor dusts are easier to define but in many ways harder to study. For example, the cartoon counterparts of the results of M1965c{N7} and M1967i{N9} would overflow with tedious and unnatural complications.

Be that as it may, the randomized Cantor cascade in Novikov & Stewart 1964 (a model of the intermittence of turbulence) arose quite independently of Berger & M 1963. It did not proceed to interpret D as a fractal dimension.

3.2 Rectilinear "determining functions" $\tau(q)$ and absolute curdling

M1974f{N15} and M1976o{N18} investigated the random Cantor dusts in greater detail, as a special limit case of general multifractality. Section 4.3 of M1974f{N15} characterizes this case drily as having a "rectilinear determining function" (the term used in that paper for the function now denoted by $-\tau(q)$). M1976o{N18} is more colorful: it describes random Cantor dusts as generated by "absolute curdling," and the more general multifractals in the bulk of Part IV as generated by "relative curdling".

3.3 The regrettable label "beta-model"

The random Cantor cascade benefited from a skillful further investigation in Frisch, Sulem & Nelkin 1978; in this subsection, let that paper be called FSN. I am grateful to one of the authors for an aside in his review of M1982F{FGN}; indeed Nelkin 1984 took the trouble of observing that FSN "made Mandelbrot's idea accessible to a wider audience ... added nothing new, but only explained what he had done."

FSN also relabelled an already familiar cascade as "beta-model". This label became accepted, despite objections that deserve to be restated. In E-dimensional Euclidean space, $\beta = N/b^E = b^{D-E}$ combines two meaningless quantities, and does not enter explicitly in any physics; not even in the formulas in Frisch & Parisi 1985 which are reproduced in Section 5.5.2.

Hence β does not deserve any prominence and is misleading when used to identify the unifractal model. Higher levels of mislabeling are reached by "generalizing the β-model."

4. THE MULTIFRACTAL FORMALISM, WITH COMMENTS AND A FEW NEW WRINKLES

4.1 Notions of self-affinity and ratio-self-similarity for measures

The notions of self-similarity and self-affinity become unavoidably more complicated when applied to measures rather than sets. The complication is best described on the basis of the binomial measure.

The measure or mass in an interval $[t_1, t_2]$ shall be denoted by $\mu([t_1, t_2])$ or $\mu(t_1, t_2)$. Divide the interval $[0,1]$ of t recursively by binary splits and select p satisfying $0 < p < 1/2$. The mass $\mu(0, 1)$ is subdivided unevenly into $\mu(0, 1/2) = p\mu(0, 1)$ and $\mu(1/2, 0) = (1-p)\mu(0, 1)$. After the same process is continued recursively, $\mu(0, 1/2)$ is obtained from $\mu(0, 1)$

by a *t*-axis reduction of ratio 1/2, but also a *μ*-axis reduction of ratio *p*. Each transformation is an affinity, not a similarity.

But the same construction can also be expressed in the form of a "ratio-self-similarity." Take an interval to be called *large*, and subintervals to be called *small*. Consider a semi-group of reductions *ρ* that transform the large interval into *ρ(large)* and the small interval into *ρ(small)*. When the measure is binomial and the reduction ratios are of the form 2^{-k}, the ratio *μ[ρ(small)]/μ[ρ(large)]* is independant of *t*. Whenever this identity holds, the measure will be called ratio self-similar, short for self-similar with respect to relative measures.

4.2 Sketch of the exponent function τ(*q*), as defined through population or sample moments

The multifractal formalism is familiar to most readers and repeated in this book. But it is worth restating here, with a few unfamiliar wrinkles. It centers on two mutually related functions τ(*q*) and *f*(α). Each can be introduced in several ways, which are closely related but non-equivalent.

To characterize a multifractal measure (short of a rigorous definition), denote by Δ*μ* the multifractal measure in an interval Δ*t*. The expression

$$\frac{\log\left[\text{ typical value }(\Delta\mu)^q\right]}{\log \Delta t}$$

was long implicit in works by several authors. When it tends to a limit as Δ*t* → 0, in particular when it is independent of Δ*t*, it defines a function of *q*. It was denoted by −*f*(*h*) in the originals of M1974f and M1974c, but by τ(*q*) in Hentschel & Procaccia 1983, which made it known in statistical physics. As a compromise, the old papers in this book were edited to denote *f*(*q*) as $\overline{\tau}(q)$.

There are several ways of choosing the interval Δ*t*, freely or in a lattice, and several ways of defining the word "typical value." The resulting difficulties seem to exemplify mathematical hair-splitting but turn out to matter very much. For example, τ(*q*) is defined as a limit; but that limit may fail to exist. Or the limit may exist but allow several interpretations that yield different expressions.

Those difficulties are not raised in Halsey et al. 1986. There, *narrow* multifractals will be defined as being measures for which the expression $\chi(q, dt) = \Sigma \mu_j^q(dt)$, often called "partition function," satisfies, *for every value of q*, an analytic scaling relation of the form $\chi(q, dt) = dt^{\tau(q)}$.

The empirical issue is whether or not this narrow definition is useful in characterizing interesting problems in nature. The mathematical question is whether or not it is satisfied by interesting mathematical constructs. It is indeed satisfied by a number of simple nonrandom constructions, for example the binomial measure and the slight generalization described at the beginning of Hentschel & Procaccia 1983.

Scaling holds under wider conditions when q is near 1. However, the remainder of this book and M1998L show that many explicit and simple constructions yield multifractals – both random and nonrandom – for which analytic scaling *fails* to hold when q is sufficiently larger that 1, or sufficiently smaller (e.g., negative). This failure is not an unwelcome complication but important fact with extensive theoretical and empirical consequences.

Two distinct forms of the function $\tau(q)$, as defined for populations and from sample moments. In M1974f{N15} and M1974c{N16}, the function $\bar{\tau}(q) = -\tau(q)$ enters as the exponent of the population (ensemble) expectation $\langle \chi(q, dt) \rangle$. Therefore, $\tau(q)$ is defined only when (a) this expectation is finite, which *need not* be the case and (b) this expectation takes the form $dt^{\tau(q)}$, which *need not* be the case, either.

The mechanical use of Halsey et al 1986 defines $\tau(q)$ by the partition function of a *sample average*.

In the 1960s, I would have thought that this distinction between ensemble and sample could not possibly make any practical difference. In fact, M1969b{N13}, M1972j{N14}, M1974f{N15} and M1974c{N16} revealed that moments of multiplicative multifractals raise very tricky questions. Cases of central importance later examined in M1990d and M1991k have the property that analytical scaling *does* hold for the expectation $\langle \mu^q(dt) \rangle$ but *fails* for the sample partition function.

Physics often brings out the essentials by taking bold and clever shortcuts that appear clumsy even to mathematicians who (as I am) are far from being formal extremists. But the study of multifractals gives physical content to diverse mathematical cases where $L_\infty = \lim_{n \to \infty} L_n$, but a property of L_∞ is *not* the limit of the corresponding properties of the L_n. Of course, computer programs always extract a sample function $\tau(q)$ from empirical data. But when the theoretical $\tau(q)$ fails to be defined, the interpretation of the sample $\tau(q)$ demands great caution.

4.3 The function $f(\alpha)$: the Legendre transforms and their justification

Starting with $\tau(q)$, the inverse Legendre transform defines $f(\alpha)$ as the envelope of the straight lines of equation $\varphi(\alpha) = q\alpha - \tau(q)$. This transform can be justified in distinct ways.

• *Cramèr theory of large deviations.* This is the path I followed in M1974c{N16} and continue to favor because one understands multifractals best by knowing their intimate connexions with the Cramèr theory of large deviations in probability theory. Here, the Legendre transform is a method for *evaluating* a renormalized logarithm of a probability density. Details were not explored fully until M1989g.

• *Lagrange multipliers.* The textbooks I used injected the Legendre transform in thermodynamics via the "Lagrange multipliers" formalism of Gibbs. This approach is limited to the multinomial case, but is strongly recommended in a first lecture on multifractals, because it is by far the simplest formally and removes all the mystery out of multifractals. It is developed and described in M1988c and M1989g.

• *Darwin-Fowler "saddle points" or "steepest descents."* Frisch & Parisi 1985 and Halsey et al. 1986 follow the first heuristic stages of the Darwin-Fowler approach to statistical thermodynamics. That approach was developed in the 1920s to justify in full rigor the results obtained by Lagrange multipliers. This method is rarely taught today or referenced in papers that introduce the "thermodynamical interpretation of multifractals."

Non-equivalence of alternative approaches to the multifractal formalism. In summary, as the multifractals went on to generate an enormous literature, several distinct approaches emerged. They are equivalent only in the simplest cases. Even for measures that are geometrically scaling, some forms of analytic scaling simply fail to hold. Broad multifractals *need not* be singular measures and do *not* demand either $\alpha_{min} > 0$ or $\alpha_{max} < \infty$. They allow $f(\alpha) < 0$ and even $\alpha < 0$, and include discontinous measures.

4.4 The scale factor $\sigma(\tau) = [1 + \tau(q)]/q$ and other linear transforms of $\tau(q)$, including the "generalized dimension"

The Legendre transform construction of $f(\alpha)$ defines $-\tau(q)$ as the ordinate of the intercept of the axis $\alpha = 0$ by the tangent of slope q of the graph of $f(\alpha)$. Let us add a remark that may be at least partly new.

Much about the substance of $\tau(q)$ is described by the points where a tangent to the graph of $f(\alpha)$ intersects the line $f = 1$. This intercept has the abscissa $[\tau(q) + 1]/q \geq 0$, to be denoted by $\sigma(q)$. When $f(\alpha)$ is introduced via the distribution of the measure $d\mu$ $\sigma(q)$ plays a central role as the exponent of the scale factor $\langle (d\mu)^q \rangle^{1/q}$, considered as function of dq. It determines the extent to which the distributions of $d\mu$ for different values of dt fail to "collapse," that is, to superpose by rescaling. This topic is discussed in detail in Chapter E6 of M1997E.

Both the abscissa and the ordinate of the intercept by $f = \alpha$ are $\tau(q)/(q-1) \geq 0$, a quantity denoted by $D(q)$. It is discussed in Section 5.6 and the term "generalized dimension" for it is challenged.

5. DEVELOPMENT OF THE NOTION OF MULTIFRACTAL; DOCUMENTS ON KOLMOGOROV AND LEGENDRE TRANSFORMS

Several constructions one may call "proto-multifractal" are analogous in spirit to the "proto-fractal" constructions exemplified by the original Cantor dust, the Koch and Sierpinski curves and a few others. Their relation to the notion of fractal was compared in the Preface to the relation of addition and rotation to the general mathematical notion of group.

From today's vantage point, the history of multifractals subdivides into stages, some so short that they can be called steps. Those steps, and the most important or best-known contributions, will be listed in Section 5.1, which acts as Table of Contents. Then the remainder of the Section, following a roughly chronological order, will take up several documents whose historical importance combines with brevity.

5.1 Sketch of prehistory and history

Prehistory in esoteric pure mathematics: Hölder, Besicovitch and $\tau'(1)$. The variable α in the function $f(\alpha)$ concerns the behavior of a measure near a point. See Section 5.2.1.

For the multinomial measures, the quantity $\tau'(1)$, often called "information dimension," was well-known to Besicovitch. Multinomial measures had even earlier roots. See Section 5.2.2.

Surprisingly, $1/f$ noises had no root in mathematics. See Section 5.3.

The binomial measure in an esoteric corner of mining engineering science: de Wijs 1951-53. To my knowledge, the earliest correct statement of a binomial cascade as tool of model-making is found in de Wijs 1951-53. A quote in M1974F{FGN}, p 376, describes the context of that model in ore distributions. However, de Wijs did not proceed to multifractals.

1962. The transition from prehistory to history is largely arbitrary. Obukhov and Kolmogorov published near-simultaneously in 1941 and 1962. It is fair to say that history begins with Kolmogorov 1962, a flawed work, but one that set a vibrant direction; see Section 5.4.

Around 1970: the function $\tau(q)$. The origin of $\tau(q)$ is obscure. For multifractals, it was evaluated in Kolmogorov 1962, but only in the self-contradictory lognormal context. Kolmogorov implied that $\tau(q)$ is "universal" and represented by a parabola. One can argue that the study of multifractals started from the awareness that the correct expression for $\tau(q)$ is, in fact, *not* universal.

The form of $\tau(q)$ was evaluated correctly in M1969b{N13} and Novikov 1969, but those independent and simultaneous discoveries differ on a fundamental point. Both authors observed that the correct non-universal $\tau(q)$ contradicts lognormality. However, Novikov did not resolve the contradiction, while I did, by showing that in the study of multifractals the central limit theorem is far from being sufficient. That is, the distribution of $\Delta\mu$ is *not* "approximately lognormal," in the sense that its approximation by the lognormal is *not* significant. $\tau(q)$ was rediscovered in Hentschel & Procaccia 1983 and evaluated in the context of chaotic dynamics.

The M1972 and M1974 models and the sources of $f(\alpha)$. See Section 5.5.

Step into a dead-end: the expression $D(q) = \tau(q)/(q-1)$. See Section 5.6.

A functional equation that generalizes Lévy stability. See Section 2.5 of Chapter N3.

5.2 Proto-multifractals within pure mathematics

5.2.1 The exponent α; reasons for calling it "Hölder exponent" rather than "pointwise dimension."
In Chapter 39 of M1982F{FGN}, pp. 373-4, heuristic manipulations that concern the exponent α are called "Lipschitz-Hölder heuristics." However, the needs of physics proved more diverse than those of mathematics, and required many successive generalizations. Rudolf Otto Lipschitz (1832-1903) and Ludwig Otto Hölder (1859-1937) mostly considered an exponent that characterizes a function in an interval and is denoted in current terminology by $\alpha_{\min}$. Without being able to weigh the merits in an ancient dispute, I felt that Hölder's name has the asset of beginning with the same letter as that of Harold Edwin Hurst (1880-1978), who introduced into hydrology an exponent that represents roughly the same idea. Combining those two traditions made it necessary

to generalize and split the classical mathematical definition into many separate notions: coarse versus fine, global versus local, and others.

Be that as it may, the concept that M1982F{FGN} handled in that entry on "heuristics" was called "singularity" in Frisch & Parisi 1985 and "dimension" in Halsey et al. 1986. Other authors call it a "pointwise dimension." A dimension satisfying $\alpha > 1$ is strange on the line but "pointwise dimension" would be tolerable if it were enlightening. I believe it is confusing.

5.2.2 Besicovitch and his students. The binomial (or Bernoulli) measure was familiar to H. Lebesgue, early in this century. But this and the multinomial measure were not studied systematically until Abram Samoilovich Besicovitch (1891-1970) and his students at the University of Cambridge, UK. (This is why M1982F{FGN} proposed for them the term *Besicovitch measures*, which was not accepted.) Besicovitch studied sets defined by number-theoretical considerations, but from a concrete point of view he studied geometrically scaling (self-similar) measures obtained by nonrandom multiplicative cascades.

Incidentally, the term "Hausdorff-Besicovitch dimension" deserves being explained. An original definition due to Hausdorff demanded that the Hausdorff measure for the dimension D_{BH} be positive and finite. This requirement excludes most random sets, that is, the sets of greatest significances in science. Besicovitch redefined D_{BH} as a crossover.

The physicist Cyril Domb knew Besicovitch as a student, then as lecturer at Cambridge. In Domb 1989, we read that Besicovitch "described himself as an expert in the 'pathology' of mathematics. If someone put forward a conjecture that [he] suspected to be untrue, he would keep worrying until he had produced a counter-example... Neither Hausdorff nor Besicovitch dreamt that their ideas on the fractional dimension of sets would be of practical use in the real world; likewise for the work of Weierstrass on curves which are continuous but not differentiable at all points. It is a remarkable achievement of Mandelbrot to have identified so many practical applications of these concepts and of the abstract mathematics which flew around them ... 'The *real* mathematics of the *real* mathematicians of Fermat and Euler and Gauss and Abel and Riemann is almost wholly *useless*.' Thus wrote Hardy in his famous book *A Mathematician's Apology*, Hardy 1940. As usual, he was being provocative and stimulating, with the aim of discouraging study for materialistic purposes. But the counter-examples of the practical applications of the 'real' mathematics of Weierstrass, Hausdorff, Besicovitch, Boole and Hilbert must surely demolish his thesis. However abstract and remote from reality the mathematics in which you are engaged, there is no escape from the conclusion that it may some day be put to practical use."

The most striking example concerns a strong relation between fractal dimension and the entropy-information $-\Sigma p\log p$ that follows from the role of $\tau'(1)$. In Besicovitch's papers from the 1920s and 1930s this relation involved an inequality. His student I. J. Good pointed out the thermodynamic analogy, and his later student T. Eggleston proved equality. The Eggleston theorem is discussed in Billingsley 1967, a book I found helpful.

5.2.3 Before Besicovitch. Brodén.
A rather general multifractal measure hides obscurely but unquestionably in Brodén 1897, behind an early but forgotten example of continuous function having nowhere a positive and finite derivative. The Weierstrass function oscillates up and down, but Brodén's function is nondecreasing. Built on the same principle as a cumulative binomial or multinomial measure, it is very general and the discussion takes many pages. The reference survived in a footnote in Knopp 1918. Apparently, Brodén fell victim to the belief that virtue, fame or promotion reside in generality.

A century before Besicovitch: Bolzano. An example of multifractal oscillating cartoon, in the sense of Chapter N1, stars in a manuscript on "The Theory of Functions" written by Bolzano in 1838. After it was actually published in the 1920s, it was criticized for excessive complication and replaced by simplified variants, all of which happen to be unifractal. Bernhardt Bolzano (1781-1848) is (very belatedly!) attracting a great deal of attention, and his collected papers are being published. I intend to comment on his work on a more suitable occasion.

5.2.4 A "Cinderella field" called "geometric measure theory".
The Preface ends by describing the search for a definition of the word *fractal* as not very interesting. To the contrary, the precise historical background is interesting, and deserves elaboration.

Fractal geometry embraces concrete and physical reality in its messiest forms, while the dominant mood of mathematics from 1875 to 1975 was to flee from reality and feed only on itself. Thus, in reviewing M1977F, F.J. Dyson commented that I had shown that Nature has "played a joke on the mathematicians." Lately, the mood within mathematics has changed and it is again politically correct to acknowledge that mathematics can find interesting problems both within itself and outside.

Bandt 1997, a review by a mathematician of a mathematical treatise on fractals, observes that "Mathematical theories tend to deviate far from their origin. If you want to learn about measure theory and you look into a contemporary course, or into the American Mathematical Society Subject Classification, you will hardly recognize that this field started with the determination of length, area and volume of sets in ordinary space. Yet there is a subfield called *geometric measure theory* which concentrates on the measurement of Borel sets in R^n – sometimes even of compact sets in the plane – and nevertheless embraces some of the deep problems and theorems of modern analysis,...

"For a long time, discoveries concerning ordinary space were considered rather as curiosities, and geometric measure theory played the role of Cinderella beside her big sisters abstract measure and integration theory, who laid foundations for analysis and probability. Later, the measure-theoretic structure of sets in R^n turned out to be fundamental for analysis, too – both for the treatment of boundary value problems with singularities and for the study of "exceptional sets" in various contexts. Still, these were rather delicate and exclusive topics.

"Meanwhile, complicated plane sets are commonly known as fractals due to the efforts of Mandelbrot and to the expanding facilities of computer graphics, and it is generally accepted that such sets appear 'everywhere' in mathematics as well as in reality."

Comparing a science to Cinderella is charming. It was no lesser for being kept in a corner, but one may indeed say that it was first invited to the Ball by fractal geometry.

To add perspective to Bandt's words, I was acquainted with the work of Besicovitch since the early 1960s. I knew Jean-Pierre Kahane since we were students and collaborated with him on Kahane & M1965{N11}. However, it was Frederick J. Almgren Jr. (1933 - 1997; I met him shortly before the publication of M1975O) who informed me that (to quote Bandt) "there is a subfield called geometric measure theory." So far, its impact on fractal geometry has so far been lesser than the converse impact.

5.3 Absence of a systematic "proto-fractal" study of 1/*f* noises within pure mathematics

Absence of news is a form of news. 1/*f* noises, made my mind wander from physics to the fundamental "counter-examples" of analysis. The continuous but nondifferentiable functions of Weierstrass (1872) are 1/*f* noises motivated by the needs of mathematics rather than physical observation. But this property was not deemed to deserve attention.

Later, around 1900, when describing several basic proto-fractal sets, mathematicians encountered additional functions with an $1/f$ spectrum. They came close to paying attention when studying the devil staircase constructed on the original triadic Cantor dust, of dimension $D = \log2/\log3$. That set suggested a periodic function $C(t)$ of period 2 made of an ordinary ascending Cantor staircase and a descending Cantor staircase. $C'(t)$ is a generalized function whose spectrum is by and large proportional to f^{-D}, but the mathematicians focussed on a sequence of exceptional frequencies that correspond to "spikes" in the spectrum. As confirmed by a verbal communication from J.-P. Kahane, the $1/f$ background received no attention.

Well-meaning anecdotes take it for granted that the Cantor dusts was contrived to be a counter-example one may call an "anomalous" set. The name of George Cantor (1845-1918) being indissolubly linked with set theory, a field rife with counter-examples, this expectation is natural but historically invalid. Cantor began by studying trigonometric series, and defined S as being a *set of multiplicity*, if a trigonometric series that converges to zero outside of S may fail to converge to zero on S. All the other sets are called *sets of unicity*. Intervals are sets of multiplicity and finite sets are sets of unicity, and Cantor defined the triadic dust while searching for a set of multiplicity whose measure is zero. The tools needed to prove whether S is a set of unicity or multiplicity were not available, and Cantor moved on to "invent" set theory.

Only much later (as described in Kahane & Salem 1963) was it discovered that S is a set of multiplicity if, and only if, it carries a measure whose Fourier spectrum is (in a technical sense) close to being $1/f$. Unicity corresponds to an $1/f$ background combined with an infinity of spikes. Kahane & M1965{N11} showed that, with probability one, the Lévy dusts are *sets of multiplicity*. Being far simpler than its many predecessors, this solution led straight to what is known today as the "natural method" of dealing with an ancient problem.

5.4 Documents relative to Kolmogorov 1962 and Kolmogorov's students

Kolmogorov 1962 was the work of a great mind but hastily thought-through and casually written. The lognormal model it described is self-contradictory. My resolution of the contradiction is described in Part IV. It is surprising, therefore worth observing, that Kolmogorov 1962 or the authors who followed him closely (most of them from Moscow) make no reference to singular measures and to the "prehistoric" results from the Besicovitch school.

A dedication paraphrased from M1991k. A token of the greatness of Andrei Nikolaievitch Kolmogorov is that his work spanned fields far removed from one another, both within mathematics and outside. I profess special admiration for his several brief forays in highly specialized fields where he could not be more than a visitor. Kolmogorov 1941 was spurned when I was a student, but is now a classic. His second foray into turbulence was extremely brief. Kolmogorov 1962 may seem at first sight to be an expository report on Obukhov 1962, but in fact puts forward a bold 'third hypothesis' of exact lognormality. Careful reading by several authors has revealed profound flaws in this hypothesis. When a giant stumbles, it is safe to expect subtle issues to be involved. We must be grateful to Kolmogorov for having pointed out a path he did not choose to follow and for spurring much hard and rewarding work.

Comments on the "impact of carelessness." The preceding dedication refers to an issue eloquently discussed in Kahane 1991b, p. 293. "It is a fact that some mistakes, because of the personality of their author or the subtlety of the arguments they involve, play a stimulating role. Anyone interested finds this factor is not to be neglected. [For example, Lebesgue had commented on the double chance he had had] of finding but also of making big mistakes that became starting points... in the field of trigonometric series... Careless statements and mistakes by eminent mathematicians became the points of departure of eminent investigations."

The contribution of A.M. Yaglom. Elaborating on Kolmogorov 1982, the eminent fluid dynamics expert A.M. Yaglom contributed several papers, some of them referenced in M1974f{N15}. Nowhere does he question lognormality and he misses multifractality altogether. M1974f{N15} credits the precise cascade argument to Yaglom but a fresh reading of Yaglom 1965 shows that a more tenuous credit is due.

A statement in A. N. Kolmogorov's "Selected Works." Kolmogorov's stumble may be unique in his published work, and to many readers it continues to appear inconceivable. However, its being genuine is confirmed in Kolmogorov 1985-1991, Vol 1. The English translation, pp. 497-8, includes the following annotation by A. M. Yaglom. "A number of experimental works [attempted] ... to verify the hypothesis ... of lognormality .... In all cases it was observed that the corresponding distribution function is in good agreement with the lognormal distribution for a wide-range of moderate values of the argument, but on the tails... it deviates from the lognormal [distribution]. The deviations turned out to influence substantially the moments of higher orders of the probability distributions,.... In this connection, [M1972j{N14}] noted that the fact that the probability distribution of [locally averaged dissipation] tends asymptotically to the lognormal with increasing Reynolds number Re by no means implies that the moments of [dissipation] must be close to the moments of this limit distribution for large values of Re."

These words confirm the above comment about Yaglom 1965; they acknowledge the *empirical inadequacy* of the lognormal "third hypothesis," but fail to face its *theoretical impossibility*. My argument is reported very incompletely with no hint of the subtle theoretical difficulties that soon developed into a theory of multifractals. Those difficulties concern the central limit theorem invoked in Obukhov 1962 and Yaglom 1965 to justify lognormality. Once again, the central limit behavior is largely irrelevant in the case of multifractals and interest must focus on the distribution tails given by the large deviations theorem of H. Cramèr.

5.5 Sources of $f(\alpha)$

5.5.1 Photographic excerpts from the original M1974c. See Figure N2-1.

5.5.2 Excerpt from Frisch & Parisi 1985. The low quality of printing in the original made resetting unavoidable. This was an opportunity to use the reference style of this book and come closer to current notation: h was replaced by α, p by q, and $d(h)$ by $f(\alpha)$, and ζ_q by $\tilde{\tau}(q) = \tau(q) - 3$.

"[Examine the]" singularities of the Euler equations considered as limit of the Navier-Stokes equations as the viscosity tends to zero.... M1976o{N18} ...and Frisch et al. 1978 considered models with singularities concentrated on a set having noninteger (fractal) Hausdorff dimension. We shall here show that the data suggest the existence of a hierarchy of such sets (a multifractal) ... The velocity field at a given time $v(x)$ is said to have a singularity of order $\alpha > 0$ at the point x if

$$\overline{\lim_{x \to y}} |v(x) - v(y)| / |x - y|^\alpha \neq 0. ...$$

"We call $S(\alpha)$ the set of points for which the velocity field has a singularity of order α [and] denote by $f(\alpha)$ the Hausdorff dimension of $S(\alpha)$ ... different kind of singularities are associated with sets having different Hausdorff dimensions ... To connect $f(\alpha)$ with the exponents ζ_q which control the asymptotic behavior of the longitudinal structure functions. We can try to rephrase the previous statements on the Hausdorff

CALCUL DES PROBABILITÉS. — *Multiplications aléatoires itérées et distributions invariantes par moyenne pondérée aléatoire.* Note (*) de M. **Benoit Mandelbrot**, présentée par M. Szolem Mandelbrojt.

1. CONSTRUCTION. — Soit une suite de « poids », des v. a. indépendantes et identiquement distribuées (i. i. d.). C étant un entier > 1 donné, les C premiers poids seront désignés par $W(i_1)$, $0 \leq i_1 \leq C-1$; les C^2 suivants par $W(i_1, i_2)$, etc. Soit dans l'intervalle $]0, 1]$, un réel t développé dans la base C sous la forme $t = 0, i_1, i_2, \ldots$. Partant de $X'_0(t) \equiv 1$, la suite de mesures aléatoires $X'_n(t)$ sera définie par

$$X'_n(t) = W(i_1) W(i_1, i_2) \ldots W(i_1, i_2, \ldots, i_n).$$

Posons $X_n(t) = \int_0^t X'_n(s) \, ds$. Notre objet premier sera d'étudier $X_\infty(t) = \lim_{n \to \infty} X_n(t)$, et, si X_∞ est dégénérée, $Y_\infty(t) = \lim_{n \to \infty} Y_n(t)$, où $Y_n(t) = X_n(t)/A_n$, A_n étant une suite normalisatrice non aléatoire choisie de façon appropriée.

Nous écrirons $F(w) = \Pr\{W < w\}$. ...

10. LE CAS POSITIF. DÉFINITIONS. — Lorque $F(0) = 0$ et $EW = 1$, soient

$$\alpha = \max\{1, \sup[h : EW^h < C^{h-1}]\} \quad \text{et} \quad D = 1 - EW \log_C W.$$

Lorsque $F(0) = 0$, soit $\varphi(h) = \log_C[EW^h/C^{h-1}] = \log_C EW^h - (h-1)$. Ce $\varphi(h)$ est convexe, et lorsque $\varphi(1) = \log EW = 0$, α est en quelque sorte le deuxième zéro de $\varphi(h)$ (admettant que celui-ci dépasse le premier zéro $h = 1$). Par ailleurs, D est formellement $-\varphi'(1)$. ...

L'extension au cas de log W non gaussien passe par l'inégalité de Chernoff ([3]), laquelle exige $E(\log W) < \infty$; $EW = \infty$ est admissible. Posant $-\log A_n = T n \log C$, et négligeant quelques complications qui n'affectent pas le présent raisonnement, on obtient

$$\Pr\{\log[X'_n(kC^{-n})C^{-n}] \geq -Tn\log C\} \sim \exp[-nQ(T)]$$

avec

$$-Q(T) = \inf\{T\log C h + \log E(W/C)^h\} = -\log C + \log C \inf[\varphi(h) + hT]. \ldots$$

FIGURE N2-1. A translation of M1974c (together with annotations) is available below as Chapter N16, but it is good to reproduce a few lines of the French original that first used the "Legendre" transform in the study of multifractals. A laser-based copying machine reduced this text horizontally (by a diagonal affinity!) to fit in a narrower page.

dimensions of $S(\alpha)$ by saying that the probability of having $|v(x) - v(y)|$ of order $|x - y|^\alpha$ goes to zero like $|x - y|^{3 - f(\alpha)}$ when $|x - y| \to 0$. Thus ...

$$\langle (\delta v(l))^q \rangle \sim d\mu(\alpha) l^{(q\alpha + 3 - f(\alpha))},$$

where $d\mu(\alpha)$ is a measure concentrated on the region where $f(\alpha) > 0$.

"Using the saddle-point method, we easily find

$$\tilde{\tau}(q) = \min_\alpha [q\alpha + 3 - f(\alpha)].$$

"[This] is the Legendre transform of the codimension $c(\alpha) \equiv 3 - f(\alpha)$ of the set $S(\alpha)$. The dimensions $f(\alpha)$ can be extracted from the $\tilde{\tau}(q)$s by the inverse Legendre transform

$$f(\alpha) = 3 - \min_q (\tilde{\tau}(q) - q\alpha).$$

" $\tilde{\tau}(q)$ appears to significantly deviate from a linear function of q. The function $f(\alpha)$ is thus nontrivial and singularities of different kinds, if they exist, are concentrated on sets having different Hausdorff dimensions.

"The multifractal model is not completely consistent with the Kolmogorov 1962 lognormal model, for which $\tilde{\tau}(q) = q/3 + \mu q(3 - q)^2/18$. Indeed, with this choice of $\tilde{\tau}(q)$ we find that beyond $q_{max} = 9(2/3\mu)^{1/2}$, a negative dimension is obtained. Accurate measurements of very-high-order structure functions are required to test for a possible inconsistency of the multifractal model.

"Finally, one may wonder how the above multifractal model relates to the models of M1974f{N15}, 1976o{N18}) and 1977F. [There] a random weighting factor W appears at each stage of the cascade. The case when W has a binomial distribution (absolute curdling) corresponds to a single fractal approach... For more general W-distributions (weighted curdling) one obtains exponents ζ_q that depend nonlinearly on q like in the multifractal model. There is a single fractal for the energy dissipation, but it is conceivable that other fractals will be uncovered by investigating all possible singularities of the dissipation. Still the multifractal model appears to be somewhat more restrictive than Mandelbrot's weighted-curdling model which does include the lognormal case."

General comments on the preceding quote. Frisch & Parisi 1985 – to be denoted in these comments as F&P – deserves high credit for being the first work to interpret $f(\alpha)$ as a fractal dimension, but the role of the Legendre transform was clear in M1974c{N16}.

F&P also deserves credit for introducing the excellent term *multifractal*, and for making no attempt to define it. Their heuristic argument applies to special examples of the M1974 construction and to an unspecified broader class of measures. Finally, F&P were correct in surmising that my earlier work was of wider scopes than theirs.

Comments on references to lognormality. These references are mystifying. M1972j{N14} had shown that the lognormal model fails to be self-consistent, therefore a model that is consistent with exact lognormality would be unsound.

Comment on negative values of $f(\alpha)$. The possibility of $f(\alpha) < 0$ is mentioned by F&P in half a line and not pursued. As mentioned in Sections 1.1 and 1.7, I noticed it independently and pursued it energetically, but (as mentioned in Section 1.7) not until after 1985.

If F&P had dealt with a Hausdorff-Besicovitch dimension, the conclusion that $f(\alpha) < 0$ would have shown the heuristics to be faulty. But more is at stake. The F&P argument does not apply to the "canonical" multiplicative process with lognormal weights W, as described in M1972j{N14}, M1974f{N15} and M1974c{N16}. In that process, the measure is *not* itself lognormally distributed. However, an intrinsic part of the story of canonical multiplicative processes is that negative dimension are not absurd but to the contrary carry essential information. It is hardly necessary to repeat from Section 1.7 that the concept of negative dimension necessarily differs from the concept due to Hausdorff and Besicovitch.

5.6 Comments on "dimension," as generalized from sets to measures

I found myself in the mid-1980s in the unexpected role of vehement opponent of extending the notion of fractal dimension from sets to measures. In particular, I saw little reason to single out the expression $= \tau(q)/(q-1)$, which is mentioned in Section 4.4 and denoted by $D(q)$. Of course, the special values $D(1) = \tau'(1)$ and $D(2) = \tau(2)$ are not in question: they relabel notions well-known before 1983. The criticism was addressed to other $D(q)$s: they never enter by themselves, only enter when $\tau(q)$ is re-written as $(q-1)D(q)$. Let us elaborate.

The incidental critical role of $D(q)$ in M1974f{N15}. This role was stated for integer qs, but is readily interpolated for all qs. Take a grid of base b and a cascade based on the multiplicative random weight W. When a cascade with those b and W proceeds in a space of dimension E, one has

$$\tau_E(q) = -\log_b\langle W^q\rangle + E(q-1), \text{ hence } D_E(q) = -\frac{\log_b\langle W^q\rangle}{q-1} + E.$$

This $D_E(q)$ can be represented in the form

$$D_E(q) = D_0(q) + E,$$

where $D_0(q)$ no longer depends on E.

Suppose that the cascade is canonical in the sense of M1974f{N15} and denote by μ_E the measure it generates. A basic property of μ_E, first reported in M1974f{N19}, concerns the value q_{crit}, other than $q = 1$, that satisfies $\tau[q_{\text{crit}}(E)] = 0$. For given E, one has $\langle(\mu_E^q)\rangle < \infty$ for $q < q_{\text{crit}}(E)$, while $\langle(\mu_E^q)\rangle = \infty$ for $q \geq q_{\text{crit}}(E)$. This argument, when inverted, shows that $D_0(q)$ has the following property: For given q, one has $\langle(\mu_E^q)\rangle < \infty$ for $E > D_0(q)$ and $\langle(\mu_E^q)\rangle \infty$ for $E < D_0(q)$. Thus, $D_0(q)$ plays the role of a critical dimension, a fine result but one of limited importance and almost esoteric.

$D(q)$ as "generalized dimension". Receiving a draft of Hentschel & Procaccia 1983 made me send back a number of comments. (A) $\tau(q)$ was presented as a new notion, but in fact it had been used by many earlier writers. The text that went to print acknowledged (in Note 17) that the inequality concerning $\tau(q)$ were available in M1974f{N15}. B) One must not speak of $\tau(q)$ as referring to "a fractal," without specifying that the fractal in question is a measure and not a set. C) "Dimension" is best viewed as a property of a set, not of a measure, but the Hentschel & Procaccia preprint called $D(q) = \tau(q)/(q-1)$ a "dimension" without specifying any set. Because of Comment C, the authors changed "dimension" into "generalized dimension," which did not respond to my concerns.

The Rényi connexion. It soon became known that $D(q)$ was previously written down in Rényi 1959 and called a "dimension". The process of generalization is given generous credit in mathematics when it yields new results and/or helps organize old ones. But Lebesgue railed against "notions that are attractive, to be sure, but serve no other purpose than to be defined." I feel that Rényi 1959 gave advance moral support to a concept that we are better without.

5.7 Concluding comment

When mathematicians, physicists and others contribute to the same project, their later recollections of what happened tend to diverge. When different national "schools" are involved, misunderstandings deepen and there is no completely nice way of straightening them out. Most fortunately, the misunderstandings touched in this chapter largely belong to the past, but casual early accounts do continue to be repeated, even when no harm is intended. This is the sole reason why I resolved to put several available and unequivocal events on record in this chapter.

N3

Scaling, invariants and fixed points

✦ **Abstract.** This brief chapter's goal is to show that, next to obvious and deep differences and without actual intent or interaction, fruitful and surprising parallels exist between my scientific work from 1956 to 1972 (as exemplified in M1997E and this book) and modern statistical physics.

The latter, to be denoted in this chapter as MSP, will be understood as focusing on critical phenomena. Pioneered in Kadanoff 1966, Wilson 1972 and Wilson & Fisher 1972, it was eminently successful in explaining or predicting many experimental observations. It started with solid and unquestioned laws of physics and used power-law relations and powerful tools called scaling and renormalization, invariance and fixed points.

In many important ways my work was very different. The early stage reported in M1997E concerned phenomena in economics, whose study cannot wait for the emergence of solid and unquestioned basic laws. As to the later work reported in this book, it concerned phenomena exemplified by turbulence. Solid and unquestioned basic laws are provided by the Navier-Stokes equations and there is a rich and efffective phenomenology. Unfortunately, the link between those laws and that phenomenology remains elusive.

Granting those deep differences, two facts often come as a surprise. Firstly, my work used many of the same tools as MSP, namely, power-law relations and scaling and renormalization arguments. This happened long before the day in 1972 when I first heard of MSP and even before the period around 1965 when MSP gathered speed. Secondly, my use of those tools was extensive and (on its terms) successful. After 1972, I became increasingly influenced by MSP and worked on problems having a basis in solid and unquestioned laws. To a lesser extent, my work influenced MSP, and interesting further developments were influenced by both.

The main tradition that influenced me was the use of scaling and renormalization arguments in probability theory (as a fractured education made me interpret that discipline). The most influential data were the power-law distributions discovered in the social sciences by Pareto and Zipf. A later influence was the use of scaling in the study of turbulence.

Particularly central to the point made in this chapter is the ancient use of power-law distributions and of scaling, fixed point, and renormalization arguments in the context of the Cauchy-Polyà-Lévy's "stable probability distributions". Those uses were purely mathematical until M1956c and M1956w injected them into a corner of science. They became central to the works I devoted from 1959 to 1972 to economics/finance (as reprinted in M1997E). They are also central to the works written before 1976 and reprinted in this book. Therefore, the deep connection that Jona-Lasinio 1975 discovered between the stable distributions and MSP automatically implies a deep connection between MSP and my work; this connection greatly contributes to this chapter's thesis. A token of deep differences is that in MSP the parameters of the stable distributions follow from basic principles, while my work infers them from the data.

Inevitably, the inclusion of this brief special chapter is in part motivated by current efforts to expand the use of the methods of physics to domains where, despite long and systematic search, basic law are not available. A popular target domain is finance/economics. Every broadening is welcome, but finance had no need to borrow the basic ideas of scaling and renormalization from MSP, simply because my work had rooted them in finance even before MSP came to be. ✦

THE HISTORICAL AND IN PART AUTOBIOGRAPHICAL TASK undertaken in Chapter N2 continues in this chapter, but in totally different style. Instead of many specialized topics, this chapter is restricted to a closely-knit complex of ideas surrounding power-law relations and probabilities distributions, scaling and renormalization, with an emphasis on invariances and fixed points. Instead of the activities of a few scholars working over a short time period, this chapter concerns events that engaged several large scientific communities over long time periods. A key factor is that those communities rarely interacted and could not view themselves as contributing to a common goal.

A broad comparative study of the various "flavors" of scaling and renormalization would be an excellent big project for a historian of science. It should include biological "allometry" and the long-known but also long-neglected scaling rules found in seismology. But the present chapter has a narrower goal, largely directed to the reader acquainted with the modern statistical physics of critical phenomena (MSP). The goal is to document the following fact: years before MSP arose, my work was consistently based on power-law distributions and relations and the broadly understood notions of scaling, renormalization, fixed point and invariance.

Those surprising circumstances exemplify a rare sequence of events. A conceptual construct arose away from any application and became elaborated in finance/economics and engineering/turbulence, therefore far from the physics mainstream. Later, a parallel development was undertaken quite independently in the physics mainstream and led to MSP. But there is an enormous difference between MSP and my work exemplified in M1997E and this book. MSP can proceed to great depths because its scope is restricted to parts of physics that are based on solid and unquestioned basic laws. This is not at all the case in any of the many fields represented in my early work.

A digression concerning reductionism. Given the choice of matching Newton's achievements or Galileo's or Kepler's, every scientist will choose Newton's. However, an investigation in science must not be judged on ambition but on accomplishment and need. From the viewpoint of this historical chapter, the absence of convincing "reduction" is not a significant issue.

M1997E and this book report on work that was successful in reducing extremely messy phenomena to scaling and renormalizability, but not to more basic principles. This is the best I could (or can) do, and it must not be judged by the standard of Newtonian dreams, but by the reality of alternative models. Some consist in "mere phenomenology" that represents each set of data by a simple formula justified by its good statistical fit. At worst, the separate fits are mutually incompatible. At best, they provide no significant understanding. Other alternative models seek equations that generate distributions that fit the data. Chapter E10 of M1997E analyzes a broad class of such models. The details vary but all too often scaling is found in the output because it is (unwittingly) inserted in the input.

How MSP developed and came to my attention. A bare sketch of the early stages of MSP will suffice, because of the availability of Domb 1996, Fisher 1998, and other accounts with very extensive bibliographies. The study of critical phenomena was active until 1908. Then interest waned, to revive only in the late 1950s. Onsager's solution of the plane Ising model was poorly understood and did not suffice to revive interest. But it coalesced with other contributions at an NBS conference held in 1965, when many separate activities merged into a proper scientific endeavor. Scaling became fully understood thanks to Kadanoff 1966, and fixed points under intrinsic "renormalization" transformations followed with Wilson 1972 and Wilson & Fisher 1972.

Those developments rapidly became widely known, but had not yet reached the IBM Mathematics Department when (still in 1972) the American Physical Society asked me to be the after-dinner speaker at a regional meeting. Having presented my scaling and renormalization work on "the price of cotton and the River Nile," I returned to my dinner table to be greeted exuberantly by the person sitting next to me. He was Herbert Callen, a fellow thermodynamicist, and he gave a short but wonderful account of what was happening in physics. In the fall of 1972, I went on sabbatical to Paris and met Pierre-Gilles de Gennes; we talked about critical phenomena and I helped him add a section on Appolonian packing to a paper he was about to co-author, Bidaux et al (see M1982F, p. 176). I was at work on a book that eventually became M1975O and critical phenomena were added to this book. They formed Chapter IX, which, despite its brevity, described possible contacts between MSP and the fractal geometry that developed from my papers reprinted in M1997E and this book.

Parallels between MSP and my work that result from the essential role played by divergences criteria. Needless to say, divergences related to scaling are a mainstay of MSP. As to my work, the Preface described it as being near exclusively focussed on variability and randomness that is not *mild* but *wild*. It happens that wildness often manifests itself by divergences that are symptoms and consequences of scaling. By far the best known is *infinite* length of coastlines, but the examples most relevant to this book concern probabilistic notions. Here is a non-exclusive list:

• *infinite* expectation (it was first encountered in M1956c, which will be discussed momentarily);

• *infinite* variance (the main example concerned price variation, M1963b {E14});

• *infinite* spectral density at $f = 0$;

• reality or perhaps only appearance of *infinite* spectral energy near $f = 0$ (as in $1/f$ noises, see Section 2.4 of Chapter N2);

• *infinite* probability measure (as in Chapter N10).

Surprise and contention met each new instance of divergence. Invariably, the issue concerned a familiar statistical expression that was reputed to be finite and to define an intrinsic quantity. The issue arose in a context where the behavior of such a quantity for ordinary finite data seemed paradoxical or self-contradictory. I postulated that this quantity was in fact divergent; the paradoxes vanished immediately, interesting predictions could be made and then verified empirically.

An ironical twist of fate. As will be mentioned in Section 2.2, I first faced scaling and renormalization in M1956c and M1956w, two works concerned with an otherwise minor issue of taxonomy. At the end of Section 1, M1956c, observes that the theory about to be presented "reminds one of critical points of physics; the difference is that infinite fluctuations are inacceptable in a physical model, while they exist, and must be explained, in the data about to be investigated."

(Here is the original French wording: "...le nombre probable des espèces dans un genre est infini. La théorie rappelle celle des points critiques en physique, sauf qu'en physique, lorsqu'un modèle conduit a une fluctuation infinie, il y a des bonnes raisons de le changer, tandis qu'ici ce sont précisément ces cas que l'expérience nous conduit à étudier.")

It can be revealed that I had in mind the density fluctuations that lead to critical opalescence. But the expert I consulted was ill-informed of the developments that were going on and later coalesced into MSP (at the already-mentioned 1965 meeting at the NBS). My expert's name will not be revealed and it is forbidden, when writing history, to ask questions that begin with "if."

1. SCALING, RENORMALIZATIONS, FIXED POINTS, AND THE FORM THAT UNIVERSALITY TAKES IN PROBABILITY THEORY

In probability theory, renormalization is an old and standard tool in the statement of, and proof of, various limit theorems, and many limit theorems of probability are usefully viewed as concerning convergence to fixed points: under which conditions convergence occurs and how it proceeds. This section collects a few standard facts that will form the basis for Section 2. Cauchy in 1853, and much later Fréchet, conceived two renormalization schemes and searched for their fixed points. Much of my early work started by adapting those schemes for diverse scientific needs. In due time, I had to introduce many other renormalization schemes.

1.1 Sums of random variables and their limits and fixed points, from Cauchy 1853 to Lévy 1925; the Cauchy functional equation

Augustin Cauchy did not turn to probability theory until late in his life; yet – perhaps not unexpectedly – his contribution was striking and well

N3 ◊ ◊ SCALING, INVARIANTS AND FIXED POINTS

ahead of his time (Heyde & Seneta 1997). To deal with sums of independent random variables with $F(x) = \Pr\{X < x\}$, Cauchy innovated by introducing the Fourier transform

$$\varphi(u) = \int \exp(iux) dF(x),$$

which is now called the characteristic function, ch.f. He recognized that the ch.f. of the sum $S_n = X_1 + \cdots + X_g + \cdots + X_n$ is $\varphi_1(u) \ldots \varphi_g(u) \ldots \varphi_n(u)$. When the X_g are identically distributed, this expression becomes $\varphi^n(u)$. Therefore, the renormalized sum $S_n/A(n)$ has the characteristic function $\varphi^n[u/A(n)]$. Cauchy sought to identify ch.f's $\varphi(u)$ having the invariance property

$$\varphi^n[u/A(n)] = \varphi(u).$$

Fixed points. In present day words, Cauchy asked whether the transform $\varphi^n[u/A(n)]$ can have fixed points. The ch.f. of a random variable that is invariant under averaging is a fixed point, and Cauchy took it for granted that the converse was also true. He restricted himself to one-dimensional symmetric variables and claimed that, for the above invariance to hold, it is necessary and sufficient that $A(n) = n^{-1/\alpha}$ and $\varphi(u) = \exp(-|u|^\alpha)$, with $\alpha > 0$. The case $\alpha = 2$ yields the Gaussian distributions and the case $\alpha = 1$ yields the distribution of density inversely proportional to $1 + x^2$, which is now called Cauchy distribution (a term introduced in the 1920s by Lévy). No other value of α yields an explicit analytical density. (The first numerical tables appeared in M1963b{E14}.)

Unfortunately, the sufficiency claimed by Cauchy is incorrect. Indeed, Polyà showed that $\alpha > 2$ leads to an inverse Fourier transform of $\varphi(u)$ that is negative for certain x, while a probability density must be positive or zero for all x. In addition, Lévy completed this study by introducing non-symmetric solutions of Cauchy's equation. To all those random variables, he attached the term "stable."

Observe that the Cauchy-Polyà-Lévy renormalization can be called "base-free," because the integer $n > 1$ need not be an integer power of an integer base b. However, Polyà and Lévy also considered "base-bound" renormalization under addition. One replaces the original variables by "blocks," each made of b variables. Next, one considers "superblocks," each made of b blocks, hence b^2 variables. And so on. The challenge is to satisfy the following property: the sums carried over blocks, superblocks, etc. can be renormalized to follow the same distribution. The answer is that one must have $\varphi(u) = \exp[(-|u|^\alpha P(\log_b |u|)]$, where P is a periodic function of period $\log b$.

Fixed points as limit points. The term, "stable distribution," was unfortunate but is not changeable. It was not arbitrary, because the fixed points are also limit points under suitable renormalization. Lévy 1925 describes the addition of random variables in terms of a dynamic system that "attracts" a point representing a variable towards a stable fixed point.

To each α correspond addends that form a "domain of attraction" in the language of probabilists or a "class of universality" in the language of the physicists.

The domain of attraction is therefore determined by the behavior of $\log \varphi(u)$ near $u = 0$. For the Gaussian fixed point $\alpha = 2$, the behavior of $\log \varphi(u)$ is analytic near $u = 0$, and the domain of attraction is very wide, containing all addends having a finite variance. For each $\alpha < 2$, the behavior of $\log \varphi(u)$ near $u = 0$ is non-analytic, and the domain of attraction is narrow. Scaling and power-law distributions enter here in two ways. Firstly, as was already seen by Cauchy, the proper renormalization at the fixed point is $A(n) = n^{-1/\alpha}$. Secondly, all the non-Gaussian stable distributions follow a power-law distribution in their "tails".

In the domain of attraction, slightly more general $A(n)$ should be allowed, namely, those of the form $n^{-1/\alpha} L(n)$, where $L(n)$ may be $\log n$ or perhaps $1/\log n$, or some other function that is "slowly varying" in a technical sense due to Karamata that is discussed near the end of M1967i {N9}.

1.2 Maxima of random variables and the corresponding fixed points; the Fréchet function equation.

Replace the sum S_n by the quantity $M_n = \max(X_1, X_2, ... X_n)$. In this context, Fréchet carried out renormalization arguments parallel to Cauchy's. Since $\Pr\{M_n < x\} = [F(x)]^n$, the fixed-point functional equation is

$$\{F[x/A(n)]\}^n = F(x).$$

Again, one must have $A(n) = n^{-1/\alpha}$ and the fixed point becomes

$$F(x) = \Pr[X < x] = \exp(-x^{-\alpha}).$$

Now, the only restriction on the exponent is $\alpha > 0$.

1.3 Generalizations and a digression

Additional, non classical, renormalizations will be mentioned in Section 2.

Digression about bifurcation theory. The Cauchy and Fréchet functional equations have an analog in the theory of iterated real maps. Denote by $\phi(u)$ a map of a real interval upon itself and by $\phi_n(u)$ its $n-th$ iterate. The Cvitanovic-Feigenbaum renormalization equation

$$\phi_n[u/A(n)] = \phi(u)$$

proves immensely harder to investigate, but comparison with Cauchy and Fréchet makes its remarkable scaling aspects less surprising.

2. PROBABILITY DISTRIBUTIONS THAT ARE LIMIT AND FIXED POINTS, VIEWED AS "PRIVILEGED MODELS"

My work injected probabilistic fixed points into science by defining in several ways a concept for which I now propose the term, "privileged models." Without following a strict chronological order, I propose to show that this concept is closely related to renormalization. Its uses cut across the social sciences, both social and physical. Thus, Section 2.3 deals with numerical quantities whose definition is clearcut and not arbitrary. Examples include the bulk of the physical quantities studied in this book and some important social science quantities, most notably, the prices quoted on "transparent" financial exchanges. Section 2.2 deals with the very different cases where, for one reason or another, the definitions of the basic quantities are largely arbitrary. Section 2.1 serves as introduction.

2.1 Role of privileged statistical models in Berger & M 1963 {N6} and M1965c {N7}

Chapters N6 and N7 involve a built-in "natural" renormalization, since they investigate the intervals between individual events called errors, but also between milliseconds, seconds or longer "blocks" that include at least one error. It was very puzzling to hear that the distribution of error-free inter-block intervals is, in fact, largely independent of block length. This property suggested that the study of errors should start by trying out a model that is privileged by being invariant, a "fixed point," under the natural renormalization. Berger & M 1963 {N6} approximated this fixed point by a power-law distribution, with a single parameter interpreted as a fractal dimension. The fit was not perfect, but greatly improved upon earlier proposals.

M1965c {N7} went further and performed a painstaking exact base-free renormalization; the good news is that the privileged model that came out provided a very much improved fit to the data. This renormalization examines a set or process in an increasingly coarse-grained fashion; it does *not* consist in taking the sum (Cauchy-Polyà-Lévy) or a maximum (Fréchet).

One begins with a discrete set defined for all values of t and an integer b imposed by the ratios of lengths of alternative boxes in error. Using b as base, one constructs boxes of increasing size b^k, defines an index for each box and describes a generating rule that obtains the index of a box from the indices of its sub-boxes. Chapter N7 uses an index that takes one of two possible values: 1 if the box is occupied and 0 if not. The generating rule is that a box is occupied if, and only if, at least one of the subboxes is occupied.

2.2 Role of privileged statistical models in the argument in M1956c and M1963e {E3} – a paper titled "new methods in statistical economics"

One of my main papers in economics was M1963e{E3}, whose bold title proclaimed an idiosyncratic argument that runs as follows:

Macroscopic physical quantities such as velocity and mass are well-defined irrespectively of the observer. But much of economics and the social sciences does not have this luxury. The notion of "personal income" is not an intrinsic concept, but an inextricable combination of reality and tax code, with its loopholes, inclusions and exclusions. The notion of "city size" is an inextricable combination of human geography and the politics, often long forgotten, that set city boundaries. The notion of "firm" is affected by national and local laws. For example, the laws governing liability lead shipowners to register each "bottom" as a separate corporation. The expense and nuisance cost of paperwork and filing fees is also significant. Thus, an increase in filing fees led to a collapse in the number of firearms dealers in the U.S.A, and an explosion in the average sales volume per dealer! Similarly, in old-fashioned taxonomy before the advent of molecular biology, the definitions of genera and families, and sometimes even of species, used to be to a large extent arbitrary, therefore dependent on the taxonomist.

Disregarding this acknowledged indeterminacy, frequency distributions were plotted by many writers: Pareto did it for personal incomes as reported by the Census, Auerbach for city size as reported in the Almanach, and Willis for the number of species in a genus.

The reason for bringing those disparate cases together is that each yielded a scaling ("power law," "algebraic") distribution. As already mentioned in passing, there were (and there are) many attempts to explain the Pareto, Auerbach or Willis laws. To "explain" meant to start with basic principles and end with those laws, rather than any in a long list of alternatives, such as the lognormal. No one is against explanation. However, an idiosyncratic argument advanced in M 1963e{E3} throws serious doubt on the list of conceivable alternatives. It concedes the indeterminacy of economic quantities and brings to bear a form of renormalization.

This idiosyncratic argument begins by viewing the Almanach definition of city sizes as having been obtained from "true city sizes" by applying a filter that may vary in space and time. Even if one believes that a true city size distribution hides somewhere, it would be sensible to expect the observed city size distribution to also vary in space and time. Therefore, it is unreasonable to expect "universal" statistical regularity, unless an extremely specific situation happens to prevail. The true distribution happens to be a fixed point, an invariant under the filters that produce the observed distributions. The argument advanced in M1963e{E3} ends up by observing that the only true distributions one may ever hope to discover are the scaling ("power law") distributions. The sole alternative is a completely lawless mess.

The preceding argument claims no "explanatory" value. The sole reason for reporting it here resides in its heavy reliance on diverse forms of renormalization. Once again, Chapter E10 of M1997E expresses reservations about many available explanations of those probabilities distributions.

2.3 The variation of speculative prices: use of Lévy stable distributions in the model described in M1963b{E14}

It has become widely known that M1963b{E14} advanced the first realistic model of the variation of certain speculative prices. An earlier model was Brownian motion. It is (almost surely, almost everywhere) continuous, while price variation is characterized by sharp discontinuities. By contrast, the main thrust of the M1963 model was to account for the discontinuities while keeping the dependence between price changes for later study.

Most speculative prices lack "seasonals," that is, all time increments are equally "natural." Different financial market "charts" provide records of price when followed day by day, week by week, or year by year (today, some charts follow a price minute by minute). Therefore, I found it puzzling that all "charts" should have the reputations of "looking alike." The proper "privileged model" seemed to be one for which the probability distributions of all price increments ΔP over time increments Δt, could be made to collapse after proper renormalization. The only fixed points were the stable distributions discussed in Section 1.1. M1963b{E14} and a follow-up paper, M1967b{E15}, established that those fixed points represent many price records quite well.

2.4 The River Nile, $1/f$ noises, and the fractional Brownian model pioneered in M1965h{H}

A correlation that follows a power law is synonymous to a spectral density proportional to f^{-B}. When a process is Gaussian, such correlations are the solution of a functional equation that expresses (base-free) renormalizability. I invoked them for the first time in M1965h{H}, in my resolution of the Hurst puzzle concerning the River Nile.

2.5 The multiplicative multifractals of M1974f{N15} and M1974c{N16}; fixed points of a functional equation that generalizes that of Cauchy

M1974f{N15} and M1974c{N16} use a rule of renormalization that is described in the Annotations of Chapter N16 and had no antecedents in mathematics. This rule takes two forms, the more interesting being called "canonical." That canonical construction leads to an important functional equation that generalizes the Cauchy-Polyà-Lévy stability. For the sum $S_n = X_1 + X_2 + ... X_n$ of independent identically distributed variables, Cauchy's fixed point condition is that S_n should be renormalizable to have the same distribution as the addends. There is no gain of generality if one prescribes non-random weights W_j and considers the weighted sum $V_n = W_1 X_1 + W_2 X_2 + ... W_n X_n$. The novelty in my 1974 papers is that they take up the case when the weights are random, independent and identically distributed.

The fixed points of the resulting renormalization have been investigated with some care but would warrant additional work. Some of the many known properties play an important role in Section E6.3 of M1997E. As in the Cauchy-Polyà-Lévy "stable" case, the distribution of the fixed point is generally not known in closed form. An isolated exception due to J. Peyrière is described in the Appendix to Chapter N17.

N4

Filtering and specification of self-affinity

✦ **Abstract.** Chapter N1 argued that Ff^{-B} noises with the same exponent B can take any of many very different forms. This brief chapter takes the next step and faces the challenge of going beyond the spectrum and discriminating between those various possibilities. Two very different questions come to mind.

Is it legitimate to view the measurements performed on a finite sample as estimating an underlying Wiener-Khinchin spectrum? The key issue is whether or not the prefactor F is independent of the sample size T.

It is useful to evaluate and compare the Fourier spectra of suitably chosen non-linearly filtered functions $G[X(t)]$, exemplified by x^q. Often, the filtered function is also a f^{-B} noise, at least for low frequencies. If so, the B exponents that correspond to different filters G constitutes a (finite or infinite) "signature" of the process. This chapter focusses on three forms of $1/f$ noise: Gaussian, "dustborne," and multifractal, and shows that each has its own very distinctive signature. ✦

ALMOST LIKE A MANTRA, Chapters N1 and N2 repeat that a Fourier spectrum proportional to f^{-B} ought to be interpreted one of many possible symptoms of underlying self-affinity, and that the spectrum fails to discriminate between a broad range of very different kinds of behavior. Some are observed in nature, for example in turbulence and diverse electronic $1/f$ noises. Other are observed in mathematical constructions, therefore under fully controllable conditions. Examples include Gaussian noises (to be discussed in M 1998H), noises that will be described as "dustborne" because they only vary at time instants that belong to a fractal dust (to be discussed in Parts II and III), and multifractals (to be discussed in Part IV).

With experience, the simplest fully controllable examples become readily distinguished by the visual appearance of the sample functions. But (to invert a statement often made in the Preface) geometry is not enough. Several analytic challenges arise immediately, to which partial and separate responses are made explicitly in M1967i{N9}, implicitly in my work on multifractals and explicitly in Taqqu 1975, which reports on a Ph.D. thesis I suggested and supervised. Those challenges became more insistent due to recent work in finance reported in M1997E. This chapter has rather little to report, its main role being to draw attention to the following very important issues.

A) Strict adherence to the mathematical definition of self-affinity involves identity between graphs of functions. Therefore, it involves an infinity of equalities between probability distributions. Furthermore, each probability distribution is a function: it is in general equivalent of an infinite number of parameters. It follows that a complete confirmation of self-affinity is beyond any conceivable experimental measurement. Therefore, given an empirical $X(t)$, one would welcome down-to-earth criteria to test whether $X(t)$ can safely be taken to be self-affine.

B) A more realistic challenge follows from the physicists' need for numerically parameterized functions. This challenge splits into two: B') Set up a representative catalog of interesting and conveniently parametrized alternative behaviors compatible with a $1/f$ spectrum. B") Develop or identify objective and quantitative criteria that involve a finite number of steps and usefully distinguish between those different behaviors.

Section 1 will focus on the prefactor F and the remainder of the chapter on the exponent B.

1. The prefactor F and the discrimination between two forms of the spectrum: Wiener-Khinchin and conditional

Suppose that a sample of total length T is known for values of t multiple of ε, and that spectral algorithms applied to this sample yield a spectral density of about Ff^{-B}, where F is independent of f. The papers reprinted in Part III include a message that deserves increased attention.

According to the Wiener-Khinchin theory, F is a random variable whose expectation is independent of T, whose variance (if smoothed) tends to 0 as $T \to \infty$, and whose sample dependence can be controlled.

But none of those properties of F can be taken for granted. As a matter of fact, suppose that spectra of different data sets can be renormalized to yield a beautiful collapse on a line of slope $-B$ in doubly logarithmic coordinates. Very often, for example in turbulence, the data sets yield separate straight lines. The collapse is only obtained by forcibly translating those lines so that all go through an arbitrarily chosen point. The need for this procedure suggests that F depends on the sample length T and on the sample itself. For example, it depends on whether the sample came from a region of low or high overall level of turbulence.

It is to show how these properties can fit together that I constructed the examples described in Chapter 8 and 9, for which the spectral density has effectively the form $F T/(1 + f^B T^B)$, where the (redefined) prefactor F is still a random variable but no longer depends on T.

The presence or absence of the treacherous possibilities that hide in the prefactor may help discriminate between different possibilities.

2. The method of non-linear transforms and filters

Many effective responses to the challenges listed early in this chapter belong to a single procedure. Its most important ingredient consists in replacing t and/or x by non-linear transforms $\Gamma(t)$ and/or $G(x)$, and investigating the transformed function $G(\Gamma)$. This approach went farthest in the cases when either t or X is tranformed, but not both.

2.1 Non-linear transforms $G(X)$, while keeping $\Gamma(t) = t$, followed by a Fourier analysis of $G[X(t)]$

To an electrical engineer, the analysis of a signal $X(t)$ is not complete if not followed by the analysis of "rectified" forms of this signal. This practice is not restricted to $1/f$ noises, in fact, is ordinarily applied to noises having a bounded spectral density or spectral peaks away from $f = 0$. Furthermore, rectified functions of the form $|X(t)|$ or $X^2(t)$ were also used (independently, or so it seems) in the analysis of price changes found in Olsen 1996 and Voss 1992. This procedure deserves to be generalized.

Starting with an $1/f$ noise, the question is how do diverse transformations G affect the spectral density? In the background resides an optimistic expectation: even though (once again) self-affinity cannot be proven by any measurement, it "should" manifest itself by the fact that every $G[X(t)]$ has a spectral density that is itself represented by a power law.

Actually, this optimistic expectation often fails. But in cases where it holds, the original exponent B and the exponent $B(G)$ after transformation may be either equal or distinct. In case this $B(G)$ is defined, it is to be counted as a quantitative characteristic of $X(t)$, in addition to the original B. Since the latter corresponds to the linear transformation $G(x) = rx$, it can be called the "linear B." In the spirit of mathematical statistics, one can agree on sequence of functions $G(x)$ that will be used for successively finer tests, until an agreed-upon "stopping rule" is activated, the search is stopped, and two $1/f$ noises are declared to be either identical or different.

The quality of an investigation can be assessed by the appropriateness of the choice of $G(x)$.

2.2 The choice of G and the size of the signature

The two inputs of a non-linear transformation are the original process and the transform. Sections 3 to 5 examine how different choices of G affect each basic class of $1/f$ noises listed in Chapter N1: fractal dustborne noises, multifractals, and the Gaussian $1/f$ noise, which will be exhaustively tackled in M1997H. The simplest transformations are $G(x) = e^x$; $\log x$; $G(x) = x^q$ for integer q and $G(x) = |x|^q$ for arbitrary q. Next in complication is the "symmetrized" form of $|x|^q$ defined as $G(x) = \pm |(x)|^q$, with $+$ and $-$ being chosen at random with equal probabilities. Also very useful is the following function: $G(x) = x$ if $|x| < \varepsilon$, but $G(x) = \varepsilon$ if $x > \varepsilon$ and $G(x) = -\varepsilon$ if $x < -\varepsilon$; this function G and $|G|$ only record the times when $X(t)$ crosses zero. The Gaussian case has specific requirements, being best understood by taking for G the Hermite polynomials, which begin with x and $x^2 - 1$.

As to the signature, its size may be viewed as smallest when all the $G(B)$ are functions of one another, and largest when there is an infinite number of $G(B)$ related to one another by inequalities.

2.3 Transformation of the time parameter from the future clock time $t > 0$ to the exponential time $\theta = \log t$

This transformation applied to a self-affine function of t yields a stationary function of θ. The classic case is that of the ordinary Wiener Brownian motion $B(t)$. Before t is made into exponential time, $B(t)$ is transformed linearly into the non-stationary function $\tilde{B}(t) = t^{-1/2} B(t)$. At the point $t = 0$ (and only at that point), $\tilde{B}(t)$ is self-affine with $H = 0$. That is, for every multiplier h and collection of times t_n, the joint distributions of the functions $\tilde{B}(ht_n)$ are independent of h and identical to those of $\tilde{B}(t)_n$. Now move on to the exponential time $\theta = \log t$ and write $\tilde{B}(e^\theta) = \tilde{B}_0(\theta)$, $\log h = \Delta\theta$ and $\log t_n = \theta_n$. From the self-affinity of $B(t)$, the joint distribution of the quantities $\tilde{B}_0(\theta_n + \Delta\theta)$ is independent of $\Delta\theta$. That is, $B_0(\theta)$ is stationary.

3. Behavior of fractal dustbound 1/f noises under non-linear transformations

Once again, "dustbound" or "dustborne" noises are those which vary only for "non-trivial" time values, defined as belonging to a fractal dust. It is in the context of the dustbound noises that I first observed that the spectral exponent B can be changed by a simple transformation G. This finding was reported, well ahead of its time, in M1967i{N9}. In two of the examples studied in that work, the values of the function for different non-trivial values of t are independent. Non-trivial times are defined as belonging to a set being investigated. When that set is fractal, one needs to introduce normalizations beyond the scope of this chapter.

The function V defined in M1967i{N9} satisfies $V = 1$ for nontrivial times, and $V = 0$ elsewhere. But it can be generalized by having its values for non-trivial times be independent samples of a random variable with $EV > 0$. When the dust is a Lévy dust of dimension D, the spectral density depends on D, that is, distinguishes between different supporting dusts.

For the white function $X(t)$, one has $EX = 0$ and $EX^2 = 1$ for non-trivial ts and $X = 0$ elsewhere. The spectral density of $X(t)$) is f^{-0} independent of D. However, $|X(t)|$, $X^2(t)$ and all other positive tranforms of $X(t)$ are V-functions. Therefore, the whiteness of $X(t)$ can be described as "unstable". This fact reveals a "blind spot" of spectral analysis that is examined in Chapter E6 of M 1997E. Thus, a nonlinear transformation suffices to distinguish between unstable and stable occurrences of white spectrum.

A third example of dustborne function is the "coindicator" or coin function $W(t)$. It is defined by the following properties: $W \neq 0$ for all non-trivial t ; over every gap (defined as an interval between two points in the dust), $W =$ constant; the values of W over distinct gaps are independent samples of a random variable with $EW = 0$ and $EW^2 = 1$. When the dust is a Lévy dust of dimension D, the spectral densities of the functions $W(t)$ and $|dW(t)|$, have different exponents B. As to the transform $|dW(t)|$ itself, a calculation too delicate to include here shows that the covariance is integrable, therefore the spectrum is $\sim f^{-0}$ when f is small.

4. Behavior of multifractals under power law and logarithmic non-linear transformations

By construction of the multiplicative multifractals examined in Part IV, the corresponding expression $\log(d\mu)$ is a f^{-B} noise with $B = 1$, i.e., a proper $1/f$ noise. In other words, the intrinsic transform $G(x) = \log x$ preserves the 1/f property, but erases all other properties of the multifractal. Therefore, it is not useful.

To the contrary, the effect of power-law filters $G(x) = x^q$ on multifractals is highly non-trivial and easy to determine. It follows by simple reinterpretation from the definition of the fundamental function $\tau(q)$. Since a multifractal measure is ≥ 0, the power q may, but need not, range from $-\infty$ to ∞. The spectral exponent B is determined by $\tau(2) = D_2$ which is why D_2 is called correlation dimension of $d\mu$. "Normally," the correlation dimension $D_2(q)$ of $(d\mu)^q$ is simply

$$D_2(q) = \tau(2q).$$

Thus, the quantity $\tau(q)$, introduced for altogether different reasons, proves to be a special example of the general nonlinear transform method.

5. Behavior of pre-Gaussian 1/f noises under nonlinear transformations

Physicists believe that they never need to think about details of probability theory. In the study of mild randomness, this neglect tends to be innocuous, but not in the study of 1/f noise. The practical significance of precise versus approximate Gaussianity is a suprisingly delicate point, and the reader is directed to a Post-publication Appendix to Chapter N9.

Transforms of Gaussian 1/f noises and the findings in Taqqu 1975, 1979. Consider a Gaussian random function $Y(t)$ of spectrum proportional to f^{1-2H} and nonlinear map $G(Y)$. Taqqu 1975 reports the following facts:

• $G(Y)$ is a 1/f noise for large values of f.

• The exponent $B(G)$ of the spectrum of $G(Y)$ depends on H and on G in unexpected fashion. When $G(x) = x^q$, then (contrary to the multifractal case of Section 4) it is *not* true that $B(G)$ depends on q. In fact, the transforms $G(x) = x^q$ are no longer the most intrinsic test functions; instead, $B(G)$ depends on an integer called the "Hermite rank of the nonlinear transform $G(Y)$." This rank is defined as the order of the first non-vanishing term in the decomposition of $G(Y)$ into a Hermite series.

Best understood are ranks 1 and 2. Rank 1 leads to $B(G) = B$, independently of G; it is encountered for $G(x) = x$ (the first Hermite polynomial) and all antisymmetric Gs. Rank 2 is encountered for $G(x) = x^2 - 1$ (the second Hermite polynomial) and all symetric Gs. The value of $B(G)$ depends on H.

Composite of several published texts   **N5**

Short excerpts (1964-1984)

Table of contents of the chapter, together with references to the sources

1. Failure of the spectral analysis of turbulence (M1964w)
2. Fractal dimension and additional parameters
   needed to characterize the intermittence of turbulence (M1972f)
3. Comment distinguer entre le turbulent et le laminaire? (M1975O)
4. Objets relativement intermittents, ouverts et partout denses (M1975O)
5. Relative intermittence (M1982F)
6. Exponents of intermittent turbulence:
   fractal dimension and inequalities involving it (M1978h)
7. Decomposable dynamic systems (also known as IFS) (M1982F)
8. Some random fractal measures (M1984w)

● *Chapter foreword.* Most pieces of this chapter are intended to document early stages in the development of multifractals, starting years before that notion began to attract broad interest. I coined the word *fractal* in 1975 and the word *multifractal* was coined after the period covered by this chapter.

Books or portions of books are rarely incorporated in *Selected Works* or *Selecta*. In this case, however, it seems useful to include, as Sections 3, 4, 5 and 7, several excerpts from M1975O and M1982F *(FGN)*. The excerpt from M1975O was the first "popular" presentation of the proto-multifractals mentioned in Chapter N2 and of my work that followed in 1974.

Insertion of old unpublished material is also largely left to the editor's discretion. I felt that this volume will be improved by the inclusion of two texts (Sections 1 and 2), which are excerpts, respectively, from a Research Report and a reasonably widely circulated draft.

Finally, Sections 6 and 8 are from wide-ranging surveys written, respectively, for the physicists and the mathematicians. The remainder of those papers does not belong in this book.

Thus, the original dates of the purely expository short pieces exceed the 1962-1976 span of the main papers. The order is chronological, except that Sections 5 and 6 were exchanged, to make Section 5 follow Section 4, of which it is a translation. Deletions are marked by ellipsis points (...), except in Section 1. When the continuity of an argument must be preserved, long deletions are summarized within square brackets [ ].

Until 1975, I was the only scientist using fractals and multifractals, the usefulness or even necessity for multifractals being particularly difficult to communicate. In despair, I chose to "outflank" the opposition by crossing (without doing any harm!) into the neutral territory of geography. M 1967s investigated (fractal) dimension in the context of the now-famous question, "How Long is the Coast of Britain?" This outflanking maneuver succeeded with M1975O; however, that very early mention of multifractals (without the name) attracted the attention of few readers.

Additional short excerpts are found in some Chapter Forewords. •

1. Failure of the spectral analysis of turbulence (M1964w)

It is known that Wiener's spectral theory of the "second-order" stationary stochastic processes was expressly tailored to provide a mathematical model of incoherent colored light. It fulfills this aim outstandingly, and is also strikingly useful in many other areas. Its attempted application to the study of turbulence in fluids (G.I. Taylor, Von Karman), however, has proven very puzzling and ought to be controversial. On one hand, Kolmogorov, Obhukov, Onsager and von Weizsäcker, using purely dimensional considerations, independently predicted that, if the Reynolds number is very large, the measured spectral density $S(\omega)$ should be proportional to the power $-5/3$ of the frequency ω. Happily (and surprisingly), this "$\omega^{-5/3}$" law does indeed constitute an excellent prediction. From most other viewpoints, however, spectral analysis is unsuccessful or at least very questionable. Its most important failure [concerns] the expression $S(\omega)\,\omega^{5/3}$. [As observed in] Obukhov 1962, "although each measurement is in satisfactory agreement with a $-5/3$ power law in a certain range of scale, $S(\omega)\,\omega^{5/3}$ varies from measurement to measurement. These slow fluctuations of energy dissipation are due to the large scale processes in the observation region, or 'weather' in a general sense."

In fact, various empirical studies of turbulence have yielded so many properties that seem mutually contradictory that it may be best to begin by demonstrating that they are mutually compatible, while postponing for a while the task of relating each of these properties to the Navier-Stokes equations. I believe that this task requires a substantial generalization of the usual concepts of random variable, of stationarity,... of spectrum and even of stochastic process. On reflection, this unpleasant finding is hardly a surprise: much of the body of mathematics is made of concepts that had at one time expressed the latest knowledge about the physical world, therefore new facts may well require new concepts.

Clearly intermittent phenomena, analogous to atmospheric turbulence, are quite widespread in nature, and features uncannily parallel to those of turbulence are found in the self-similar model of the distribution of errors in communication circuits advanced in M1965c{N7}.

In studying an intermittent phenomenon, there seems to be an irresistible urge to subdivide it immediately into quiescent and active periods .... As a result, the common vocabulary embodies an informal science that often disguises the intermittent nature of a phenomenon, and the present [Research Report] led me to infer that many ... [of the] data were obtained from an intermittent function by excising the quiescent periods. From the experimentalist's viewpoint, this leads to questions such as the following: Suppose that one is presented with two wind tunnels, one which generates a steadily turbulent flow and the other which generates a very intermittent one. The usual attitude is to consider the former tunnel better; one may wonder, however, whether the latter is not really closer to free atmospheric turbulence that is not constantly restarted by energy injected through large "eddies."

Analogous censorship may be especially widespread in the case of the intermittent phenomena encountered in economics.

I conjectured that the observed erratic behavior of the spectra [of turbulence] was due to difficulty in actually separating the laminar and turbulent portions, e.g., to the difficulty of determining the actual duration of time over which the total observable turbulent energy is spread. This conjecture goes deeper than Kolmogorov's classic considerations about the self-similarity properties enjoyed by the purely turbulent regions. I believe that if turbulent flows are indeed scale-invariant, the idea of self-similarity should be modified to include laminar inserts as well. The present [Research Report] is devoted to an examination of this generalized self-similarity.

2. Fractal dimension and additional parameters needed to characterize the intermittence of turbulence (M 1972f)

Part I of this paper [titled *On Fractally Homogeneous Intermittent Dissipation*] has served to introduce the fractal dimension D, stressing the autonomous and nearly exclusive importance and explanatory power exhibited by D when a quantity describing turbulence is distributed uniformly on certain fractals. The present Part II [titled *On Fractally Non-Homogeneous Intermittent Dissipation*] goes one step further, and demonstrates the limitations of D. It describes another class of fractal patterns generated by cascade models, [on which] the distribution of stuff is non-homogeneous. A first result is that the value of D, while still important, provides a very incomplete characterization. It still suffices to determine the dimensions of the intersections. But [consider] the exponent occurring in the expression for the flatness factor, or the correction to the 5/3 law; in both cases, the value of D provides only an upper or a lower bound. As a matter of fact, the number of different parameters continues its regrettable growth. [Part II described two such parameters] (in addition to D), and now we encounter at least one more. We also describe a first and a second order approximation, which provide two very different ways of relating the various parameters to D; they are, respectively, akin to the Novikov-Stewart model, and to the Obukhov-Kolmogorov-Yaglom-Novikov log-normal model. Each is bound to be of use somewhere, but neither is more than an approximation.

3. Comment distinguer entre le turbulent et le laminaire? (M1975O)

Du point de vue fractal, l'aspect le plus nouveau de la turbulence [est] la nécessité d'étudier des objets dans l'espace à travers les coupes... D' après von Neumann 1949, la turbulence se concentre sans doute "dans un nombre asymptotiquement croissant de chocs affaiblis." Sur nos coupes unidimensionnelles, chacun desdits groupes de chocs apparaîtra comme une rafale de points.... On verra ainsi percer...] structure cantorienne... [Cependant,] dans les intermissions laminaires, l'écoulement du fluide ne s'arrête pas, mais devient simplement beaucoup plus régulier qu'ailleurs.... [Le même phénomène se rencontrerait dans le cas des erreurs de téléphone] si on regardait, non pas seulement les erreurs, mais le bruit physique qui leur est sous-jacent. Ceci suggère que l'on construise un modèle de la turbulence et du bruit [en question] en deux approximations. La première... nous ramèn[era] au schéma cantorien; la deuxième... supposera que tout cube de lespace contient au moins un peu de turbulence.... Dans ces conditions, si on néglige la turbulence là où son intensité est très faible, on se ramènera, à peu de chose près, à l'approximation [de ce chapitre].

4. Objets relativement intermittents, ouverts et partout denses (M 1975O)

En décrivant la généralisation à laquelle nous allons nous adresser [maintenant, il est bon] de se rappeler la nature d'une approximation faite dans plusieurs de nos applications. En discutant des erreurs en rafales, nous refoulions notre certitude qu'entre les erreurs, le bruit sous-jacent faiblit, mais sans cesser. En discutant des distributions stellaires, nous refoulions notre connaissance de l'existence d'une matière interstellaire, qui risque, elle aussi, d'être distribuée très irrégulièrement. Et en discutant des feuilles de turbulence, nous tombions à notre tour dans le panneau de l'a-priori, tout au moins dans notre première approximation, en admettant une image du laminaire où il ne se passe rien. Nous aurions également pu, sans avoir à introduire d'idée essentiellement nouvelle, consacrer un Chapitre à la distribution des minéraux: entre les régions où la concentration est suffisamment forte pour justifier l'exploitation minière, la teneur d'un métal tel que le cuivre devient faible, très faible même, mais on doute qu'il y ait une région qui [en] soit absolument dépourvue. Tous ces vides, il faut les remplir, en espérant qu'en ce faisant nous n'allons pas trop affecter les images [fractales!] déjà établies. Nous allons maintenant esquisser une manière de s'y prendre, qui convient lorsque les objets et les intermissions sont de même nature et ne diffèrent que de degré. Ce Chapitre sera relativement technique et sec.

L'ensemble de Besicovitch. Dans le contexte triadique original de Cantor, [laissons] chaque étape [de construction] diviser la masse initiale en parties égales, respectivement, à $p_1, p_2,$ et p_3.... Quand on aura répété la procédure à l'infini,... presque toute la masse se concentrera sur 3^{kD} intervalles de longueur 3^{-k}, où l'on a:

$$D = - \sum p_i \log_3 p_i.$$

... Formellement, D est une "entropie" au sens de la thermodynamique, ou encore une "information" au sens de Shannon (voir Billingsley 1967), mais cette analogie ne va pas bien loin. Plus important, D .... est une dimension de Hausdorff. Toutefois, l'ensemble de Besicovitch est ouvert, [et de ce fait la dimension de Hausdorff] se trouve perdre beaucoup de ses propriétés, y compris certaines qui possèdent un intérêt pratique direct, et auxquelles nous commencions à être habitués!

5. Relative intermittence (M1982F)

[Many of the discussions in Chapters 1 to 38 of *FGN* deliberately] negate some unquestionable knowledge about Nature.

For example, we forget in Chapter N9 that the noise that causes fractal errors weakens between errors but does not desist.

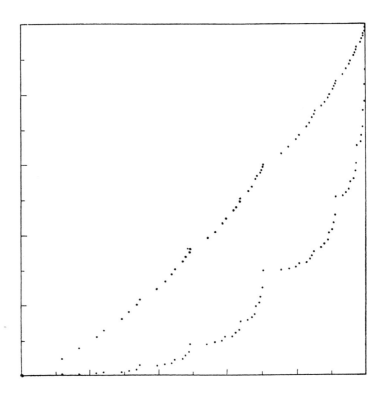

FIGURE N5-1. L'escalier de Besicovitch. C'est une contrepartie de l'escalier du diable, qui ne varie que sur un ensemble fractal partout dense... Entre les abscisses 0 et 0,7, l'ordonnée augmente, soit de 0,6 (courbe en haut), soit de 0,3 (courbe en bas)... Si l'on divise l'abcisse en un grand nombre de petits segments, une très petite proportion de ceux-ci rend compte d'une très grande proportion du déplacement vertical total.

{P.S. 1998. This is Figure 27 of the first edition of M1975O. In current terminology, the above two dotted curves represent $\mu[0, t]$ for two binomial measures. Confirming the "proto-multifractal" character of the construction, this figure is reproduced from Dubins & Savage 1965.}

We neglect in Chapter N9 our knowledge of the existence of interstellar matter. Its distribution is doubtless *at least* as irregular as the distribution of the stars.

And in Chapter N10 the pastry-like sheets of turbulent dissipation are an obviously oversimplified view of reality.

The end of Chapter N9 very briefly mentions the fractal view of the distribution of minerals. Here, the use of closed fractals implies that, between the regions where copper can be mined, the concentration of copper vanishes. In fact, it is very small in most places, but its concentration cannot be assumed to vanish anywhere.

In each case, some areas of less immediate interest were artificially emptied to make it possible to use *closed* fractal sets, but eventually these areas must be filled. This can be done using a fresh hybrid. To take an example, a mass distribution in the cosmos will be such that no portion of space is empty, but, for every set of small thresholds θ and λ, a proportion of mass at least $1-\lambda$ concentrates on a portion of space of relative volume at most θ.

6. Exponents of intermittent turbulence: the fractal dimension and inequalities involving it (M1978h)

The different physical properties of a system near a critical point are well-known (as we saw in the case of percolation) to be distinct exponents related by inequalities that become equalities under the scaling assumption. I have established that a different but somewhat analogous situation holds true for intermittent turbulence. The first difference is that in turbulence, dimension is a new exponent, which does not *in general* reduce to a combination of old exponents. Also, inequality between exponents is compatible with some form of scaling, while equality demands an especially strong form of scaling. Furthermore, my argument is *not* based upon any a priori analogy presumed to exist between turbulence and critical point behavior or any other chapter of physics. An earlier theory of intermittence due to Kolmogorov and Yaglom, can be characterized by the use of the lognormal *Ansatz*. This theory had not recognized that there is a multiplicity of distinct exponents, because it claimed to be able to deduce all the basic properties from one constant, μ.

The present section is a very sketchy overview of results given elsewhere (M1976o{N18}). First, recall that the classical picture of intermittent turbulence, as it emerged in 1962, involves several exponents.

B: The spectral density of turbulent velocity takes the form $k^{-5/3-B}$. The term B is a correction to the classic Kolmogorov exponent 5/3.

Q: The covariance of local dissipation takes the form R^{-Q}. The variance of the dissipation contained within a ball of radius R is R^{2-Q}.

M_q: The q-th moment of the dissipation contained within a ball of radius R is R^{M_q}. (In particular, $M_1 = 1$ and $M_2 = 2 - Q$.)

Expressed in terms of B and Q alone, the main result of M 1976o{N18} resides in the basic inequality

$$0 \leq 3B \leq Q.$$

In the Kolmogorov-Yaglom model, $3B$ is much smaller than Q. However, in a different model which I proposed to call *absolute curdling* (in the case of turbulence, it was pioneered by Novikov & Stewart), $3B$ is identical to Q.

To make the above inequality fully meaningful, fractal geometry adds another exponent D, which is nonclassical and can be interpreted as a similarity dimension.

D: Let a basic spatial cube of side L be paved by small cubes of side ζ, with $\zeta/L \ll 1$. Intermittent turbulence is characterized by an exponent D such that dissipation is either wholly or overwhelmingly concentrated in a very small set that can be covered by $(L/\zeta)^D$ of these small cubes. This new exponent D clearly satisfies $2 < D \leq 3$. Inserting this quantity D transforms the basic two inequalities into the three inequalities

$$0 \leq 3B \leq 3 - D \leq Q.$$

Continuing with these inequalities and the classical probabilistic inequalities between moments, we reach the following result: there must exist a cup convex function $\varphi(q)$ passing through the following points:

$$(2/3, B), \quad (1, 0), \quad (2, 2 - Q = M_1), \quad \ldots, \quad (q, M_q).$$

Equality of $3B$, $3 - D$ and Q occurs when the graph of this function $\varphi(q)$ reduces to a straight line. A consequence of equality is that all exponents can be deduced from any one of them.

The exponent D should be viewed as first among equals because it relates to the problem's basic geometric substrate. Geometrically speaking, a special and unique virtue of a straight $\varphi(q)$ is that it requires the dissipation to be uniform or *homogeneous* over the *closed* subset of space to which it is restricted. The carrier of such dissipation is akin to a cluster in Hammersley percolation in the sense that it has a well-defined boundary and is "all skin" with no significant "inside."

Every other $\varphi(q)$ implies that the bulk of dissipation is homogeneous over a small set, but some dissipation must be considered separately, in the form of two small remainders. One remainder is spread around, and another remainder is concentrated in sharp peaks. These terms are unexpectedly significant, in the sense that different moments of the dissipation turn out to be very much affected by one or by the other.

For example, if the graph of $\varphi(q)$ is not straight, its next simplest form is a parabola. This $\varphi(q)$ happens to be implicit in the lognormal *Ansatz*, and indeed the approximations that underlie the *Ansatz* happen to be equivalent to curve-fitting the above $\varphi(q)$ by a parabola. The value of B is affected inordinately by the low "density" remainder measure, and the value of Q is affected inordinately by the second remainder measure. It is difficult to avoid postulating that this influence is of limited physical significance and that the basic physics is mostly dependent upon D. {P.S. 1998. The *Annotations* comment on this last sentence.}

It may be observed that the original derivation of the above inequalities, in M1976o{N18} begins with a specific model I call *curdling*. In this model, a cup convex function $\varphi(q)$ is a basic assumption. A linear $\varphi(q)$ corresponds to *absolute curdling*. However, the preceding argument suggests that the basic distinction between closed and open carriers of dissipation can be established without invoking curdling; it is, therefore, natural to conjecture that the validity of the basic irregularities is not dependent on any specific model.

As a useful restatement of absolute curdling, Frisch et al. 1978 devised the "β-model." It is more specific on some points and less demanding on others but yields identical results. See also Siggia 1977. {P.S.1998. This issue is discussed in several places in Chapter E2.}

Universality. The D of turbulent dissipation may or may not be universal. The question is open and deserves attention.

Some other aspects of turbulence involving fractal dimensions also raise the issue of universality. For example, the boundary of a domain resulting from turbulent dispersion is a fractal (see M1977b); is its D a universal exponent?

7. Decomposable dynamical systems (also known as IFS) (M1982F)

[There are many dynamical] systems with geometrically standard attractors but interesting repellers. To invert the roles of these two sets, thus making time run backward [demands a little thinking whenever, as is the case for rivers] the path is uniquely determined in the downhill direction, but in the uphill direction each fork involves a special decision.

For example, [the map] $x \to \lambda x(1-x)$ has two possible inverses [hence] inversion requires choosing between two functions. In other examples, the number of possibilities is even larger [and] we want them to be selected by a separate process.

We demand that one of the coordinates of the state $\sigma(t)$ – call it *determining index*, and denote it by $\sigma_0(t)$ – evolve independently of the state of the other $E-1$ coordinates – call it $\tilde{\sigma}(t)$ – while the transformation from $\tilde{\sigma}(t)$ to $\tilde{\sigma}(t+1)$ is determined by both $\tilde{\sigma}(t)$ and $\sigma(t)$. In the examples I studied most, the transformation $\tilde{\sigma}(t) \to \tilde{\sigma}(t+1)$ is chosen in a finite collection of G different possibilities ... selected according to the value of some integer-valued function $g(t) = \Gamma[\sigma_0(t)]$. Thus, I studied dynamics in the product of the $\tilde{\sigma}$-space by a finite index set.

In fact, in the examples that motivate this generalization, the sequence $g(t)$ either is random or behaves as if it were .. [A] serious difficulty [is] that dynamical systems are the very model of fully deterministic behavior, hence are forbidden to accommodate randomness! However, one can [take for] $g(t)$ the value of a sufficiently mixing ergodic process.

Plates 198 and 199: attraction to fractals. {P.S.1998: See Figure N5-2 and the P.S. at the end of this section.} These two shapes illustrate long orbits of two decomposable dynamical systems. The *Pharoah's Breastplate* in Plate 199 is self-inverse..., being based upon 6 inversions selected to insure that the limit set is a collection of circles. The *San Marco dragon* in Plate 198 is self-squared..., being based upon the two inverses of $x \to 3x(1-x)$.

The determining index is chosen among 6, respectively 2, possibilities, using a pseudorandom algorithm repeated 64,000 times. The first few positions are not plotted.

The neighborhoods of cusps and self-intersections are very slow to fill.

{P.S. 1998. This book's cover and jacket illustrate Plate 199 in color, with gratefully acknowledged assistance of K. G. Monks. The algorithm, inplementing the "osculating open discs" described in Chapter 18 of M1998F {FGN} and M1983i, outlines the exact attractor much faster than the IFS. The first step is to draw an "osculating base;" in Figure N5-2, it is

FIGURE N5-2. Small-size reproductions of Plates 198 and 199 of M1982F {FGN}.

{P.S.1998. This figure also contains a diagram that explains the construction of Plate 199. The generator is made of six circles (drawn as thin lines), each tangent to four of the remaining ones. The corresponding decomposable dynamic system or IFS consists in ordinary geometric inversions with respect to those circles. Applied to a starting point, a pseudo-random sequence of inversion indexes creates a discrete orbit, which is plotted as a sequence of points and outlines the actual attractor. The completeness and precision of the outline increases with the length of the orbit. The "density" of the outline is a multifractal, possibly the first to be drawn in this fashion.}

made of the eight open discs with the following property. The circles that bound them (drawn as thick lines) are orthogonal to three of the generating circles and do not intersect the other three. Each generates an open set, called "clan," that is made of its images with respect to every sequence of inversions in the generating circles. Up to eight distinct colors are chosen and the open discs belonging to a clan are given the same color. On the cover and jacket, four of the colors are black. The basic result in the already-quoted Chapter 18 is that the attractor is made of those points in the plane that do not belong to any of the eight clans. The book's cover and jacket also show "gold threads" that create a "cloisonné" effect. These are the clans of a) the eight circles bounding the oscillating base, and the three circles that are orthogonal to four among the generating circles.}

8. Some random fractal measures (M 1984w)

Consider the following array of independent and identically distributed random variables (i.i.d.r.v.'s): b r.v.'s $W(g)$, followed by b^2 r.v.'s $W(g,h)$, then by b^3 r.v.'s $W(g,h,k)$, and so forth. Write the number $t \in [0,1]$ in the counting base b, as $t = 0, t_1, t_2, ..., t_n, ...$ Define

$$X'_n(t) = W(t_1)W(t_1, t_2)W(t_1, t_2, t_3)...W(t_1, t_2, ...t_n), \text{ and } X_n(t) = \int_0^t X'_n(s)ds.$$

Problems and conjectures: M1974c{N15} posed and partly solved many problems that concern variety of classes of W's. These problems concern the weak or strong convergence of $X_n(t)$ to a non-vanishing limit $X_\infty(t)$, the numbers of finite moments of $X(t)$, and the dimension of the set of t's on which $X(t)$ varies. *Partial answers*: Kahane & Peyrière 1976{N16} confirmed and/or extended these conjectures and theorems; for example, $E\ W \log_b W$ is a codimension.

The fixed points of related smoothing transformations of probability distributions. Multiplicative chaos. Take b i.i.d.r.v.'s W_g and b i.i.d.r.v.'s X_g having the same distribution as the $X(1)$ in the preceding paragraph. The weighted average $(1/b) \sum W_g X_g$ (with the sum from 0 to $g - 1$) has the same distribution as each X_g, meaning that $X(1)$ is a fixed point of the transform consisting in weighted averaging.

N5 ◇ ◇ SHORT EXCERPTS 129

&&&&&&&&&&& **ANNOTATIONS** &&&&&&&&&&&

Section 1. Further excerpts from M 1964w are reproduced in the *Foreword* of Chapter N10. When I first studied turbulence as a student at Caltech, spectral analysis of turbulent velocity was the primary technique of the sophisticate. No other tool was available, and spectra seemed safe, since the distribution of velocity variations seemed to be near Gaussian. In addition, it seemed that physically more meaningful "components" required for honest and safe spectral analysis could be identified, if only one could give better meaning to the notion of eddy.

A high but for many years controversial achievement of that day was the $k^{-5/3}$ spectrum of turbulent velocity, as obtained simultaneously by Kolmogorov, Obukhov, Onsager and Von Weizsäker. This was the only prediction about turbulence that one had been able to draw from pure thought (as Onsager did in his derivation of $k^{-5/3}$), or from cascades of eddies assumed to be real (as in the Kolmogorov derivation). The point of Section 1 was that a focus on spectra ceased to be natural when, before 1964, interest turned from velocity to turbulent dissipation, which is very far from Gaussian and highly intermittent.

Section 2. M1972f, a never-published draft of M1974f{N15}, survived in a publicly documented form, because a copy was sent to Uriel Frisch in Nice for comments. His copy was better filed than the original.

Sections 3 and 4. These sections in French are excerpted, respectively, from Chapters VIII and IX of M 1975O. They document that, in 1975, the notion of multifractal seemed sufficiently important to warrant a full chapter in my first book on fractals, plus mention in another chapter, but too complicated to be dwelt upon and not yet deserving a proper name.

Section 5. This is a free English translation of the first paragraph of Section 4. It first appeared in Chapter 39 of M1982F{*FGN*} as part of an entry that contains much important and not yet exhausted material. In that entry, *multifractals.* are called *non-lacunar fractals*, an overhasty and unfortunate term no one used again.

Section 6. This text documents the importance I attached in 1977 to the notion of multifractal (before the term), in particular, to the function $\tau(q)$ (before this notation). M1978h reported on an invited talk presented at the Statphys XIII Conference held in Haifa in 1977 for the IUPAP (Statistical Physics Conference of the International Union of Pure and Applied Physics.) The whole of M1978h might be reprinted in a later volume of these *Selecta*. The portion that concerns multifractals was Section 6 of M1978h (which became Section 6 of this Chapter). An erroneous sentence that contradicts Section 2 was corrected in a Post-script.

The fact that a multifractal has a power-law correlation, as noted in this Section, means that multifractals are $1/f$ noises, a fact stressed in this book's introductory chapters.

A question Section 6 asks explicitly was not faced again until 1983. Why is it that turbulent intermittence requires an infinite sequence (or "hierarchy") of exponents, while the critical phenomena of physics require only finite numbers of exponents? Much later, there was a hand-waving answer, that critical phenomena also involve additional exponents, which relate to the so-called "irrelevant" fixed points of the renormalization group. In the 1960s, no such phenomena were known, but it became widely agreed that it would be a good idea to seek them out.

Section 7. This text and figures, and other brief excerpt from M 1982F{*FGN*} is unlike this chapter's remaining sections. It is meant to motivate and explain one of the two representations of multifractals illustrated on this book's cover and jacket. A multifractal can be illustrated by a wiggly line that represents a "coarse-grained" measure. But it can also be "sampled" by using a construction I called "decomposable dynamic system". This construction was taken up and greatly developed by Michael Barnsley, and became widely known under the term used in Barnsley 1988, namely, *iterated function systems* (IFS). I make no claim of being the first to use this construction and I miss my own term for it, but IFS has by now taken root. The original caption of Plate 198 in M1982 {*FGN*} was brief to excess (and a typographical error transformed 6 into 4). The caption of Figure 2 is complete and meant to explain the colorful design on this book's jacket.

Section 8. This is an excerpt from an invited talk at the International Congress of Mathematicians held in Warsaw in 1983. Additional excerpts from M1984w might be reprinted in a later volume of these *Selecta*.

PART II: UNIFRACTAL ERRORS AND LÉVY DUSTS

This Part is primarily concerned with the properties and concrete use of the simplest grid-free unifractal model. The underlying constructions are random Cantor-like point sets and closely linked with the Lévy dusts on the line, as well as closely related discrete sets.

Chapter N6 reproduces Berger & M 1963 and describes how Cantor-like behavior was unexpectedly discovered in what might have seemed a run-of-the-mill problem of telephone engineering. As is well-known, fractal models soon became widely accepted. In particular, the reader acquainted with M 1982F {FGN} will recognize that a 3-dimensional generalization of Berger & M 1963 provided the "toy model" of fractality in the distribution of galaxies: that model is described in Chapter 32 of FGN and several papers scheduled for being either reprinted or first published in M 1998L.

Chapter N6 ends with a lengthy Post-publication Appendix *that describes additional observations motivated by Berger & M 1963. They confirmed the basic finding that the inter-event intervals follow a power law distribution. But different channels yielded different values of the fractal dimension D. Those discrepancies in the estimated dimension proved to be an early symptom of what was to develop later into the phenomenon of multifractality.*

The bulk of Chapter N7 investigates the properties of Lévy dusts that have been coarse-grained into sequences of intersected and empty boxes. The resulting theoretical inter-error distribution also has a practical aspect: it improves the fit between the theory and the data in Chapter N6.

Another part of Chapter N7 dwells on a fundamental fact: to be invariant by dilation and contraction, fractals cannot be invariant by translation. That is, stationarity must go, A good replacement is conditional stationarity. a new notion first introduced in Chapter N7.

New model for the clustering of errors on telephone circuits (Berger & M 1963)

✦ **Abstract.** This paper proposes a new mathematical model to describe the occurrence of errors in data transmission on telephone lines. We claim that the distribution of inter-error intervals can be well approximated by a Pareto distribution of exponent less than one. It follows that the relative number of errors and the information-theoretical equivocation tend to zero as the length of the message is increased. The validity of those claims is demonstrated with the aid of experimental data obtained from the German telephone network. Further consequences, refinements, and uses of the model are described. ✦

1. INTRODUCTION: COMPETING MODELS OF ERROR CLUSTERING

The model that we propose for describing the occurrence of errors is best described by contrasting it with two earlier alternatives.

Binary symmetric channels without memory. A first model postulates that the transmission of any given bit is not influenced by the correctness of the transmission of earlier parts of the message. Consider the probability $\Pr\{U = u\}$ that the inter-error interval U is equal to u. The reader will observe that random variables will always be designated by capital letters, while their possible values are denoted by the corresponding lower-case letters. The binary symmetric channel without memory yields the geometric distribution $\Pr\{U = u\} = (1-p)p^{u-1}$, which is the variant of the exponential law relative to integer variables. The number of errors in a sequence of N bits is a binomial random variable. However, many tests performed in recent years on actual channels have demonstrated that this model is grossly inadequate. Actual error data have been qualitatively described as appearing in bursts or bursts of bursts of errors, in addition to single, independent error events.

Channel having two or more states or levels of error susceptibility, with transition probabilities between these states. Within each state the occurrence of errors is taken to be described in terms similar to the binary symmetric channel, and the probability of error is assumed to depend upon the state. For example Gilbert 1961 postulates one good state with a zero probability of error, and one bad state with a large probability of error. However a good detailed agreement with the data can only be achieved by models with several states.

The model we propose in the next section is motivated by the qualitative observation that clustering of events as well as long periods without events were characteristic of certain processes governed by distributions of the Pareto type. The simplicity of these distributions warranted their being investigated for this new application.

The validity of a specific scientific claim cannot be influenced by facts concerning other disciplines. But we wish to point out that the fundamental idea and the techniques of this paper were suggested by the extensive work which one of the authors (B.B.M.) has devoted to "scaling" phenomena in many different contexts.

2. A NEW ALTERNATIVE MODEL

2.A. Statement of the fundamental postulates

Our explanation of error bunching incorporates three distinct postulates:

(A) We claim that, although the distance $T_{n+1} - t_n$ between successive errors is surely not ruled by a geometric distribution, the random variable $T_{n+1} - t_n$ is *statistically independent* of the numbers t_i which specify the positions of earlier errors ($i < n$).

(B) We assume that the distribution $F(u) = \Pr\{T_n - t_{n-1} < u\}$ *attributes to small values of t a probability very much larger* than predicted by the geometric distribution.

(C) We also assume that $F(u)$ *attributes to large values of u a probability very much larger* than predicted by the geometric distribution. As a result, medium-sized t will have an unusually small probability.

A simple way of satisfying (B) and (C) is to assume that $T_{n+1} - T_n$ is a scaling random variable, which means that

(B') $P(u) = 1 - F(u) = u^{-D'}$ for small u;
(C') $P(u) = 1 - F(u) = Pu^{-D}$ for large u .

In general, D' and D could be two different positive numbers. Things greatly simplify when $D' = D$ (implying that $P = 1$), but the general idea of our model is valid if $D' \neq D$. In fact, our different assumptions need not be satisfied with equal degrees of precision. (A referee has pointed out that our Assumption (C') has been anticipated. However, Mertz 1961 appears to share the general opinion on error clustering, and avoids anything resembling our Assumption (B'), by taking the law of Pareto under the truncated form $P(u) = (1 + u/h)^{-D}$. Note also that many of the statistical expression that he uses (e.g., long-term averages) are misleading because their population values are finite.)

2.B. Motivation of the fundamental postulates by the random walk process

Postulates (A) and (C') were suggested by the striking qualitative resemblance we saw between the empirical records of inter-error intervals and the sequences of returns to the origin in the classical game of tossing a fair coin. Let us restate the rules of that game. Henry and Thomas began to play (circa 1700 A.D.) with infinite fortunes; whenever their coin fell on "heads," Henry wins a cent from Thomas, and whenever it falls on "tails," it is Thomas who wins a cent from Henry. The behavior of $G(m)$, Henry's gain after m coin tosses, is well known to mathematicians and to some professional gamblers to be totally contrary to what is sometimes referred to as "intuition." Examine indeed our Figure 1.

By definition, the intervals between successive roots of the equation $G(m) = 0$ are independent random variables. There is no question, however, that the roots appear to be grouped in clusters and that there are violent fluctuations in the lengths of inter-root intervals. The complications of the sequence of roots is extreme, which is why most books on probability, particularly the older ones, contain extensive studies of coin tossing. A useful modern reference is Feller 1957.

The probability for the distance between roots is already given in Cournot 1843. Emile Borel wrote many popular books that discuss coin-tossing. His opponents were Parisian philosophers of 1900, but their "common sense" arguments are spontaneously re-invented by every American engineer of 1963, when he first approaches these problems. Let us quote from p. 48 of Borel 1914: "Suppose that the 2,000,000 adults living in Paris associate themselves in teams of two, and begin tomorrow morning to play at heads or tails until the winnings of both return to zero. If they work very fast, they may go through a play per second, that is, through 10,000,000 plays per year. Well, one must predict that after 10 years, 100 couples will still be playing, and that, if the players entrust their interests to their heirs, 10 games or so will still be continuing after 1000 years."

The fact that error clustering and the violent fluctuations in the bit-error rate of telephone lines recall the zeros of coin tossing suggests that

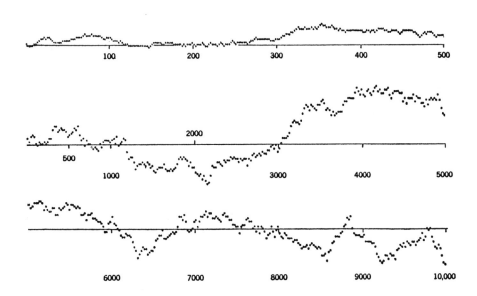

FIGURE N6-1. Record of Thomas's winnings in a coin-tossing game played with an actual coin, presumed to be fair. Zero-crossings appear to be strongly clustered, although the intervals between them are obviously statistically independent. In order to appreciate fully this Figure, one must note that the unit of time used on the second and third lines equals 20 plays. Hence, the second and third lines lack detail and each of the corresponding zero-crossings is actually a cluster or a cluster of clusters. For example, the details of the clusters around time 200 can be clearly read on line 1, which uses a unit of time equal 2. This illustration is adapted from Figure III, Feller 1957.

they *need not* be due to dependence between the inter-error intervals; both may be compatible with *independent* inter-error intervals, if they follow a suitable distribution.

Returning to the distance between successive roots of $G(m) = 0$, it is well known that its probability is nonzero only if m is even and is then equal to

$$\binom{1/2}{m/2}(-1)^{(m/2)-1}.$$

For larger m, this quantity is proportional to $m^{-3/2}$, which asymptotically follows Pareto's law $P(u) \sim u^{-D}$ with $D = 1/2$. We took this scaling behavior seriously, but replaced $D = 1/2$ by a more general D, and we were thus led to our assumption (C'). As to assumption (B'), it was added in order to account for the possibility that transmission errors may be even more clustered than the roots of the function $G(m)$ of coin tossing.

The novelty of our suggestions is that *they generate apparent patterns of "contagion" by a suitably chosen process of independent events. That is, they generate non-stationary sequences with the help of stationary processes* (for many similar examples, see the references in the last paragraph of Section 2A).

2.C. Further illustration of the clustering properties of scaling distributions

Our suggestion is further motivated by Figure 2, which is more fully explained in M 1960i. Consider three successive errors and suppose that the positions t_{n-1} and t_{n+1} are known, while T_n is unknown; the distribution of the random T_n will be studied under three basic assumptions.

When the inter-error distances are *geometrically* distributed, the position of the random T_n is *uniformly* distributed between the known t_{n-1} and t_{n+1}, so that the actual instants t_i are neither uniform nor bunched.

If the inter-error intervals follow a *binomial* law, a situation not encountered in practice, their distribution could be approximated by a Gaussian, and T_n would approximately be a Gaussian variable, having $(1/2)(t_{n-1} + t_{n+1})$ as its mean value. Errors would tend to be almost uniformly distributed.

But in our case $T_{n+1} - T_n$ is a scaling variable. The resulting density of $U' = T_n - T_{n-1} = u'$, conditioned by $U = U' + U'' = T_{n+1} - T_{n-1} = u$, is closely approximated by

$$\frac{D^2 u'^{-(D+1)}(u-u')^{-(D+1)}}{2Du^{-(D+1)}} \sim (D/2)u^{-(D+1)}\left[\frac{u'}{u}\left(1-\frac{u'}{u}\right)\right]^{-(D+1)}.$$

The main feature of this probability of T_n is that it has *two very sharp peaks*, located near the instants of time $t_{n-1}+1$ and $t_{n+1}-1$, and having equal amplitudes. [P.S.1998: this probability attains its minimum for $u' \sim u/2$, where it is very small]. As a result, the middle error tends to be close to either of the end errors and the three errors separate into an isolated one and a cluster of two.

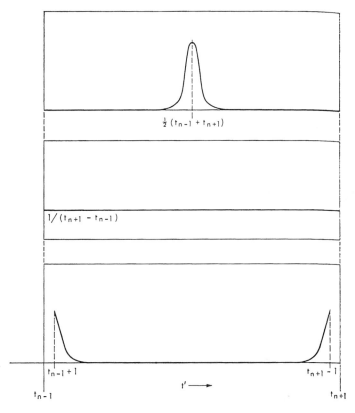

FIGURE N6-2. Three cases of the distribution of the inter-error interval $U' = T_n - T_{n-1}$, when successive intervals $T_n - T_{n-1}$ and $T_{n+1} - T_n$ are statistically independent, and when the value of u of $U = T_{n+1} - T_{n-1}$ is known. In each case, the horizontal scale is that of u', the vertical scale is that of the probability of u'. (For the sake of legibility, this probability is interpolated by continuous lines; the reader should recall that the geometric distribution is nothing but a discrete form of the exponential.)

Similarly, in the case of a large number N of errors, a sizeable portion of the total sample length is the contribution of a few of the longest error-intervals, say of L of them, thus creating a pattern in which errors are mostly grouped in L clusters.

It will naturally be desirable to explain the preceding phenomena, by reducing them to more elementary physical facts. We shall not attempt to do so in this paper. Our approach is "phenomenological," in the sense generally used to describe classical thermodynamics.

3. FIRST EXPERIMENTAL VERIFICATION OF THE NEW MODEL

3.A. Doubly logarithmic plots

The standard method of checking the law of Pareto for $T_{n+1} - T_n$ consists in plotting $\log \Pr\{T_{n+1} - T_n \geq u\}$ as a function of $\log u$, and examine how closely the resulting graph is approximated by a straight line. Unfortunately, this procedure has been neglected by statisticians, so that no small-sample tests are available and valid conclusions can be drawn only in the case of very long samples. The corresponding test of the independence of $T_{n+1} - T_n$ with respect to the past consists in plotting $\log \Pr\{T_{n+1} - T_n \geq u\}$ *for every pattern of past errors*. Unfortunately, again, the need for very large samples has limited us to the first-order transitions and some of higher order. Therefore, in varying u' from 1 to 100,000, we have performed the following tests.

We studied the function

$$\log \Pr\{T_{n+1} - T_n \geq u'', \text{ if } T_n - T_{n-1} = u'\},$$

which depends upon *three* successive errors:

Then the function $\log \Pr\{T_{n+1} - T_n \geq u''\}$, was studied in the following cases, which depends upon *five* successive errors:

either $T_n - T_{n-1} = u'$, $T_{n-1} - T_{n-2} = 1$ and $T_{n-2} - T_{n-3} = 1$,
or $T_n - T_{n-1} = u'$, $T_{n-1} - T_{n-2} = 1$ and $T_{n-2} - T_{n-3} = 2$,
or $T_n - T_{n-1} = u'$, $T_{n-1} - T_{n-2} = 2$ and $T_{n-2} - T_{n-3} = 1$,

Finally, we studied the distribution of the sum of *three* successive inter-error intervals, $T_{n+1} - T_{n-2}$, and compared it with the distribution of the sum of three independent random variables having the same distribution as $T_{n+1} - T_n$. It can be shown that, when the three variables are independent and scaling, the tails of the doubly logarithmic graphs of each addend and of the whole are identical, except for a vertical translation by an amount equal to log 3 (see for example M 1960i).

The plots of these three functions are given in Figures 3 through 6. While a more refined study of the model and the supporting empirical evidence will be made in Section 5, some general conclusions can be drawn without delay. Note that data are based on those described in Appendix I that correspond to the transmit level (in decibels) of –22 dbm. The approximate over-all sample size is indicated by reference to Table I of that Appendix I, which describes similar tests. Thus, approximately 8×10^7 bits were transmitted at this level and our sample of error-bit events is of the order of 10^4. Therefore, the sample size for the marginal distributions shown in Figures 3 and 4 is of the order of 10^4.

The plots of the marginal distribution $\Pr\{T_{n+1} - T_n \geq u''\}$ constitute the first test of the model. The fact that a straight line with slope of about 0.4 fits this curve very well over approximately four decades of inter-error intervals is a definite first-order verification of that portion of the model. The behavior in the last decade is not meaningful, as seen in Section 4.

The plots of the function listed $\Pr\{T_{n+1} - T_n \geq u'' \text{ if } T_n - T_{n-1} = u'\}$ are given in Figures 3 through 6 for various values of u'. The sample size is a decreasing function of u' and is of the order of 5000 events for $u' = 1$, 2000 for $u' = 2$, 1000 for $u' = 3$, 500 for $u' = 4$, 500 for u' between 10 and 20, 100 for u' between 50 and 60, and 300 for u' between 100 and 200. If the inter-error intervals were independent, we would find the same distribution for all values of u'. The shapes of the curves are indeed strikingly similar. Apart from the case of $u' = 1$, which is discussed further in Section 5, they yield at least a first-order verification of the hypothesis of independence. The ordering of the curves for $u' = 2, 3, 4$ does not appear significant at this time.

The plots of our second function are similar to those of the first, except that they single out some specific previous patterns which correspond to typical "bursts" of errors as discussed in the literature. The plots in Figures 5 and 6 are necessarily based on small samples of the order of a hundred events. Since they exhibit the same characteristics as the other curves, they further establish the hypothesis of independence and our thesis that clusters *per se* or "bursts" do not have a separate intrinsic meaning and identity.

The function $\Pr\{T_{n+3} - T_n \geq u''\}$, plotted in Figure 3, is based on a sample of several thousand events. Consistent with the assumption of independence, the vertical displacement by $\log 3$ yields good agreement.

These results, taken together, appear to establish quite definitely the main features of the proposed model. The geometric distribution, which is plotted in Figure 7 for several values of its parameter p, is most certainly inapplicable.

3.B. Joint distribution of successive inter-error intervals

Pareto's graphical method has been criticized on various grounds. Actually, if the alternative hypothesis is the geometric distribution, and if D is small, say, is less than 3, this procedure is much better than is believed (see Figure 7 and M 1963e). It seems useful, however, to present another form of the evidence, originating in M 1963e and Gerstein & M 1964.

Let us begin by considering continuous variables, U' and U'', such that $\Pr\{T' > u'\} = u'^{-D}$ for $u' > 1$ and $\Pr\{U'' > u''\} = u''^{-D}$ for $u'' > 1$. If U' and U'' are independent, their joint probability density is the product of the marginal densities:

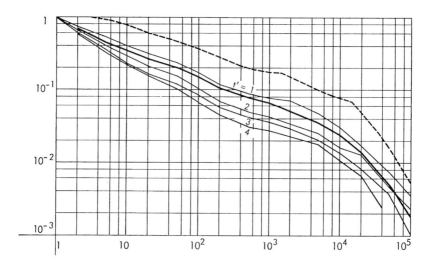

FIGURE N6-3. Cumulated doubly logarithmic plots of empirical inter-error distributions at the transmit level of -22 dbm. Bold line: marginal (unconditioned) frequencies $\Pr\{T_n + 1 - T_n \geq u''\}$. Dashed line: marginal frequencies $\Pr\{T_{n+3} - T_n \geq u''\}$. Thin lines (looking from the top down): conditioned frequencies $\Pr\{T_{n+1} - T_n \geq u''$, when $T_n - T_{n-1} = u'\}$, for the following values: $u' = 1$, $u' = 2$, $u' = 3$, $u' = 4$. The peculiar behavior for $u' = 1$ is discussed in the body of the paper.

$$Du'^{-(D+1)} Du''^{-(D+1)} = D^2 (u'u'')^{-(D+1)}.$$

Hence, the lines of equal density are of the form $u'u'' = constant$. They are hyperbolas truncated to the region where $u' > 1$ and $u'' > 1$.

This is to be contrasted with the two usual cases. When U' and U'' are both exponential, the lines of equal probability are straight and parallel to the line $u' + u'' = 0$. When U' and U'' are Gaussian, the lines of equal probability are circles. The contrast between these three curvature is so sharp, that no elaborate goodness of fit testing is needed when the sample is very large.

The cases of integer-valued variables U' and U'' are very similar but more complicated to write down analytically.

Figure 8 reproduces a three-dimensional model of the joint probability distribution. The display clearly favors the scaling hypothesis, viewed as being tested against the Gaussian and exponential alternatives.

Section 5 describes higher-order complications necessary to account fully for the data. This will not change the general implications of the model, but will make them somewhat less clear-cut. It is therefore proper to interrupt the examination of the empirical verification at this stage, to draw some implications in the clearest case.

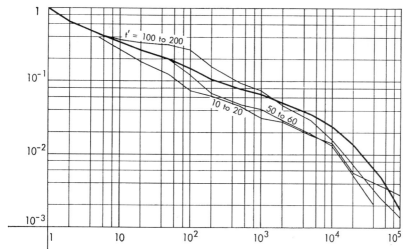

FIGURE N6-4. Cumulated doubly logarithmic plots of empirical inter-error distribution at the transmit level of -22 dbm. Bold line: same as in Figure 3. Thin lines (looking from the top down, in the region of $u'' = 100$): conditioned frequencies corresponding to the following ranges of values of $u'':u'$ between 100 and 200, u' between 50 and 60, u' between 10 and 20.

4. SOME CONSEQUENCES OF THE NEW MODEL

4.A. Higher population moments of the inter-error interval are infinite

The bulk of statistics concerns random variables that possess finite moments of all orders, or at least possess a finite variance. However, a discrete scaling variable has a kth population moment given by

$$D \sum_{u=1}^{\infty} u^{k-1-D}.$$

When $0 < D < 1$ (as for the coin tossing), all integer-order population moments are infinite. When $1 < D < 2$, the first integer moment is finite. At the same time, when the sample is finite the *sample moments* of all orders are finite. The useful consequences of infinite population moments are therefore those relative to the *variation* of sample moments with sample size (Appendix 2). For example, the theory predicts that, in the case of scaling random variables with $0 < D < 1$, the first sample moment

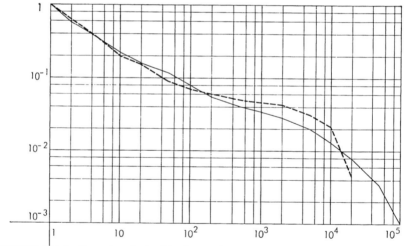

FIGURE N6-5. Cumulated doubly logarithmic graphs of the following empirical inter-error distributions: thin line: $\Pr\{T_{n+1} - T_n \geq u''$, when $T_n - T_{n-1} = 3$, irrespectively of the positions of still earlier errors$\}$: dashed line: $\Pr\{T_{n+1} - T_n \geq u''$, when $T_n - T_{n-1} = 3$ and either of the following is true: $T_{n-1} - T_{n-2} = 1$ and $T_{n-2} - T_{n-3} = 1$, or $T_{n-1} - T_{n-2} = 1$ and $T_{n-2} - T_{n-3} = 2$, or $T_{n-1} - T_{n-2} = 2$ and $T_{n-2} - T_{n-3} = 1\}$. Granted that the second curve is based upon a much smaller sample, the difference is negligible. Extremely similar results hold for $T_n - T_{n-1} = 2$ or 4.

varies quite erratically with sample size and, on the whole, it scales like (sample size)$^{-1+1/D}$.

All this makes it very important to determine D. Judging from error-intervals of up to 10,000, D was small (1/3 to 1/2), and even tended to decrease with $1/u$. But the local slope of the Pareto graph suddenly increases when u passes 10,000, and it may even exceed 1 when u equals 100,000. This fact did not, however, lead us to abandon the conjecture that D is the same for all t and of the order of 0.4. We viewed it as very likely that, for some connections at least, our new mechanism is mixed with independent errors having a very small probability of occurrence. If so, the observed curve should be some average of a straight line and of the curve corresponding to the geometric distribution on a doubly logarithmic graph; this is indeed how the empirical curves look.

Recently, however, our lack of concern with the final sections of our curves has been sanctioned in a far better way. We found indeed that many runs including fewer than 5 errors had failed to be reported on our tape, because they were felt to be statistically irrelevant. This means that the actual data have "fatter tails" than the curves which we have plotted. We conjecture that the correct value of D corresponds to the linear extrapolation of the flattest part of our curves, i.e., that it is closer to 1/4. A larger D would then be required to represent short inter-error intervals.

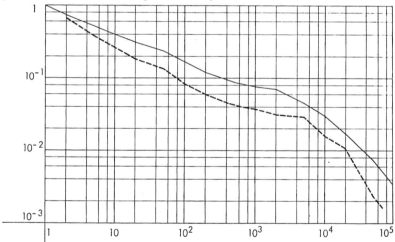

FIGURE N6-6. The data used in this figure are similar to those of Figure 5, except that the value of $T_n - T_{n-1}$ is 1 instead of 3. The lower curve is very close to the three curves of Figure 3 that correspond to $u' = 2$, 3 or 4, and therefore to both curves of Figure 5. This seems to mean that there is no need here for the second-order correction which is suggested by the special behavior of the line $u' = 1$ of Figure 3.

In any event, we have confidence in our conclusion that $D < 1$, and proceed to describe some of its consequences.

4.B. Probability distribution of the number of errors per m symbols, N_m

It has long been known in the theory of coin-tossing (see Feller 1957, p. 83) that if a play lasts over m tosses, the number N_m of roots of the equation $G(m) = 0$ satisfies

$$\Pr\{N_m < xm^{1/2}\} = (2/\pi)^{1/2} \int_0^x f\exp(-1/2s^2)ds.$$

Roughly speaking, N_m will increase as $m^{1/2}$, even though the most probable single value for N_m always remains equal to 1. More generally, the following was proved in Feller 1949. Let the known quantity m be the sum of an unknown number N_m of identically distributed random variables, each following the same law of Pareto with $0 < D < 1$. Then, for large m, $N_m m^{-D}$ will tend towards a nondegenerate limit: the mean of N_m is $[\sin(D\pi)] \cdot (D\pi)^{-1}m^D$, the standard deviation of N_m also increases as m^D, and the relative number of errors N_m/m will almost surely tend to zero with $1/m$.

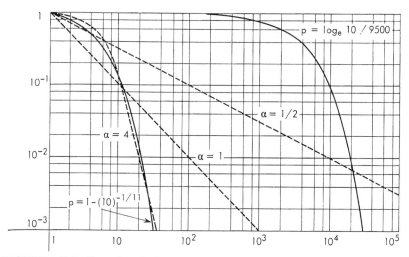

FIGURE N6-7. Cumulated doubly logarithmic plots of two geometric inter-error distributions, corresponding respectively to a high and low value for the probability p of the mutually independent errors. The dashed lines give two scaling laws of exponent, 1/2 and 1, as well as a law that becomes scaling with exponent 4 when u'' exceeds 10.

FIGURE N6-8. Three-dimensional wire model of the joint probability distribution of $U'' = T_{n+1} - T_n$ and $U' = T_n - T_{n-1}$ for several transmit levels mixed together.

Choice of coordinate scales. To achieve a reasonable size display while preserving legibility close to the origin we used variable coordinate scales: units from 1 to 49, tens from 50 to 99, hundreds from 100 to 999, etc.... To each change of scale corresponds a "spike" devoid of intrinsic meaning.

Applying this result to our problem, we see that *the average number of digits in error should be expected to tend to zero as the length m of the message increases to infinity.*

It follows that one of the central notions of information theory, called equivocation, vanishes for our channel. Indeed, denoting by U_m and V_m an emitted and received sequence of m symbols, Shannon has defined the equivocation per sequence as being equal to $-\log \Pr\{U_m|V_m\}$. The precise definition of this conditional probability is difficult in the case of channels with infinite memory. In the present special case, however, equivocation is simply the nonaveraged information required to specify the positions of the digits in error, i.e., to specify the distances between the beginning of the sequence and the first error and the distances between succeeding errors.

Writing $p(u) = \Pr\{\text{inter-error interval} = u\}$, we see that the specification of an inter-error interval requires on the average the information

$$\sum_{u=1}^{\infty} p(u)\log_2 p(u) \sim \log_2 D - (D+1)\sum_{u=1}^{\infty} u^{-(D+1)}\log_2 u, = Q,$$

which is finite. Moreover, the information required to specify the position of the first error is surely contained between 1 and $\log_2 m$. Hence, averaging the equivocation with respect to all possible positions of the errors, we find it to be between $(N_m - 1)Q$ and $(N_m - 1)Q + \log_2 m$. Finally, averaging with respect to values of N_m, we find for large m that the averaged equivocation is of the form "(constant prefactor) $m^D Q$." The residual dependence to be examined in Section 5 only affects this constant prefactor.

Hence, the equivocation per symbol tends to zero with $1/m$. That is, despite the presence of an unbounded number of errors, *the channel has a capacity equal to one.*

The limit theorems of the theory of information are therefore inapplicable to binary transmission over actual telephone channels.

Generally speaking, the mathematical theory of coding considers proposed "precorrecting codes," and compares them with an ideal of performance associated with certain probability limit theorems. The theory of information was divided by Shannon into two parts, according to the absence or the presence of noise.

Actually, this division is an oversimplification, because the theory of noiseless transmission is *not* the limit of the theory of transmission in the presence of vanishingly small noise. For example, since the capacity of the circuit described in Section 3 is equal to one, there seems to be no need for error correction of any kind. Actually, as the word length increases, the limit of the best error-precorrecting code is not identical to the best code corresponding to the limit capacity of one. This anomalous limit behavior is frequently observed in physics and engineering, and is referred to as being a "singular perturbation;" its classical prototype arises in the comparison of nonviscous fluids with fluids of very small viscosity. (An earlier example in which a singular perturbation enters into information theory concerns is the *limitation* of the propagation of errors on slightly noisy channels; see M 1955b).

Note that the inefficiency of error correction for our channels has already been pointed out in the literature; Fontaine & Gallagher 1961 and Fontaine 1961.

4.C. On the practical utility of the new model of errors

It is a simple matter to program a computer to generate sequences of errors according to our first-approximation model or to our second approximation. Therefore, these models make it a simple matter to compare codes by the Monte Carlo method. While this method is semi-empirical, it surely compares favorably with the completely empirical approach based upon the observation of actual performance in long taped sequences of errors. Our error model may also be used in analyzing queues in a communication link, in studying networks of such data circuits, and in finding optimum block-length in error-detection and decision-feedback systems.

However, at least until statisticians develop adequate small-sample techniques, the parameters of our model are hard to estimate.

To find out how various precorrection schemes perform, when the noise follows our model, a variety of problems of probability theory will have to be faced and solved.

5. FURTHER EXPERIMENTAL AND THEORETICAL CONSIDERATIONS

5.A. Other experiments and some amendments to our model

A single parameter D does not account for the detailed structure of our various Figures. True, the conditioned distributions of Figures 4 and 5 differ from the marginal distribution by a factor of two at most, while the geometric distribution is off by a factor of over 1000. However, it would be good to account for those variations and also to account for the fact that the density of errors seems to increase as the level of the signal decreases. We suggest that the data can be better represented by either of the following two-parameter laws.

- The first law supposes that the inter-error intervals are such that

$$\Pr\{T_{n+1} - T_n = 1\} = 1 - p$$
$$\Pr\{T_{n+1} - T_n \geq u, \text{ knowing that } u \geq 2\} = p(u/2)^{-D}.$$

Here the additional parameter is the probability that the inter-error interval be equal to 1, and it may be made dependent upon past errors.

- The second possibility is to write

$$\Pr\{T_{n+1} - T_n \geq u\} = \left(\frac{u - V}{1 - V}\right)^{-D}.$$

This factor V was extensively used in the theory of word frequencies, M 1961b. It could also depend upon the positions of past errors.

Indeed, tests similar to those of Section 3 have been performed in the case of transmit levels of −10 dbm, −16 dbm and −28 dbm (see Appendix 1). Some of the results are plotted in Figures 9 through 12. The general appearance of these graphs is the same as for the transmit level of −22 dbm. However, Figures 11 and 12 suggest that the parameters D, p or V must be made to depend upon the outside conditions. We shall not study the dependence of $T_{n+1} - T_n$ on transmit level any further. We shall, however, examine in greater detail the assertions concerning $T_{n+1} - T_n$ made in Section 3. First, we plot the variation of the function

$$\Pr\{T_{n+1} - T_n = 1 \text{ when } T_n - T_{n-1} = u'\},$$

when its argument u' varies from 1 to 10,000. In Figure 13, based upon a sample of several thousand events, we observe that this function is practically independent of u', with the possible exception of $u' = 2$, and certainly with the exception of $u' = 1$, where the function's value is much smaller than elsewhere. This means that *sequences of three successive errors are actually markedly less frequent than predicted by our first-approximation model*. In other words, *our model is too successful in predicting the extent of clustering of errors*, even when the inter-error intervals are statistically independent. The number of "clusters" of the form error-correct-error-error is also somewhat overestimated.

The test conditions have one disturbing implication, due to the fact that the tests were performed on loops. Depending on the length of the dialed-up connections and the character of the transmission facilities (microwave or loaded cable), there is a high probability of encountering a clustered-error correlation time equal to the total transit time around the loop. It follows that a disturbance (e.g., dropout) at the transmitting end is likely also to disturb the currently received baud. This loop test condition is atypical of normal transmissions. It is the reverse of the error trend underlined in the preceding paragraph.

In contrast, the curve corresponding to $T_{n+1} - T_n = 2$, which is shown in Figure 13 and is based on a smaller sample, indicates no systematic deviation from independence with respect to u'.

In the same vein, it is illuminating to consider the distributions of $T_{n+1} - T_n$, conditioned upon a given value of $u' = T_n - T_{n-1}$, and given that $T_{n+1} - T_n$ is *not* equal to 1. For that, it is sufficient to move all the conditional distributions up, until they go through the point of abscissa 2 and ordinate 1. All the various curves become practically superposable, winding around each other without a clear pattern.

In other words, our earlier model may be amended, preserving a value of D independent of $T_n - T_{n-1} = u'$, while the quantity p or V introduced at the beginning of this section takes different values for $u' = 1$ and for $u' > 1$.

Our first approximation model may alternatively be improved in another way if one considers "loose clusters" for which $T_n - T_{n-1}$ is between 3 and 10. Our various Figures show that, after such a cluster, and for u'' ranging from 50 to 10,000, $\Pr\{T_{n+1} - T_n \geq u''\}$ is *smaller* than if $T_n - T_{n-1}$ equals 1 or 100. Hence, the observation of a "loose cluster" seems to decrease the probability that the ensuing error-interval be *very* long, even though it has very little effect for $T_{n+1} - T_n$ around 10.

5.B. Generation of "pathological" random variables as mixtures of non-pathological ones

When the "pathological" behavior sketched in this paper is observed, statisticians tend to conjecture that the pathology is due to mixing of data, implying of course that nonmixed data may be non-pathological. Even if correct, this argument would not be useful, because it would introduce so

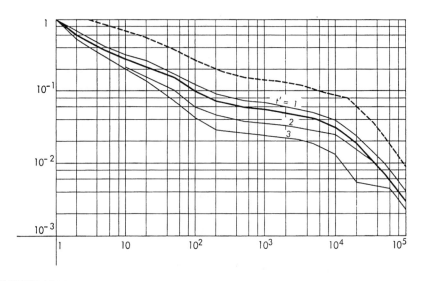

FIGURE N6-9. Data similar to those of Figure 3, but the transmit level was −28 dbm, and only three lines are given: $u' = 1$, $u' = 2$, and $u' = 3$.

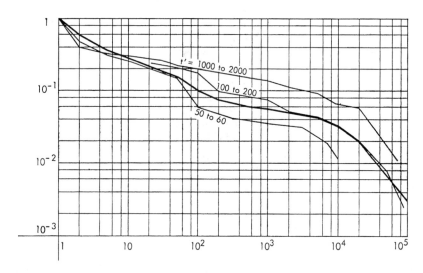

FIGURE N6-10. Same explanation as for Figure 4, but the transmit level was −28 dbm. Looking from the top down (in the region of $u'' = 100$), the three lines correspond to the following ranges: u' between 1000 and 2000 u', between 100 and 200, and u' between 50 and 60.

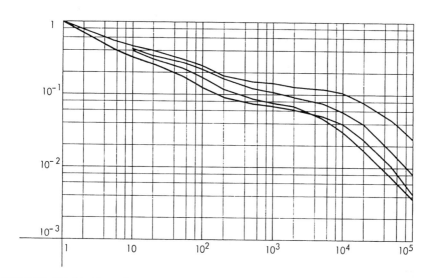

FIGURE N6-11. Cumulated doubly logarithmic plots of the distribution $\Pr\{T_{n+1} - T_n \geq u''$, given that $T_n - T_{n-1} = 1\}$, for four transmit levels. Looking from the top down (in the region around $u'' = 100$), the four lines correspond to the transmit levels of -28 dbm, -22 dbm, -16 dbm, and -10 dbm.

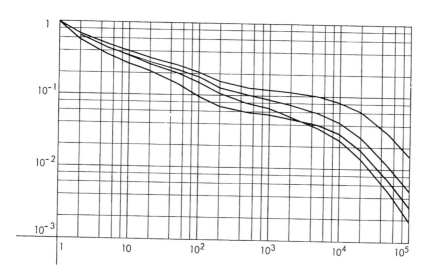

FIGURE N6-12. Data similar to those of Figure 11, except that the four curves correspond to marginal (unconditioned) distributions of $\Pr\{T_{n+1} - T_n \geq u''\}$.

many parameters that reliable estimation and prediction would be impossible anyway. Moreover, the law of Pareto has some very attractive properties relative to its behavior under mixture, as seen in M 1963e. Finally, by looking at small subsamples of our data, we find (see Figure 14), that differences between them are slight

5.C. On a possible role of the model of "quality states" on a more macroscopic level

By comparing successively long stretches of data concerning errors (that is, stretches lasting several hours), we found further effects that our model seems unable to explain. Therefore, even if our model represents fully the data on shorter sequences of errors, the fundamental parameters may vary in time, after all.

Acknowledgement

The authors wish to express their gratitude to their colleague J. Christian of the Advanced Systems Development Division of IBM, who wrote the programs used to test our theory and generally was in charge of experimental verification.

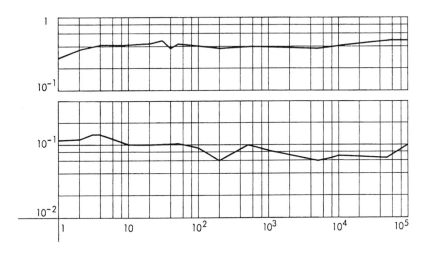

FIGURE N6-13. Variation of $\Pr\{T_{n+1} - T_n = 1$ when $T_n - T_{n-1} = u'\}$, upper curve, and of $\Pr\{T_{n+1} - T_n = 2$ when $T_n - T_{n-1} = u'\}$, lower curve, for a range of values of u'.

The theoretical work reported here was supported in part by the Office of Naval Research under Contract Nonr 3775(00), NR 047040. The authors are grateful to W. Hoffman of IBM Germany for supplying the data, and to the German Postal Administration for permission to publish them.

APPENDIX 1. ERROR RATES IN PHASE MODULATION TESTS

Our data was obtained from tests on the German public network performed jointly by IBM Germany and the German Postal Administration. The tests we used were performed on both rented and dialed-up connections using frequency modulation operating at 1200 bauds. Annex 12 of "Study Group Special A – Contribution No. 15" to the International Telephone and Telegraph Consultative Committee (CCITT) by the Federal Republic of Germany dated September 25, 1961 describes related phase modulation tests. However, the same connections and procedures were followed in the frequency modulation tests.

Briefly, four different dial-up connections were tested, each at four different transmit levels, i.e., –10 dbm, –16 dbm, –22 dbm and –28 dbm. Each level was tested continuously for 15 minutes (1.08×10^6 bits transmitted) using a periodic pseudo randomly generated 256-bit pattern. After

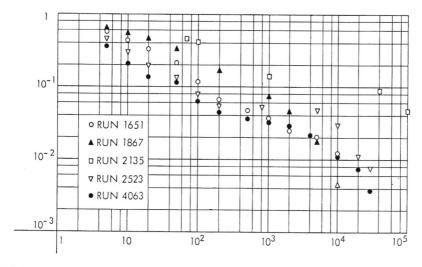

FIGURE N6-14. Cumulated double logarithmic plots for several telephonic connections of short duration, considered separately.

each of the four levels had been tested, the connection was redialed so that a total of 19 hours were used in testing each of the four connections (i.e., 8.2×10^7 bits transmitted per connection). The tests were all performed on a loop basis. The returned message was compared bit-by-bit with the transmitted message and each bit error, together with its distance from the previous error, was recorded on magnetic tape. This information was subsequently reduced to punched cards except for 91 of the 15-minute tests in which five or fewer bit errors occurred. The absence of these tests from our data is, if anything, a further verification of the models, since including them would decrease the curvature in the plotted distribution for very large u''.

The parameter most used for describing the quality of a communication facility is the average error rate: the quotient of the total number of bit errors and the total number of bits transmitted during the test. This error rate will generally vary widely with the connection and the transmit level used. The degree of variation is indicated in the published results of the phase modulation tests which are repeated in Table 1.

Analogous results for the frequency modulation data that we have used exhibit an even larger variation from level to level and between connections. It is, therefore, of considerable interest that the empirical distribution that was originally obtained from the total data should act as an equally good first approximation, independent of the transmit level and connection, for the limited samples of a single 15-minute test.

Table N6-1. Total number of bits in error in phase modulation test. For each connection, the total number of bits transmitted per transmit level was $2.05 \; 10^7$.

Connection	*Number of bits in error transmit level*			
	-10 dbm	-16 dbm	-22 dbm	-28 dbm
IV	162	617	2175	8430
V	513	2272	6137	15,880
VI	21	29	28	62
VII	79	417	3293	14,155

&&&&&&&&&&& **ANNOTATIONS** &&&&&&&&&&&

Deletion. The original appendix 2 of Berger & M 1963 was largely an excerpt from M 1963e {E3} devoted to the properties of distributions with infinite population moments. It was too concise to be useful, therefore was deleted.

Changes of notation. The original denoted both time instants and increments by diverse forms of t. For clarity and consistency with later Chapters, the increments were changed to diverse forms of u.

How this paper came to be written. My research, hence my whole life, was changed by accidental circumstances that led to Berger & M 1963. See Chapter N2, Section 2.3.

&&&&& **POST-PUBLICATION APPENDIX** &&&&&

ADDITIONAL TESTS ON CLUSTERING

✦ **Abstract.** This Appendix written especially for this volume reports on some tests performed starting in 1963. Those tests support the model of Berger & M 1963{N5}, as generalized in M 1965c{N7}. As already mentioned in Chapter 2, they point out interesting phenomena that were later incorporated in the notion of multifractal. ✦

1. Introduction

In November 1962, I took a day off from the Harvard economists to describe the preprint of Berger & M1963b to a group of electrical engineers from MIT and Lincoln Laboratory. The audience did not know about scaling but included engineers properly trained to try and control Nature, even when its ways are not yet understood. The overall response was very positive. In particular some of those who attend immediately replotted the data they had prepared for other purposes, and provided useful confirmation of Berger & M 1963. These discussions led very rapidly to a consensus, and I stopped further work in telephone engineering. But in my mind this consensus combined with the intermittence of turbulence and eventually led me to *multifractals*.

Recall the principal finding in Berger & M 1963: within a wide range of time spans the inter-error interval lengths are approximately independent and follow an approximately scaling distribution with an exponent D between 0 and 1. But for very long inter-error interval lengths, the distribution exhibits what physicists now call a "crossover to roughly exponential tail behavior."

The studies that followed Berger & M 1963 went beyond this approximation: successive inter-error intervals fail to be statistically independent, except in the first approximation. And the kinks observed on the graphs of Berger & M 1963 kept reappearing in everyone else's graphs.

While this evidence was accumulating, I was hard at work on an improved model described in M 1965c{N7}. The new model predicts interdependence, and replaces

$$P(u) = u^{-D} \quad \text{by} \quad P^*(u) = u^{1-D} - (u-1)^{1-D}.$$

The two expressions are compared in Figure N7-1, and it is clear that $P^*(u)$ fits the evidence better that $P(u)$. One may say, therefore, that the tests in this Appendix concern Chapter N7 as well as the text of Berger & M 1963 which is reproduced in the body of this Chapter.

Beyond comparatively minor improvements of Berger & M, the new empirical studies sketched in this Appendix extended to physically different channels. They showed how an increase in transmission signal-to-noise ratio changes the distribution of inter-error intervals. While all channels do indeed exhibit a fractal range, it became immediately obvious that the value of D is not universal. An increase in signal-to-noise ratio ought to decrease the frequency of errors, and it was found that this decrease is mainly associated with a decrease in the fractal dimension D. Other effects of a change in signal to noise ratio are to displace point of crossover, and to change the shape of the exponential tail.

These interesting findings were reported in several papers that are impossibly hard to get hold of. This Appendix reproduces the main findings and illustrations of the papers kept in my files.

2. The work of B. J. Moriarty (1963); Figure 15

Moriarty 1963, an internal report distributed to a dozen-odd individuals, essentially consists in one figure, together with a caption and a brief comment. An advantage of the self-similar model in M 1964c{N7} is that it applies as well to blocks. This is most fortunate because the data collected by most agencies are very seldom complete enough to check any inference relative to symbols, but allow checks relative to long blocks. Moriarty's Lincoln Laboratory investigations performed at my suggestion indeed showed that interblock intervals follow the law $P^*(u)$ described in Section 1, with the right kink in the right place.

An even more striking feature of our Figure 15 is the wide scatter of the values of D between channels.

3. The work of S. M. Sussman (1963); Figure 16

Sussman 1963 is a careful reexamination, using engineering methods, of "one particular model which has attracted great interest recently, [the model] proposed by Berger & M." The paper concludes that the scaling distribution "provides an excellent representation of the error statistics. The distribution's single parameter D [a fractal dimension], is sufficient to predict the measured statistics with surprising accuracy.

On the other hand, Sussman notes that the two sets of data that he analyzed yield clearly distinct values of D. Therefore, he observed that D is *not* a universal constant, a fact that was eventually to become very important to the theory of multifractals.

4. The work of P. A. W. Lewis and D. R. Cox (1966)

The authors of Lewis & Cox 1966 are noted experts in the statistical analysis of point processes. Therefore, instead of using the rough graphical methods used in the body of this chapter, they analyze the same data with the help of statistical contingency tables. The resulting analysis confirms the major claims of Berger and M. In the middle range of inter-error intervals, we are close enough to statistical independent. However, Lewis and Cox confirm that there are clear-cut discrepancies for very long inter-error intervals. First (in present-day terminology), the cross over to an exponential tail *is real:* The systematic downward plunge that was observed on the figures in the body of this chapter *does not*, after all, disappear with the addition of "lost" inter-error intervals. That is, the outer cutoff of the scaling region is finite rather than infinite. The resulting limitation of the fractal model is of course familiar to physicists, but it was as disturbing to the statisticians in 1966 as it is today.

A second finding by Lewis & Cox concerns a qualitative observation we had *failed* to make – although a more thorough examination of Figure N6-8 should have led to it, as will be said momentarily. They find that cases where two very long error-free intervals follow each other, are 50% more frequent than is expected from the assumption of independence that is made in the body of this chapter and is confirmed by Lewis & Cox for average length intervals. There is an excessive number of isolated errors, i.e. errors that are not part of a cluster but are contained between two long error-free intervals.

This second finding led me to take a fresh look at the bakelite-and-wire model photographed on Figure N6-8. The conclusion is that it confirms – after the fact – the finding of Lewis and Cox. In the low values corner, the iso-lines of equal frequency are of extremely large curvature.

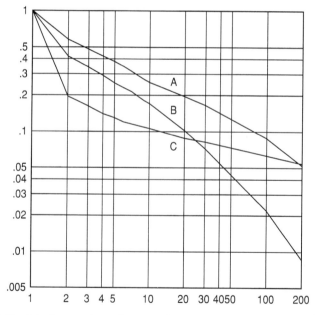

FIGURE N6-15. (Moriarty's Figure). Distribution of inter-error intervals for block lengths of 255 bits, over the following communication channels:

    A: Bermuda-Riverhead HF radio teletype test.

    B: Johannesburg-New York HF radio teletype test,

    C: High speed data telephone circuit.

This was to be expected, since the equation of the iso-probability lines, i.e., lines of equal density are expected to be dominated by the scaling density Du^{-D-1}. Their equation is, therefore, $Du^{-D-1} \times Dv^{-D-1} = $ constant, i.e., $uv = $ constant. In the high values corner, however, the iso-probability lines are expected to be dominated by the exponential tails: their equation is

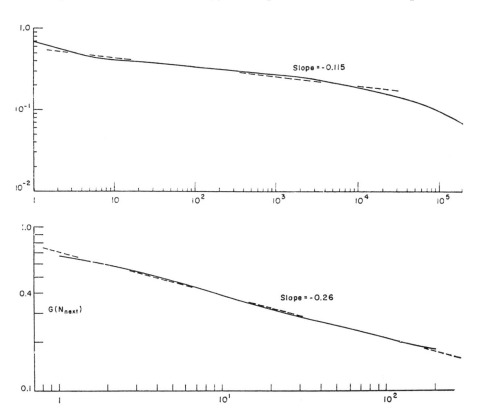

FIGURE N6-16. *Above.* (Sussman's Figure 1). Distribution of the intervals between successive errors. Estimation of D yields $D \sim 0.115$; more precisely, four alternative estimation procedures yield the closely bunched values 0.115; 0.12; 0.11 and 0.115. This figure replots data for "All Calls" from Figure 30 of Alexander, Gryb & Nast 1960; the original plot did not allow any conclusion to be reached.

Below. (Sussman's Figure 3). Distribution of the intervals between successive code words in error (Figure 4 of Fontaine & Gallager 1961). This figure yields $D \sim 0.26$; more precisely, 0.26; 0.25; 0.26; 0.29 and 0.30. The data are especially close to those to which the body of this chapter is addressed. Therefore it is noteworthy that the vertical intercept is read as about 0.75 on Figure 3, in agreement with the corresponding $1-D$ predicted from the observed D.

exp($-av$) exp($-bu$) = constant, i.e., $u + v$ = constant, and they are no longer of high curvature. The visual appearance of the data is in agreement.

It is easy to conjecture the existence of an intimate link between the presence of roughly exponential tails, and the dependence between successive long inter-error intervals. Assume indeed that the observed errors are mostly due to the fractal process described in the body of this chapter, but are contaminated by a low-frequency standard Poisson noise. Under this assumption, short clean "fractal intervals" are not much affected by the contaminator, but a very long clean fractal interval has a high probability of being "sliced" into many shorter intervals separated by single errors, rather than by bursts or clusters.

In mathematics, a set S is called perfect if every point in S is the limit of at least one sequence of points in S. Fractal errors are a concrete approximation to this "perfection," but Poisson contamination destroys it.

5. The work of M. E. Barron (1965); Figure 17

Barron 1965 was a M.S. thesis that does not appear to have been published. A noise threshold and a time threshold being given, Barron studied the distribution of the time intervals of duration > time threshold, during which the noise $X(t)$ satisfies $|X(t)|$ < noise threshold. Figure 17 reproduces one of Barron's figures: the underlying data are very different from those in the body of this chapter, and more directly physical, yet the result is very similar.

Here are excerpts from Section VI of Barron's text: "In nearly all cases, the experimental curves can be well approximated by the scaling. The fit is especially good in the tails of the distribution, in which the empirical curves are almost always asymptotically linear on a log-log plot, with small negative slopes.... The Poisson model fails completely by comparison and will always provide a poor fit in the tails. Furthermore, the tail is the most significant portion of the distribution, since these longer quiet intervals occupy a proportionately longer part of the total observation time.... The chi-squared goodness-of-fit test is not useful. In practically every case, this test would reject our hypothesis with a large value of chi-square."

The data in these figures are based on about 7000 quiet intervals at -40 dbm, and the theoretical curves marked ML, MM and LS correspond to several simple mathematical expressions. Each curve was "best fitted" to the whole body of data, using mechanical ("statistical cookbook") formula. It happens that none of the fitted curves bears even a general resemblance to the empirical curve. Therefore, there is no need to define the parameters A and m, except to point out the sharp disagreement between the following estimates.

ML *(maximum likelihood)* yields $A = 3.405$ msec and $m = 0.968$;

MM *(methods of moments)* yields $A = 2.585$ msec and $m = 2.011$,

and LS *(least squares)* yields $A = 1.428$ msec and $m = .0655$.

Barron's use of cookbook formulas based on the whole set of data flies on the face of something every student of fractals knows today: one should expect fractal behavior to hold over a range that is cut off on both sides. The fact that one *must* limit the fit to a middle range is a fact I had learned when studying the word frequency distributions in the early 1950's.

6. The work of A. Ephremides & R. O. Snyder (1982); Figure 18

Ephremides & Snyder 1982 considers ultra high frequency (UHF) and very high frequency (VHF) high-error rate wideband data channels, especially a 25 million bit VHF channel.

7. The work of B. W. Stuck & B. Kleiner (1974); Figure 19

Figure 19 reproduces the two figures in Stuck and Kleiner 1974 that bear closest on to the issues in the body of this chapter. In the top of our

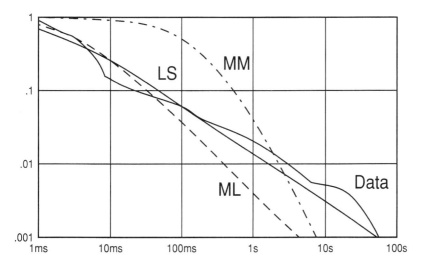

FIGURE N6-17. (Barron's Figure 4). Fraction of quiet intervals.

Figure, the most important feature resides in the shape of the curves of equal probability. In the high probability corner of the bottom left, they bear striking resemblance to hyperbolas, and the density of the points recalls strikingly the pattern of wire lengths in the bakelite-and-wire model of Figure N6-8.

8. The work of T. Aulin (1981)

Aulin 1981 is an analysis of signals from a transmitting antenna to a mobile vehicle in the city of Malmö, Sweden. Aulin's Figure 8(6) is extremely similar to those in the body of this chapter, with only a trace of exponential tail. An amusing but non-significant fact is that a brute-force statistical analysis carried out by Aulin shows statistically significant deviations from scaling. In fact, the samples he has studied are so huge, that the statistical fluctuations should be almost imperceptible. We know, however, that the scaling distribution is not expected to fit the whole body of data.

A second amusing and revealing fact is that Aulin gives a plot of consecutive gap lengths, and then asserts that "it can be seen that this process is with high probability not stationary."

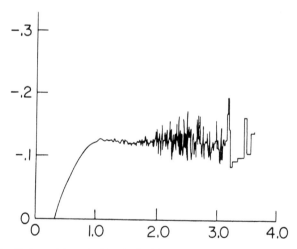

FIGURE N6-18. (Ephremides & Snyder's Figure 2). The local slope of the tail probability function plotted on doubly logarithmic coordinates, i. e., the slope fitted to short segments of the graph of that function. Beyond an initial transient that was fully expected, this slope fluctuates up and down rather roughly, but remains firmly near a horizontal trend line, meaning that this figure gives remarkable confirmation of the model in the body of this chapter.

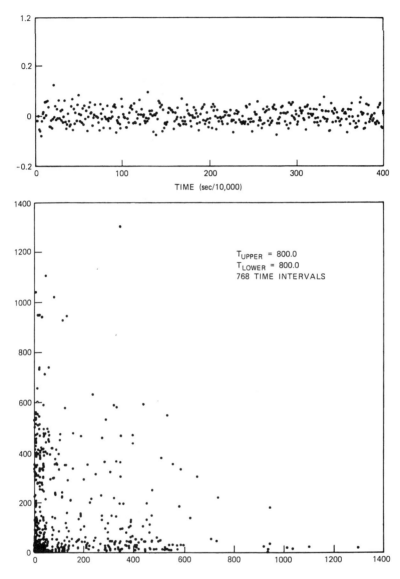

FIGURE N6-19. *Above*. (Stuck & Kleiner's Figure 28). Scatter plot of the duration of 768 time intervals between communication error bursts: in abscissa is the duration of the k-th interval, and in ordinate is the duration of the $(k-1)$st interval. This figure involves far fewer data than in the body of this chapter.

Below. (Stuck & Kleiner's Figure 29). Empirical sample autocorrelation of 1000 time intervals between bursts. Times are measured in 10^{-4} sec. In this figure, the most important feature resides in the fact that there is no obvious dependence of the correlation on the lag.

9. The work of P. Mertz (1965); Figure 20

On p.118 of Mertz 1965, we read the following: "As the test proceeds, at some given hour a large burst of errors occurs. This suddenly raises the average. For a number of hours following, the error rate may be near usual or below. The new contributions do not increase the cumulated total much, but the test time grows. Thus, the cumulated average error drops along a hyperbola (which plots as a straight line on the log-log paper). The pattern formed is somewhat that of a ripsaw tooth, with one side steep and the other sloping more gently. Stabilization of the average does not readily occur. Some traces suggest a possibility of stabilization, but examination of the other traces shows that this is not dependable.... The range between maximum and minimum error rates reported for each given test runs from a little over one to more than five decades."

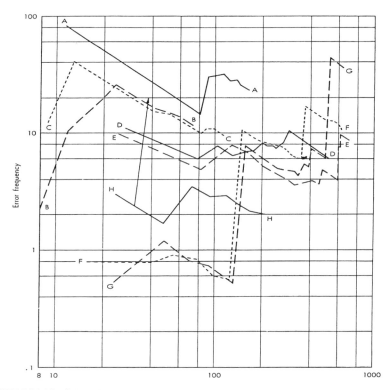

FIGURE N6-20. (Mertz's Figure 45). Cumulated averages of error incidences against time in hours, plotted in doubly logarithmic coordinates. These averages' erratic behavior is in full agreement with what Figure N6-15 makes one expect when the population expectation is infinite.

10. The work of P. V. Dimock (1960); Figure 21

Dimock 1960 is entirely made of words and graphs, without a single formula and without any quantitative conclusion. It was brought into the discussion in 1960 because its topic is close to that of this and the next chapters, but with two major differences. It concerned TV transmission, which perhaps did not justify a study as demanding as communication between computing and it concerned the duration of "hits," as opposed to the duration of hit free intervals. "The general procedure followed was to transmit a data signal continuously 24 hours a day over a period of several days or weeks on each of the circuits tested. A detailed log was kept of the number of bit errors that occurred during each individual disturbance. The number of errors was obtained either by direct observations of an electronic counter or from the automatic print-out of errors. Periodic measurements of noise, net loss, loss-frequency characteristic and delay-frequency characteristic were made during the course of the tests.... Curves showing the number of disturbances as function of duration expressed in terms of bit errors are plotted in Figure 5, 6 and y. In these figures the horizontal scale, labeled "Size of Hits," indicates the number of disturbances *per 100 hours per 1000 circuit miles* that equaled or exceeded the size of hit indicated on the horizontal scale." We reproduce the last Figure of Dimock 1960; the others are remarkably similar in appearance.

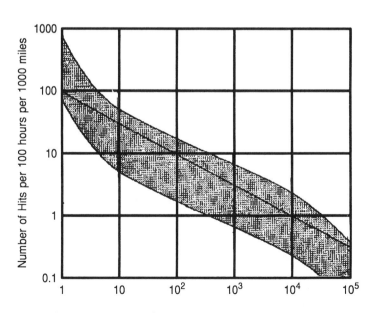

FIGURE N6-21. (Dimock's Figure 8)

Self-similarity and conditional stationarity

• *Chapter foreword.* This paper's original goal was to improve upon Berger & M 1963{N6}, therefore the original title was "Self-similar error clusters in communications systems, and the concept of conditional stationarity." But this original goal gradually paled in comparison to other motivations that are broader but were hard to express to engineers in pre-fractal times. A number of editorial changes make this aspect easier to appreciate. The original title was replaced by words taken from the abstract, and the term "burst," motivated by a specific issue of engineering, was replaced throughout by "box," which is of wider significance. In 1964, editors viewed fractal dimension as an unacceptably exotic notion; they accepted the scaling probability distribution, but asked asked me *not* to identify its exponent as being a dimension.

In this reprint, the notation was changed to conform to the rest of the book: θ became D, T became U_{max}, T^* became $\tilde{u}$, τ^* became $\tilde{\tau}$, Q^* became $\tilde{Q}$, and ε^* became $\tilde{\varepsilon}$.

The original material that duplicated Chapter N6 (including the figures) was erased and this text refers to the Figures in Chapter N6. Figure 1 compares $P(u)$ and $P^*(u)$; it did not make it to the original, but was added to this reprint.

This foreword continues by describing the goals of this paper in two modern ways: fractal terms, and the terms of a "renormalization group" argument, which may help some readers through the lengthy calculations.

Problems of observation and measurement that arise for Lévy dusts. From the mathematical viewpoint, this paper tackles a process of "error generating" events that produces a *Lévy dust* on the line. This is a fractal set S of zero length (that is, Lebesgue measure). It cannot be actually observed and measured, unless it is first "coarse-grained" within some

N7 ◇ ◇ SELF-SIMILARITY AND CONDITIONAL STATIONARITY

prescribed $\varepsilon > 0$. Then one studies the structure of the thickened set $S(\varepsilon)$, and the dependence of its characteristics upon the value of ε. Today's students of fractals know that coarse-graining can be achieved in many different ways.

One standard procedure is to draw a "Minkowski sausage," which replaces S by the set $M(\varepsilon)$ of all the points within a distance $\leq \varepsilon$ of a point in S. The present Chapter is devoted to another, even more standard, procedure. We take a grid of ε- boxes of side ε that pave the space in which S is embedded; next, we mark as "occupied" or "in error" those ε-boxes that intersect S; and we define $S(\varepsilon)$ as the union (sum) of the occupied boxes, and we coin the word "inter-box" to denote a string of empty boxes. This paper provides a number of probability distributions which describe the alternating sequence of boxes in error and inter-boxes.

A probabilistic "renormalization group" argument. Inspired by Berger & M 1963, we start with first stage boxes – call them 1-boxes – with the property that the interval between successive boxes satisfies

$$\Pr\{U \geq u\} = P(u) = u^{-D}.$$

We also assume that first stage inter-boxes – call them 1-inter-boxes – are statistically independent. If U had been continuous (a positive real variable), this distribution would have been strictly self-similar. But in the Berger & M model, U is an integer, hence its distribution is only *asymptotically* self-similar.

The next step taken in this paper is common in "renormalization group theory," and proceeds as follows: form 2-boxes (each made of 2 neighboring 1-boxes), then 2^2-boxes, then 2^k-boxes etc.., and call a 2^k-box "occupied," if at least one of the 2^{k-1}-boxes it contains is occupied. Our goal is to investigate these k-boxes as $k \to \infty$. For that, it suffices to examine the distribution of the length of the inter-boxes divided by the box length 2^k, and to determine the rules of interdependence between successive asymptotic inter-boxes. This, in effect, is the task accomplished in this paper. As $k \to \infty$, the distribution converges to a formula illustrated in Figure N7-1, namely,

$$\Pr\{U \geq u\} = P^*(u) = u^{1-D} - (u-1)^{1-D}.$$

This limit is invariant under the relevant box renormalization. Between this "fixed point" expression and $P(u) = u^{-D}$, there are two clearly visible differences:

- For large u, there is a prefactor: $P^*(u) \sim (1 - D) u^{-D} = (1 - D)P(u)$.

- For small u, $P^*(u)$ has a strong non-linear transient, not present for $P(u)$. Compare with the figures in Chapter N6. •

✦ **Abstract.** The purpose of this paper is twofold. From the viewpoint of engineering, it presents a model of certain random perturbations that appear to come in clusters, or bursts. This is achieved by introducing the concept of a "self-similar stochastic point process in continuous time."

From the mathematical viewpoint, the resulting mechanism presents fascinating peculiarities. In order to make them more palatable, as well as to help in the search for further developments, the basic concept of "con-

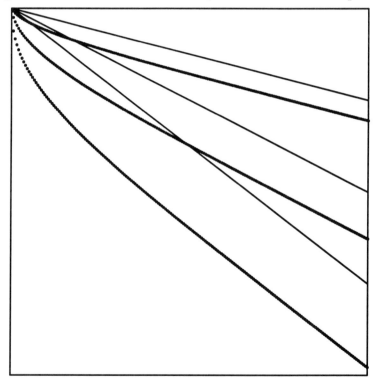

FIGURE N7-1. Illustration of the Chapter foreword. Bold lines: doubly logarithmic plots of $P^*(u) = \Pr(U > u) = u^{1-D} - (u - 1)^{1-D}$, for $D = 0.2, 0.4$, and 0.6. This tail probability renormalizes exactly. Thin lines: plots of the corresponding $P(u) = u^{-D}$ from Chapter N6.

ditional stationarity" is discussed in greater detail than would be strictly necessary from the viewpoint of engineering.

The point of departure is a model proposed by J.M. Berger and the author. The logical structure has been streamlined and a number of fresh consequences have been derived. The empirical fit has been improved, and *ad hoc* corrective terms have become unnecessary. ✦

I. INTRODUCTION

I-A. Early statistical models of communication channels

In order to make the present paper approximately self-contained, let me first recall two early descriptions of discrete channels, and present the model proposed in Berger & M 1963{N5}.

The basic model of a communication circuit is the *binary symmetric channel without memory*, which assumes that the transmission of any given symbol is not influenced by the correctness of the transmission of earlier parts of the message. If the error probability is p, one shows that the inter-error intervals U follow the geometric distribution, namely, $\Pr\{U = u\} = (1 - p)^{u-1}p$, which is the exponential distribution for integer variables. The number of errors in a sequence of N symbols is a binomial random variable.

Unfortunately, this model is incapable of accounting for the actual behavior of high-quality, or low-noise, communication circuits. Indeed, although isolated errors are occasionally encountered, most actual error records appear to be very strongly clustered. That is, *errors appear to be grouped in bursts, which are in turn grouped in bursts of bursts, and so on*. In attempt to account for this fact, it is tempting to postulate that there is some kind of "contagion" between the occurrences of errors in neighboring symbols. For example, the channel may have more than a single "state" or level of "error susceptibility." Within each state, the occurrence of errors would be described in terms similar to the binary symmetric channel, but the probability of error at any given time would depend upon the current level of susceptibility, the transitions between states being themselves ruled by a kind of "master" stochastic process.

For example, Gilbert 1961 suggested that channels have a single "good" state (with a small probability of error) and a single "bad" state (with a large probability of error), with Markovian transition probabilities for changes from the good to the bad state and vice versa. Each incursion through the bad state is assumed to give rise to a cluster of errors.

This approach does indeed give rise to certain qualitative features ascribed to the data; but quantitatively the fit is poor, as was shown especially clearly in Mertz 1961. The most striking difficulty is to be found in the fact that "first-order" bursts are clustered in "second-order" bursts, and so on. One readily imagines that any finite set of empirical data can be accounted for with arbitrary accuracy, if one agrees to introduce a large enough number of hierarchical levels and hence of independent parameters. However, such models are analytically unmanageable, and include explicitly in their input all the features that they hope to obtain in their output.

I-B. The Berger-Mandelbrot approach

As an alternative to the approaches sketched in Section I-A, Berger & M 1963{N5} proposed the following assumptions:

1) We suggested that inter-error intervals U may very well continue to be statistically independent random variables.

2) We suggested that each inter-error interval is a scaling (or "Paretian") random variable, which means that there exists a constant D, with $0 < D < 1$, such that

$$P(u) = \Pr\{U \geq u\} = u^{-D}.$$

The inter-error interval is a positive integer, and $P(1) = 1$, as expected.

The predictions based upon the combination of 1) and 2) were shown in Berger & M 1963 to be much superior to earlier models, both in mathematical simplicity and in the quality of empirical fit. We were, however, careful not to overestimate their success. Indeed, the dependence between successive U's is unquestionable. But it can be called small: when certain facts are known concerning past errors, the conditional probability of the next inter-error interval has only to be multiplied by a factor contained between, say, 3/4 and 3/2. Similarly, the marginal distribution of the inter-error interval differs from 2) when u is small; even though the discrepancy is not big, it has turned out to be systematic.

II. SELF-SIMILAR POINT PROCESSES: DEFINITION AND SUMMARY OF THEIR PROPERTIES

II-A. Statement of goals and summary

The purpose of the present paper is to improve, develop, and generalize the Berger & M model. Therefore let me first stress that the principal feature of that work, and the key to its success, was that its intrinsic entities were no longer the error bursts but the individual errors themselves. We found a striking qualitative resemblance between the empirical records of inter-error intervals and the sequences of returns to the origin in the classical game of tossing a fair coin (see Berger & M 1963{N5}, Section 2.B). This resemblance made us account for the obvious visual clustering of the errors as being the way in which a human observer will inevitably classify the sequences of events generated by certain processes in which no clustering mechanism was willfully built-in. In other words, we viewed the appearance of "contagion between errors" as being a perceptual construct.

It was unfortunate that the model of Berger & M continued to be based upon irreducible units, namely transmission symbols. This feature made it very difficult to attack a very important problem: the comparison of errors at different transmission speeds. It is not obvious, however, that any intrinsic meaning should be attributed to the scale of time that happens to separate two successive symbols when a channel is to be used with any given bandwidth. Indeed, such a unit would be intrinsic only if the main sources of error were to be found in some features of the transmitting or receiving mechanisms.

An equally plausible thought, a priori, is that errors are only a visible manifestation of certain "error-generating" events localizable at a finer level. If so, a transmission symbol would occupy, with respect to these "error-generators," the position that a box of symbols occupies with respect to a single symbol. In other words, it is tempting to interpolate the model of Berger & M without bound. Postponing the critical analysis and the empirical justification of my approach, the present essay conjectures the following. Whatever the duration of a symbol, it will be "in error," if and only if, one or more "error-generators" are encountered within its duration. To avoid a multitude of terminologies, the duration of the symbol could be referred to as a "block." {PS.1998: as mentioned, this reprint replaces *block* by the current fractal term, *box*, which emphasizes that the argument is of wide generality.} The only constraint upon the law of errors is that it must yield theoretically meaningful and empirically correct results concerning the "boxes" and the error-clusters (as defined below). We shall describe the structure of the error-generators by a mechanism that may at first appear most strange, because it involves two formally simple but fundamental generalizations of the usual theory of stationary point processes.

Section II-B will define my process, Sections II-C and II-D will discuss its more bizarre implications and will describe the reasons why it was selected, and Sections II-F and II-G will summarize its observable properties. In this way, it will be shown that the law $P(u) = \Pr\{U \geq u\} = u^{-D}$ is necessary both to achieve clustering and to fit the facts. Finally Section II-H will show that the law $P(u) = u^{-D}$ is also sufficient to describe the qualitative appearance of the data. Details will be found in Sections III and IV.

II-B. Definition of self-similar, recurrent event processes

Time will be assumed continuous, and T_k designate the moments of occurrence of certain basic "events" that are recurrent in the sense of Feller 1950-57-68. This means that the intervals U between successive T_k are statistically independent. *Self-similar, recurrent events* will be defined as being such that the intervals U follow a scaling distribution of the form $P(u) = \Pr\{U \geq u\} = u^{-D}$, where the exponent D is a parameter in the range from zero to one, and the variable u can vary continuously from zero to infinity. The behavior of this u^{-D} as $u \to 0$ or $u \to \infty$ will have fascinating and unusual implications, but in practice $P(u) = u^{-D}$ will break down at both ends. For example, $P(u)$ may become exponential as $u \to \infty$. Both limits of validity of u^{-D}, however, are at best hard to estimate; hence no problem can be considered "well-posed" unless it depends negligibly upon those limits and upon the behavior of $P(u)$ for $0 < u < \tilde{\varepsilon}$ and for $\tilde{u} < u < \infty$, where $\tilde{\varepsilon}$ and $\tilde{u}$ are some constants. From the viewpoint of such problems, $P(u)$ may be considered as "proportional" to u^{-D} as soon as proportionality holds over a finite interval $0 < \varepsilon < u < T < \infty$, where $\varepsilon < \tilde{\varepsilon}$ and $U_{max} > \tilde{u}$ are also some constants. For example, one may assume that $P(u) = 1$ for $u < \varepsilon$, $P(u) = (u/\varepsilon)^{-D}$ for $\varepsilon < u < U_{max}$, and $P(u) = 0$ for $u > U_{max}$; this implies a positive probability for $U = U_{max}$.

Notations: s and r will be used as "dummy" integration variables. Capital letters will in principle designate random variables, whose values are the corresponding lower case letters. However, U_{max} and $\tilde{u}$ are not random. The reader is advised to guard against confusion on this account.

II-C. The number of errors in a box in error tends to infinity as $\varepsilon \to 0$; the concept of "self-similarity," and its possible role as axiom

A perplexing feature of the law $P(u) = u^{-D}$ is that it assumes short interevent intervals to be so numerous that the "total probability" $P(0)$ is infinite. The key to a correct handling of this phenomenon is that, as in the last analysis, we shall never deal with u directly, but shall restrict it in various ways to make it into a bona fide random variable. The basic method of conditioning assumes that $u \geq \varepsilon > 0$, and it leads to the two-parameter family of quite proper variables U_ε, satisfying

$$\Pr\{U_\varepsilon \geq u\} = (u/\varepsilon)^{-D} \quad \text{if } \geq \varepsilon$$

$$\Pr\{U_\varepsilon \geq u\} = 1 \quad \text{if } < \varepsilon.$$

One immediately notes that the members of this family that correspond to different values of ε can be deduced from each other by scaling u appropriately. This property, which has led to the term "self-similar," will now be made into an axiom or principle, and we shall prove that, when it is appropriately specified, self-similarity characterizes the law $P(u) = u^{-D}$ with $0 < D < 1$. Let me proceed in several steps.

1) Assume that, for any $h' > 0$ and $h'' > 0$, the conditioned laws

$$\Pr\{U \geq u \mid U > h'\} \text{ and } \Pr\{U \geq u \mid U > h''\}$$

differ only in scale, meaning that

$$\Pr\{U/h' \geq u \mid U/h' > 1\} = \Pr\{U/h'' \geq u \mid U/h'' > 1\}.$$

For this equality to hold, $P(u) = \Pr\{U \geq u\}$ must satisfy the condition $P(uh')/P(h') = P(uh'')/P(h'')$. Consider $R = \log P$ as a function of $v = \log u$, and write $v' = \log h'$ and $v'' = \log h''$. Our assumption is equivalent to the functional equation $R(v' + v) - R(v') = R(v'' + v) - R(v'')$ for all choices of v, v', and v''. This requires that R be a linear function of v. Hence, $\Pr\{U \geq u\}$ must be a function of u of the form $\Pr\{U \geq u\} = (u/\varepsilon)^{-D}$, where ε is any positive number less than h' and h''.

2) More precisely, the meaningful conditioned distributions $\Pr\{U \geq u \mid U > h\} = (u/h)^{-D}$ can be computed as if the unconditioned U satisfied $\Pr\{U \geq u\} = Pu^{-D}$ for $0 < u < \infty$.

The trouble is that this function satisfies $\Pr\{U \geq 0\} = \infty$. But this divergence ceases to be shocking if one recalls that the concept of a "probability" is just a weighted expected frequency. The common procedure is to introduce relative frequencies by dividing absolute frequencies by the total number of outcomes. In the self-similar case, however, the total number of T_k per unit time is either zero, or extremely large and tending to infinity as $\varepsilon \to 0$ (as we shall see in Section IV-A). As a result, there is no meaningful way of weighting them, and the best that one can do is leave an absolute frequency with a wholly unspecified multiplier. Any expression such as $P(u) = Pu^{-D}$ will do equally well, and $P(u)$ can be normalized to equal u^{-D}.

3) D must be in the range $0 < D < 1$ in order to insure that, although the number of very short inter-event intervals is infinite, those having a length less than ε lead to no trouble. This turns out to require that $\int_0^\varepsilon P(s)ds < \infty$, which is satisfied only if $0 < D < 1$.

Incidentally, the limit case $D = 0$ of the self-similar law is the "uniform distribution" over the unbounded interval from 0 to ∞. Again, this limit is not a proper probability distribution, but becomes meaningful whenever u is restricted to a time interval of finite duration. As a result, the uniform law over $(0, \infty)$ is well known to be useful when one wants to represent certain random phenomena for which the u scale is not intrinsic, but introduced by the method of observation. *The unweighted u^{-D} is the most general law that shares this property and possesses no time scale.* This will be its major asset as well as its defining property.

II-D. The average duration of the inter-event interval is infinite; the need for a new "conditional" concept of stationarity and of a random process

To discuss certain consequences of the behavior of $P(u) = u^{-D}$ as $u \to \infty$, we shall work with the conditioned random variable U_ε.

Very long inter-event intervals are so long in the case of U_ε that

$$E(U_\varepsilon) = \varepsilon^D \int_\varepsilon^\infty sDs^{-D-1}ds = D\varepsilon^D \int_\varepsilon^\infty s^{-D}ds = \infty.$$

The notation $E(u)$, is standard in mathematics, and is adopted here, even though physicists prefer $\langle u \rangle$.

Pure mathematicians have long known infinite means to be quite proper in the case of random variables. Let us now turn to random processes. It is well known (see also Section II-A) that in coin tossing, the distribution of the intervals between successive returns to zero, is asymptotically scaling with $D = 1/2$. This inter-event law must therefore be acceptable in a random function. And yet it happens that *stochastic processes have been so defined by mathematicians that they fail to encompass the random sequences* $\{T_k\}$ *such that* $\Pr\{U \geq u\} = u^{-D}$, *where* $0 < D < 1$.

This circumstance can be described using the following notation. Consider a set of finite intervals (t'_h, t''_h) where $1 < h < H$; these intervals may overlap, and the possibility that $t'_h = t''_h$ is not excluded.. Define the *indicator function* $V(t'_h, t''_h)$ as equal to one if there is at least one T_k in the interval $(t'_h < s \leq t''_h)$; and as equal to zero otherwise. Inspired by earlier chance phenomena, probabilists define a stochastic point process by all the joint probabilities of the form

$$\Pr\{V(t'_h, t''_h) = v_h \text{ for every } 1 \leq h \leq H\}.$$

Moreover, such a process is said to be stationary if the above probabilities are unchanged when any $\delta \neq 0$ is added to every t'_h and t''_h.

Let us now see what happens in the case of a recurrent process $\{T_k\}$ for which the inter-event interval has an infinite expectation $E(U)$. When the interval $(t'_1 t''_1)$ is chosen at random, the probability that it contains an event is equal to $|t''_1 - t'_1|$ divided by $E(U)$. Therefore, $\Pr\{V(t'_1, t''_1) = 1\} = 0$ and $\Pr\{V(t'_1, t''_1) = 0\} = 1$. Similarly, examine the higher order probabilities $\Pr\{V(t'_h, t''_h) = v_h$ for every $1 \leq h \leq H\}$. If all the v_h's are zero, then the probabilities equal one. All other values of v_k have probabilities identically equal to zero. Note, incidentally, that all the above probabilities are left unchanged if any displacement $\delta \neq 0$ is added to all the t'_h and t''_h.

These degenerate results mark the limits of applicability of the usual definition of a stochastic process, and reflect the fact that both pure and applied mathematics has so far concentrated upon chance phenomena such that something is happening much of the time, in the sense that $\Pr\{|X(t)| > 0\}$ is non-zero. Moreover, different portions cut from the same function are somewhat similar in appearance to each other.

It now appears that adherence to these properties excessively restricts the practical scope of the theory of random functions. It excludes many well known mathematical examples which can in no way be considered pathological, and it also appears to *exclude the ill-understood processes that underlie communication errors.*

To help broaden the existing theory, one may recall that, in its most general and loosest sense, a random function is simply an object chosen at random out of any prescribed collection of ordinary functions. If so and the objects in that collection are labeled by an appropriate parameter λ, a random function is simply a function of two variables t and λ. The labeled collection is stationary if, whenever it contains a function, it also contains all its temporal translates, and if it attributes the same probability to two functions that differ only by translation.

To this excessively loose description, certain measurability conditions are added to make it manageable, and to distract attention from irrelevant complications that can only present themselves on sets of λ's of measure zero.

Unfortunately, the resulting framework of probability theory fails to provide a universal prescription. Indeed, for recurrent events with $E(U) = \infty$, the interesting cases are thus thrown into the dustbin. Faced with one of these cases, one can no longer rely upon finite-dimensional distributions and upon the usual measurability arguments. The problems that arise are like the problems one would encounter if one had to use the three-dimensional Euclidean measure to investigate phenomena in which all the probability is concentrated upon a one-dimensional line, or upon a surface. There is, however, a simple and familiar way to avoid such a meaningless answer, namely, to examine only the distributions conditioned by the assumption that one finds oneself upon the line or surface in question.

This suggests the following broader definition of *a stochastic process* and the following *conditional concept of stationarity*. Assume, not only that the value of the function $V(t'_0, t''_0)$ is known for some interval (t'_0, t''_0), but that it equals one rather than zero. The sequence $\{T_k\}$ will then be defined by the conditional distribution of the various finite sets of functions $V(t'_h, t''_h)$.

Examples: The condition $t'_0 = t'_1 = t''_0 = t''_1$ expresses that the position of one of the events T_k is known exactly. The condition $t'_0 \leq \min t'_h$ and $t''_0 \geq \max t''_h$ expresses that at least one of the events T_k is located within the interval (t'_0, t''_0).

Moreover, $\{T_k\}$ will be called *conditionally stationary*, if the conditioned distribution of the indicator function V remains unchanged when all the t'_h and t''_h, are displaced by some fixed δ, but V remains constrained by the condition $V(t'_0, t''_0) = 1$.

The last step in this definition may seem innocuous, but is not. Conditioning may be the key to the necessary task of describing the structure of many empirical intermittent phenomena.

The exclusive resort to conditioning entails a reduction of generality as well as a complication. Permit me, however, to sketch a tongue-in-cheek argument which suggests that, after all, few concrete problems risk being eliminated. This argument is somewhat philosophical in nature, and introduces a distinction between "new" and "old" phenomena. Suppose that Dr. K. sets up to investigate the properties of some kind of event first noted by his colleague Dr. F., in which intervals between occurrences follow a self-similar law. One could think in terms of atmospheric fluctuations in a given area (a phenomenon which in M 1964i{N9}, will formally relate to the self-similar theory). It is obvious that the very fact that Dr. K. was not the first to look for this phenomenon immediately implies that the time period during which he carried on his investigation was *not* selected at random. He may have set his experiment during a period δ of (say) one hour chosen at random during a period of (say) two years after the observation of Dr. F., and a period τ of (say) a month since the last confirmed report of a similar observation. Weak as it may seem, this additional information concerning a self-similar process suffices to change it into a process with a time scale, and it is just sufficient to transform most questions so that they yield nonvanishing probabilities. For example, Section IV will show that the probability of an event occurring τ time units after a previous occurrence (which need not be the latest one) decreases proportionately to $\delta \tau^{D-1}$. This does not vanish, even though it becomes extremely small as the latest reported occurrence of the event recedes into an increasingly distant and dim past.

But there is a zero probability of observing a phenomenon never before mentioned or noticed. There is nothing shocking to intuition, in suggesting that if an apparatus is set at random, the probability of catching such a phenomenon is indeed minute. The folklore of science suggests, however, that once caught, a new phenomenon will soon be observed again and again; this is exactly what the self-similar model would predict.

In the case of communication circuits, the preceding argument unfortunately encounters a major difficulty, since our Drs. F. and K. do not actually observe the same "phenomenon." If one deals with telephones, the number of distinct circuits is finite but enormous, hence the particular connection used by Dr. K. may well never have been used before. One may perhaps assume in such a case that the process of connecting the circuit creates electric transients that necessarily lead to incorrect communication and "trigger" the error process.

An alternative method was carefully kept in reserve. For most problems relating to processes in progress, it would be best to assume that the process was triggered due to a finite U_{max}, but then to simplify the analysis by setting $U_{max} = \infty$.

II-E. Comment upon the method used in the present paper

It is important to stress the contrast between the self-similar recurrent event processes and the bulk of the processes that were classically used as models of natural chance phenomena. While "statistical stationarity" expresses the invariance of the generating mechanism with respect to the *addition* of a constant to time, the concept of self-similarity expresses the invariance of the generating mechanism with respect to the *multiplication* of time by a constant.

Formally, the stochastic processes invariant by multiplication of time are simple, but many usual procedures of probability theory are inapplicable, while other procedures require unusual care. However, not wanting to devote too much time to formal difficulties, I shall instead restrict the value u of the inter-error interval U to some domain $0 < \varepsilon < u < U_{max} < \infty$. In this way, all the usual techniques again become applicable. Moreover, the empirical phenomena, to which the present theory may apply, can also be represented by the theory based on the assumption that $1/\varepsilon$ and/or U_{max} is large but finite.

$E(U)$ is thus made finite but very large. If the interval (t'_1, t''_1) has been chosen at random, its probability of containing an event T_k is small but nonzero and is highly dependent upon U_{max}. The geometric image of points in space restricted to be on a line or on a surface must be replaced by an image of points restricted to be on a thread or on a veil. All absolute probabilities will depend upon the thickness of this thread or veil, but conditional probabilities will not.

After the theory is built upon the assumptions $\varepsilon > 0$ and $U_{max} < \infty$, let $\varepsilon \to 0$ and $U_{max} \to \infty$, and concentrate upon problems whose solutions tend to finite nonvanishing limits, continuously and regularly. The limits of the solutions will be used as definitions of the solutions relative to the limits of the problems. Elsewhere, these limits will be rederived both directly and rigorously by generalizing the original problems. But this is a subsidiary issue for present purposes. One can indeed safely affirm that when a practical problem involves a mathematical expression of the form $\lim_{s \to 0} X(s)$, one is not really interested in the limit itself, but in the value of X for some small s. The usual reason for ever invoking a limit at all is that it is often observed that limits are simpler to write down and easier to manipulate. In the present state of knowledge and intuition concerning self-similar processes, the limit is not easy to manipulate and it is therefore better to avoid studying it directly. It should, however, be recalled that the spirit of the present theory is to consider that a problem is "well posed" if its answer depends little upon ε and U_{max}, and "ill posed" if it depends a great deal upon ε or U_{max}. Later stages of the theory may well change the meaning of "well posed." ("Well posed" is a concept introduced by Jacques Hadamard in the context of the Cauchy problem for partial differential equations.)

II-F. Principal properties of the inter-boxes

The basis of the empirical verification of my model is that it is impossible to observe individual error-generators directly. One indirect method of observation cuts time into equal chunks of duration $\tilde{\tau}$, called "boxes," and views a box as being in error if it contains at least one error-generator. If so, the meaningful quantity is the "inter-box" between successive boxes in error; its properties will now be summarized, and will be fully derived in Sections III and IV.

Let the inter-error intervals follow the distribution $P(u) = (u/\varepsilon)^{-D}$ if $u > \varepsilon$, and $P(u) = 1$ if $u < \varepsilon$. The value of ε will eventually tend to zero, and the scale of u can therefore be chosen at will. To simplify notation, this scale will be chosen so that $\tilde{\tau} = 1$. As $\varepsilon \to 0$, the distribution of the inter-boxes tends towards

$$\Pr\{U \geq u\} = u^{1-D} - (u-1)^{1-D},$$ a function to be denoted by $P^*(u)$.

This $P^*(u)$ is illustrated in the Figure N7-1 added to this reprint. For large u, $P^*(u) \sim (1-D)u^{-D}$, which differs from the Berger & M expression $P(u) = u^{-D}$ only by the multiplicative factor $1 - D$. This factor happens to improve quite markedly the fit achieved in Berger & M, even though (most fortunately) u^{-D} was an entirely admissible first approximation.

What if the expression $P^*(u) = u^{1-D} - (u-1)^{1-D}$ for the probabilities of long interblock intervals gives too small a value? If these probabilities are of the form Qu^{-D}, where $Q > 1 - D$, their form may be accounted for by assuming that the inter-error intervals continue (to the end of the problem) to be restricted to being larger than a nonvanishing bound $\varepsilon > 0$. If, in particular, the case in question is $\varepsilon = \tilde{\tau} = 1$, it becomes necessary to fall back on the earlier model of Berger & M. The resulting process is self-similar asymptotically, but no longer uniformly for all u.

It will also be shown that inter-boxes are *not* statistically independent, again improving the experimental fit beyond the model of Berger & M. The probability that the first of two successive intervals be at least equal to u' and the second at least equal to u'' is given by

$$\frac{(1-D)}{\pi} \sin(D\pi) \int\int (x+u'-1)^{-D}(y+u''-1)^{-D}(1-x-y)^{D-1} dxdy,$$

where the double integral is carried over the triangle $x > 0, x + y < 1, y \geq 0$.

For large u' and u'', the dependence between successive interblock intervals can be expressed as follows: one finds

$$\frac{\Pr\{\text{first interblock} = u' \text{ and second interblock} = u''\}}{\Pr\{\text{first interblock} = u'\} \Pr\{\text{second interblock} = u''\}} = \frac{\sin(D\pi)}{D(1-D)(1+D)},$$

which is a quantity, to be denoted by $\tilde{Q}$, that is independent of u' and u''.

II.G. Principal properties of the intervals between clusters

Instead of using boxes, the observation of a self-similar sequence can be based upon "clusters." One assumes that two successive error-generating events can be distinguished from each other only if they are separated by at least a duration $\tilde{\tau}$, and one defines a cluster or burst as follows: it is a succession of errors that includes no gap of length $> \tilde{\tau}$, but is bounded by two gaps of length $> \tilde{\tau}$.

If $\varepsilon = 0$, there is a probability equal to one that a message will be an alternating sequence of clusters and of inter-error intervals longer than $\tilde{\tau}$. Incidentally, the duration of the inter-cluster intervals satisfies $\Pr\{U > u\} = (u/\tilde{\tau})^{-D}$. This behavior was first observed in Mertz 1961. In this sense, one can say that the inter-cluster durations are proportional to $\tilde{\tau}$.

The duration of the clusters themselves is also proportional to $\tilde{\tau}$, multiplied by a random variable that will be studied elsewhere. Let us only note that it has finite moments of all orders, was first considered (in an entirely different context) in Darling 1952, and is specified by the fact that its characteristic function is

$$\frac{1}{1 - D \int_0^1 (e^{i\zeta s} - 1) s^{-D-1} ds}.$$

One readily deduces that the mean duration of a burst is equal to

$$\frac{\tilde{\tau} D}{1 - D}.$$

Assuming henceforth that time units were chosen so that $\tilde{\tau} = 1$, we see that the distance between two cluster ends is the sum of two independent random variables, one of which has a scaling tail, while the other is very short-tailed. The sum is a random variable whose density takes the form of any asymmetric continuous "bell" followed by a long tail.

II-H. The mechanism of cluster generation in the self-similar model

Consider three successive errors and suppose that the values of T_{n-1} and T_{n+1} are known, while T_n is not. The distribution of the random T_n will be studied under three basic assumptions, as illustrated in Figure N5-2.

When the inter-error distances are geometrically distributed, T_k is uniformly distributed between T_{n-1} and T_{n+1}, so that the actual instants T_k are neither uniform nor bunched.

If the inter-error intervals followed a binomial law (a situation which is not encountered in practice), their density could be approximated by a Gaussian, and T_n would be a Gaussian variable centered on $(1/2)(t_{n-1} + t_{n+1})$. Errors would tend to be almost uniformly distributed.

Finally, let $P(u) = (u/\varepsilon)^{-D}$. In that case, the conditional probability density of $U' = T_n - T_{n-1}$, given the value u of $U = T_{n+1} - T_{n-1}$, is closely approximated by the incomplete beta density

$$\frac{\varepsilon^D D^2 u'^{-(D+1)} (u - u')^{-(D+1)}}{2 D u^{-(D+1)}} \sim \frac{D \varepsilon^D}{2} u^{-(D+1)} \left[\frac{u'}{u} \left(1 - \frac{u'}{u} \right) \right]^{-(D+1)}.$$

In other words, the distribution of T_n has two sharp peaks, located at the instants of time $t_{n-1} + \varepsilon$ and $t_{n+1} - \varepsilon$; their amplitudes are equal to $D/2\varepsilon$ and independent of u. As a result, the middle error "huddles" close to either of the end errors, thus creating a cluster of two.

Consider similarly the case of *four* successive errors, and construct an equilateral triangle having a height equal to the known value $t_{n+3} - t_n$ of $T_{n+3} - T_n$. The distances between the three sides of the triangle and an inside point P are known to add to a quantity independent of P, and equal to the height of the triangle. Therefore, their distances can represent the three quantities $T_{n+3} - T_{n+2}$, $T_{n+2} - T_{n+1}$ and $T_{n+1} - T_n$, as constrained by the known value of $T_{n+3} - T_n$.

If the inter-error intervals were Gaussian, most of the probability density of P would be concentrated near the center of the triangle.

If the inter-error intervals were exponential, the probability density of P would be uniformly distributed over the triangle.

If the inter-error intervals are self-similar, the probability density is very much concentrated near the three corners, so that the four errors will mostly give rise to either one of the following patterns: "e..e..e......e," or "e..e......e..e," or "e......e..e..e," where ".." denotes few intervening correct digits, and "......" denotes many such correct digits. This means that the four errors in question will usually be clustered in two bursts of random sizes.

In the case of longer sequences of $L > 3$ errors, the inter-error interval can be represented by the L-dimensional geometric images made familiar by information theory. Their volume is increasingly concentrated near the corners as L increases. In the measure generated by self-similar distributions, this concentration is even more accentuated. A sizeable portion of the total sample length will be found in a few of the longest error intervals, thus creating a pattern in which errors are mostly grouped in clusters.

Another image of clustering can be deduced from the argument of Section II-G. Replace the original $\tilde{\tau}$ by a new cluster scale $r\tilde{\tau}$. A proportion $1 - r^{-D}$ of the *number* of old inter-cluster intervals will now be contained within the new clusters. However, assuming the total message length is much longer than $\tilde{\tau}$, a proportion close to one of the *length* of the old inter-cluster intervals will remain in the new inter-clusters.

II-I. Phenomenological approach to the study of excess noise

It would naturally be desirable to "explain" the preceding phenomena by reducing them to seemingly more elementary physical facts. I shall not attempt to do so in this essay and shall only check the model and develop its consequences. My approach is therefore "phenomenological" in the sense that is generally used to describe classical thermodynamics as opposed to the kinetic theory of gases.

The resort to such an approach does not imply any desire to berate the potential value of arguments analogous to those of kinetic theory. While we all await the emergence of a convincing argument of that type, the next stage of the present theory should perhaps consist in examining the following problems:

a) The "$1/f$" noise and other kinds of "excess noise" are also self-similar in one or several senses of the word; their descriptions should be worked out in detail (see M 1965b{N8}).

b) One should relate various kinds of excess noise to each other, without necessarily having to reduce either of them to more elementary facts.

II-J. Generation of pathological distributions through mixtures of nonpathological random variables

Extreme deviations from Markovian dependence (and from Gaussian distributions) also occur in various forms and with various intensities in the other subject matters treated in my publications in economics listed in the bibliography. In many cases, my phenomenological approach to such "pathological" behavior has elicited the counter-proposal that all these difficulties could perhaps be attributed to the mixing of data of various origins or characters. The argument proceeds by suggesting that unmixed data may be nonpathological. It is clear for example that the $P(u) = u^{-D}$ law for inter-error intervals can be rewritten as

$$u^{-D} = \frac{1}{\Gamma(D)} \int_0^\infty e^{-ub} b^{D-1} db,$$

which expresses u^{-D} as a mixture of exponentials e^{-ub}, weighted by the distribution b^{D-1}. Such a representation unfortunately "explains" nothing because the form of the "input" b^{D-1} is in one-to-one relation with that of the "output" u^{-D}. Such a decomposition would moreover be somewhat illogical even if it were successful, because it introduces a scale of time while the result that it seeks to derive contains no such scale. For finite samples, the continuous mixture could be replaced by a finite mixture with appropriate weights; but this procedure would not be very useful because it introduces so many parameters that reliable estimation and prediction would be impossible anyway (see Section I-A).

The reader who is not interested in the details of derivations is advised to proceed to Sections III-F and IV-F, where the results announced in Section II-F are verified.

III. THE MARGINAL DISTRIBUTION OF THE INTER-BOXES

III-A. Distance between a randomly chosen moment t_0 and the next error T_{next}, when the inter-error intervals are independent and bounded

Time being discrete, consider a long message made up of L successive inter-error intervals U, statistically independent of each other such that $\Pr\{U \geq u\} = P(u)$. In this preliminary subsection, it will be necessary to assume that U is bounded by $U_{max} < \infty$. Let T_{next} be the next error after the moment t_0 (if t_0 is itself in error, one will write $T_{next} = t_0$), and choose the moment t_0 at random, with equal probabilities, among the $LE(U) = L\sum_{1}^{\infty} P(s)$ instants contained within the long message under consideration. The desired equality $T_{next} - t_0 = u$ is satisfied only if t_0 happens to be the uth instant before the end of an inter-error interval of duration greater than u. The expected number of such points being equal to $LP(u + 1)$, we have

$$\Pr\{T_{next} - t_0 = u\} = \frac{P(u+1)}{\sum_{0}^{\infty} P(s)}.$$

The sum of these probabilities over u ranging from zero to ∞ equals one, as it should.

Similarly, in continuous time,

$$d\Pr\{T_{next} - t_0 < u\} = \frac{P(u)dt}{\int_0^{\infty} P(s)ds},$$

which means that *the density of the variable $T_{next} - t_0$ is equal to the cumulative probability function of the inter-error interval.*

As expected this last expression satisfies the identity

$$-\int X(h) \frac{dP(u+h)}{P(h)} = X(u),$$

which combines the following two facts: 1) By symmetry, the same law $X(h)$ represents the density of the two time increments $T_{next} - t_0$ and $t_0 - T_{last}$. 2) If the value of h is known, the density of $T_{next} - t_0$ is $-dP(u+h)/P(h)$. It is indeed easy to verify that

$$-\int_0^\infty \frac{dP(u+h)}{P(h)} \frac{P(h)}{\int_0^\infty P(s)ds} = -\frac{\int_0^\infty dP(u+h)}{\int_0^\infty P(s)ds} = \frac{P(u)}{\int_0^\infty P(s)ds}.$$

III-B. Continuation of Section III-A, assuming that $T_{next} - t_0$ is further conditioned by $0 < \tilde{\varepsilon} < T_{next} - t_0 < \tilde{u} < \infty$

Suppose now that t_0 was not chosen entirely arbitrarily, but was known in advance to be such that $0 < \tilde{\varepsilon} < T_{next} - t_0 < \tilde{u} < \infty$. Then, by ordinary conditioning, one finds that in discrete time

$$\Pr\{T_{next} - t_0 = u \mid \tilde{\varepsilon} < T_{next} - t_0 \le \tilde{u}\} = \frac{P(u+1)}{\sum_{\tilde{\varepsilon}+1}^{\tilde{u}} P(s)}.$$

In continuous time, $T_{next} - t_0$ has the conditional density

$$\frac{P(u)}{\int_{\tilde{\varepsilon}}^{\tilde{u}} P(s)ds}.$$

These conditioned distributions will be so important in the sequel that it is useful to give an alternative direct derivation. Consider again, in discrete time, a long sequence of L inter-error intervals following the law $P(u)$. The condition $\tilde{\varepsilon} < T_{next} - t_0 < \tilde{u}$ is satisfied by $\tilde{u} - \tilde{\varepsilon}$ instants out of every inter-error interval of length at least $\tilde{u}$. On the average, this contributes $L(\tilde{u} - \tilde{\varepsilon})P(\tilde{u})$ possibilities. The condition $\tilde{\varepsilon} < T_{next} - t_0 < \tilde{u}$ also holds for $s - \tilde{\varepsilon}$ instants of every inter-error interval of length $\tilde{\varepsilon} < s < \tilde{u}$. On the average, this contributes

$$L \sum_{\tilde{\varepsilon}}^{\tilde{u}-1} (s - \tilde{\varepsilon})[P(s) - P(s+1)]$$

possibilities. Altogether, $\tilde{\varepsilon} < T_{next} - t_0 < \tilde{u}$ holds for a total number of points equal to

$$L(\tilde{u} - \tilde{\varepsilon})P(\tilde{u}) + L\sum_{\tilde{\varepsilon}}^{\tilde{u}-1} s[P(s) - P(s+1)] - L\tilde{\varepsilon}[P(\tilde{\varepsilon}) - P(\tilde{u})]$$

$$= L\tilde{u}P(\tilde{u}) - L\tilde{\varepsilon}P(\tilde{u}) + L\tilde{\varepsilon}P(\tilde{\varepsilon}) + L\sum_{\tilde{\varepsilon}+1}^{\tilde{u}-1} P(s) - L(\tilde{u}-1)P(\tilde{u})$$

$$- L\tilde{\varepsilon}P(\tilde{\varepsilon}) + L\tilde{\varepsilon}P(\tilde{u}),$$

a sum that reduces to $L\sum_{\tilde{\varepsilon}+1}^{\tilde{u}} P(s)$.

The more stringent condition $T_{\text{next}} - t_0 = u$ is satisfied by only a single point of each of those inter-error intervals that have a length of at least $u + 1$; this means that the number of points favorable to this condition is $LP(u + 1)$. As expected, the ratio $P(u + 1)[\sum_{\tilde{\varepsilon}+1}^{\tilde{u}} P(s)]^{-1}$ is equal to the result obtained earlier through conditioning.

III-C. Marginal distribution of the distance between two successive boxes of duration T^* that contain at least one error each: the case when U is bounded

Continuing to assume that U is bounded by $U_{\max} < \infty$, assume time to be continuous, and choose an interval of length $\tilde{\tau}$ that contains at least one error-generating event T_k. Let X be the distance between the end point T_{next} of the interval and the last event t_0 it contains. Granted that $0 < X < \tilde{\tau}$, Section III-A shows that the probability density of X is

$$\frac{P(x)}{\int_0^{\tilde{\tau}} P(s)ds}.$$

If X is known, the probability that the next error will be in the interval from $\tilde{\tau}(u-1)$ to ∞ is

$$\frac{P[x + \tilde{\tau}(u-1)]}{P(x)}.$$

Consider finally the probability that two successive intervals in error be separated by $u\tilde{\tau}$ at least, so that the end of the first interval and the beginning of the second be separated by $\tilde{\tau}(u-1)$ at least. This probability is obtained by forming the conditional density

$$\frac{P(x)}{\int_0^{\tilde{T}} P(s)ds} \frac{P[x + \tilde{\tau}(u-1)]}{P(x)}$$

and integrating over x. If so, $P(x)$ cancels out, and one obtains

$$P^*(u) = \frac{\int_0^{\tilde{T}} P[s + \tilde{\tau}(u-1)]ds}{\int_0^{\tilde{T}} P(s)ds} = \frac{\int_{\tilde{\tau}(u-1)}^{\tilde{T}} P(s)ds}{\int_0^{\tilde{T}} P(s)ds}.$$

As expected, this expression equals one for $u = 1$. For large u, the numerator is about $P(u\tilde{\tau})\tilde{\tau}$. Therefore, $P^*(u)$ is proportional to $P(u\tilde{\tau})$.

III-D. Distance between a randomly chosen t_0 and the error T_{next}: the self-similar case

The conditional density in Section III-B, namely

$$\frac{P(u)}{\int_{\tilde{\varepsilon}}^{\tilde{u}} P(s)ds}$$

can be applied without difficulty, if $P(u) = 1$ when $u < \varepsilon$, $P(u) = (u/\varepsilon)^{-D}$ when $\varepsilon < u < U_{max}$, and $P(u) = 0$ when $u > U_{max}$, where $\varepsilon < \tilde{\varepsilon} < \tilde{u} < U_{max}$. It follows that

$$\frac{P(u)}{\int_{\tilde{\varepsilon}}^{\tilde{u}} P(s)ds} = \frac{u^{-D}}{(1-D)^{-1}[\tilde{u}^{(1-D)} - \tilde{\varepsilon}^{(1-D)}]}.$$

Suppose now that $\tilde{\varepsilon} \to 0$ (which requires that $\varepsilon \to 0$). The *conditional* density of $T_{next} - t_0$ then tends to a well-defined, finite, and nonvanishing limit, namely

$$\lim \frac{P(u)du}{\int_{\varepsilon}^{\tilde{u}} P(s)ds} = (1-D)\left(\frac{u}{\tilde{u}}\right)^{-D} d\left(\frac{u}{\tilde{u}}\right).$$

The expected position of the first error tends to the limit

$$\int_{0}^{\tilde{u}} (1-D)s^{-D+1}\tilde{u}^{D-1}ds = \frac{1-D}{2-D}\tilde{u},$$

which is proportional to $\tilde{u}$.

As $\tilde{u} \to \infty$, the probability density

$$(1-D)u^{-D}\tilde{u}^{D-1}$$

tends to zero for every u. Suppose, then, that a subinterval of unit length is chosen at random from a time span $(0, \tilde{u})$ constrained to contain at least one error. As $\tilde{u} \to \infty$, this constraint becomes decreasingly stringent. As a result, the probability for a random unit subinterval to contain an error will tend to zero as $\tilde{u} \to \infty$. This brings us back to essentially the same situation as in Section II-D.

However, errors do occur. The situation is reminiscent, with lesser severity, of the following problem. Consider a sequence of $\tilde{u} > 15$ digits and suppose that one marks at random 15 of them. There is the probability $15/\tilde{u}$ of marking any digit chosen in advance. As $\tilde{u} \to \infty$, this probability tends to zero, but one remains certain that *some* 15 of the digits will be marked. Similarly, in the present case, the probability of hitting a digit in error decreases proportionately to $\tilde{u}^{D-1}$, more slowly than $1/\tilde{u}$; but one remains certain (by assumption) that out of $\tilde{u}$ digits one or more will be in error.

III-E. Marginal distribution of the distance between successive boxes of duration 1 that contain at least one error each: the self-similar case

In Section III-C, the probability density of $T_{next} - t_0$ is evaluated only under the assumption that this quantity is conditioned to be in a finite interval. The argument of Section III-D shows therefore that Section III-C applies when $P(u)$ is proportional to u^{-D}, and is such that $P(u) = (u/\varepsilon)^{-D}$ for $u > \varepsilon$, and $P(u) = 1$ for $u < \varepsilon$. The value of $P^*(u)$ will naturally depend on the value of $\varepsilon/\tilde{T}$. This ratio is particularly simple to interpret if the "boxes" are identical to communication symbols. In that case, $\varepsilon/\tilde{T}$ is indeed the ratio between the bandwidth $1/\tilde{T}$ of the signal and the bandwidth $1/\varepsilon$ of noise. To simplify, let the unit of time be such that $\tilde{T} = 1$.

The principal zone of values of ε is $\varepsilon < \tilde{T} = 1$. Then

$$\int_0^1 P(s)ds = \varepsilon + \int_\varepsilon^1 (u/\varepsilon)^{-D} du$$

$$= \varepsilon + \frac{\varepsilon^D}{1-D}[1 - \varepsilon^{1-D}] = \frac{\varepsilon^D}{1-D}[1 - D\varepsilon^{1-D}].$$

Moreover, except when $u = 1$,

$$\int_{u-1}^u P(s)ds = \frac{\varepsilon^D}{1-D}[u^{1-D} - (u-1)^{1-D}].$$

We have thus established the following result: $P^*(1) = 1$ and for $u > 1$:

$$P^*(u) = \frac{u^{1-D} - (u-1)^{1-D}}{1 - D\varepsilon^{1-D}}$$

If u is very large, one obtains the approximation

$$P^*(u) \sim \frac{1-D}{1 - D\varepsilon^{1-D}} u^{-D}.$$

For theoretical purposes, the most important case is that of $\varepsilon \to 0$. At the limit, $P(u) = (u/\varepsilon)^{-D}$ becomes meaningless. However, for all u between one and ∞,

$$P^*(u) \to u^{1-D} - (u-1)^{1-D}.$$

This formula is one of the principal results of this paper.

Now, for the sake of completeness, let $\varepsilon > \tilde{T} = 1$; it follows that

$$\int_0^1 P(s)ds = 1$$

and the expression for $\int_{u-1}^{u} P(s)ds$ continues unchanged if $(u-1) > \varepsilon$. This means that, if u is not too small

$$P^*(u) = \varepsilon^{-D} u^{-D};$$

a result which is, in a way, obvious, since each error overlaps ε boxes.

To sum up, if u is large, one can always write

$$P^*(u) = Q(\varepsilon) u^{-D},$$

where the function Q is defined as follows:

$$\text{if } \varepsilon < 1, Q = \frac{1-D}{1 - D\varepsilon^{1-D}} < 1$$

$$\text{if } \varepsilon > 1, Q = \varepsilon^{-D} > 1.$$

It will follow that, by adjusting ε, one can use the self-similar theory to represent quite a variety of different behaviors for $P^*(u)$. However, the consequences of the above form of Q should be interpreted with great caution, because the whole model was constructed by extrapolating from experimental data for which ε is presumably quite small.

III-F. Empirical verifications

Insofar as the law $\Pr\{U \geq u\} = (u/\varepsilon)^{-D}$ is concerned, the standard method of testing its validity consists in plotting the logarithm of the frequency of $U \geq u$ as a function of the logarithm of u. The law is satisfied if the resulting graph is well approximated by a straight line. This procedure, unfortunately, has been neglected by statisticians, so that no small-sample tests are available. Valid conclusions can, however, be drawn in the case of the long samples that are available in problems of electrical communications. For example, the sample size yielding the marginal distributions shown in Figures N5-3 and N5-4 is of the order of 10^4. Moreover, as seen on Figure N5-7, there is very little danger of confusion between the distribution u^{-D} with $0 < D < 1$ and the exponential law that is suggested by all short-memory models. (Confusion is very likely when the exponent D is of the order of four or more.)

Consider therefore the bold solid lines of Figures N5-3 to N5-6, which represent the marginal distribution $\Pr\{T_{n+1} - T_n \geq u\}$. The fact that straight lines fit these curves very well over approximately four decades of inter-error intervals is a definite first-order verification of the Berger-M model. But the behavior of $P(u)$ for very small and very large values of u was puzzling. Now it can be fully accounted for.

To begin with, the "local slope" of the graph suddenly increases when u passes 10,000 and it may even exceed one when u equals 100,000. This fact did not, however, lead Berger & M to abandon outright the conjecture that D is the same for all u and is of the order of 1/3. We argued indeed that it is very likely that, for some connections at least, our new mechanism is combined with independent errors having a very small probability of occurrence. If so, the observed curve should be expected to be some average of a straight line, and of the curve corresponding to the geometric distribution as plotted on a doubly logarithmic graph. This method of eliminating the difficulties due to large values of U was mentioned in Section II-B, and the Berger & M figures would seem to imply that it has to be resorted to at an early stage. In fact, however, the problem raised by the final sections of our curves was eliminated in another very gratifying way. We found, indeed, that a few runs including fewer than five errors, were not represented on the tape used in our tests because they had been felt to be statistically irrelevant. This means that the actual records contain more data in the "tails" than suggested by the curves which we have plotted. On the basis of incomplete exploratory tests, we conjectured that the correct value of the parameter D can be obtained by linear extrapolation of the flattest part of our curves.

The next test of this conjecture was provided by Sussman's analysis of published Bell Telephone data. As predicted by the Berger & M model, the doubly logarithmic graphs of Sussman 1963 go straight through to the highest plotted values of u.

Once the predictions for u varying from 5 to 10,000 have been shown to be of such high quality, the discrepancy for small u's becomes very striking. It is clear that if one forms the linear interpolates of the data that are best in the range $5 < u < 10,000$, the resulting curves fail to go through the origin in the figures borrowed from either Berger & M 1963{N5}, or Sussman 1963. Therefore, Berger & M conjectured that a larger D may be required to represent short inter-error intervals.

It is here that the improved formula $P^*(u) = u^{1-D} - (u-1)^{1-D}$ proves its value, because its departure form u^{-D} has precisely the form required by data when D is small.

Figures N5-9 to N5-12 are similar to Figures N5-5, N5-7, N5-8, and N5-9, except that the transmit level is lower.

IV. INTERDEPENDENCE BETWEEN SUCCESSIVE INTER-BOXES

IV-A. Expected number of errors between time t_0 and time $t_0 + \tau$

Case when there is an error at time t_0. Time being discrete, suppose that the intervals between errors are mutually independent and such that $\Pr\{T_{n+1} - T_n \geq u\} = P(u)$. Now let $N^*(\tau)$ be the expected number of errors within the time span of length τ after an error. $\Delta N^*(\tau) = N^*(\tau) - N^*(\tau - 1)$ is then the probability that the instant τ be the location of an error (that need not be the first to occur between times 0 and t).

Chapter XIII of William Feller's 1957 book shows that the Laplace transform $G^*(b)$ of $\Delta N^*(\tau)$ can be deduced from the generating function $G(b)$ of $\Pr\{U = u\}$ by the relation $G^*(b) = [1 - G(b)]^{-1}$. Since $G(0) = 1$, the sum $G^*(0)$ of the probabilities ΔN^* is infinite, which means that $N^*(\infty) = \infty$.

In the models usually considered, $\Delta N^*(\tau)$ tends for large τ to the unconditional probability of encountering an error; i.e., it tends to the inverse of the inter-error interval $E(U)$. As a result, one has $G^*(b) \sim [bE(U)]^{-1}$ as $b \to 0$, which can also be established by noting that $G(b)$ behaves for small b like $1 - bE(U)$.

Now let Feller's relation be applied to the law of the Berger-M model, in which $P(u) = u^{-D}$ with integer u's. The characteristic function $G(b)$ behaves for small b like

$$\exp[D\Gamma(-D)b^D] = \exp[-\Gamma(1-D)b^D]$$

so that

$$G^*(b) \sim b^{-D}[\Gamma(1-D)]^{-1}$$

and

$$\Delta N^*(\tau) \sim \frac{\tau^{D-1}}{\Gamma(1-D)\Gamma(D)} = \frac{\sin(\pi D)}{\pi} \tau^{D-1}.$$

As $\tau \to \infty$, $\Delta N^*(\tau)$ tends to zero (a limit that is, duly, the inverse of $E(U)$).

Note also that, if it were meaningful to let T tend to zero, T^{D-1} would tend to ∞. But T is at least equal to one, so that the proportionality between $\Delta N^*(T)$ and T^{D-1} must break down for small T in such a way that $\Delta N^*(T)$ is always at most equal to one. In any event, Feller 1949 has shown that, for large T,

$$N^*(T) = \frac{\sin(\pi D)}{D\pi} T^D + \text{a corrective term.}$$

The corrective term alone depends upon the behavior of $N^*(T)$ for small T.

Now let us change the process that generates errors to read that inter-error intervals are multiples of some $\varepsilon > 0$. Then

$$N^*(T) = \frac{\sin(\pi D)}{D\pi} (T/\varepsilon)^D + \text{a corrective term.}$$

Again, the corrective term becomes negligible in relative value as (T/ε) tends to ∞. Thus, for fixed T, it becomes negligible in relative value as ε decreases.

The fact that $N^*(T)$ increases to ∞ as $\varepsilon \to 0$ proves the assertion made in Section II-C. If the number of error-generating events is nonzero, it tends to ∞ as $\varepsilon \to 0$.

If ε is very small, time becomes continuous and the event that an error occurs within a time dT near instant T has the probability

$$\frac{\sin(\pi D)}{D\pi} T^{D-1} \varepsilon^{-D} dT.$$

Case when there is at least one error within the span t_0 to $t_0 + T$, but none at the instant t_0. Knowing the position s of the first error, the expected number of errors becomes

$$\frac{\sin \pi D}{TD} (T-s)^D \varepsilon^{-D}.$$

It was seen in Section III-D that, for small ε, the probability density of s is given by $(1-D)s^D T^{D-1}$. The expected number of errors therefore becomes

$$\frac{\sin \pi D}{\pi D} (1-D) \int_0^\tau s^{-D} \tau^{D-1} (\tau - s)^D \varepsilon^{-D} ds$$

$$= \frac{\sin \pi D}{\pi D} (1-D) \tau^D \varepsilon^{-D} \int_0^1 x^{-D} (1-x)^D dx$$

$$= \frac{\sin \pi D}{\pi D} (1-D) \tau^D \varepsilon^{-D} \frac{\pi D}{\sin \pi D} = (1-D)(\tau/\varepsilon)^D.$$

This is another fundamental result of this paper. It follows that, if one of the τ instants between t and $t + \tau$ is chosen at random, the probability of hitting an error is equal to $(1-D)\varepsilon^{-D}\tau^{D-1}$. This tends to zero as $\tau \to \infty$, yielding yet another way of expressing the fact that, as $\tau \to \infty$, less and less information is provided by the assumption that a time increment of length τ contains at least one error.

IV-B. Digression concerning the inapplicability of Shannon's information theory

From the τ^D growth of the expected number of errors in a time span τ, it follows that the equivocation of our channel is zero. Indeed, W_τ and V_τ being an emitted and received sequence of τ symbols, Shannon defines the equivocation per sequence as equal to $-\log \Pr \{W_\tau / V_\tau\}$. In general, the precise definition of this conditional probability raises difficulties in the case of channels with infinite memory. In the present case, however, equivocation is simply the nonaveraged information required to specify the positions of the digits in error, i.e., the distances between the beginning of the sequence and the first error, and the distances between succeeding errors.

Using ε as unit of time, we see that the specification of an inter-error interval requires the following information on the average: C being some constant

$$-\sum_{s=1}^\infty |\Delta P(s)| \log_2 |\Delta P(s)| \sim C - (D+1) \sum_{s=1}^\infty s^{-(D+1)} \log_2 s,$$

which is a finite quantity J. Moreover, the information required to specify the position of the first error is contained between 1 and $\log_2 \tau$. Hence, averaging the equivocation with respect to all possible positions of the errors, we find it to be in the range from $[N^*(\tau) - 1]J$ to $[N^*(\tau) - 1]J + \log_2 \tau$.

Finally, averaging with respect to values of $N^*(\tau)$, we find for large τ, that the averaged equivocation per symbol is proportional to τ^{D-1}. This expression tends to zero with $1/\tau$. It follows that, despite the presence of an unbounded number of errors, the channel which we have defined has a capacity equal to one.

Conclusion: The limit theorems of the theory of information are inapplicable to binary transmission over actual telephone channels. Generally speaking, the mathematical theory of coding, that is, the theory of information as understood in the strictest sense, consists in evaluating the various "precorrecting codes" suggested by their inventors, and in comparing them with an ideal of performance associated with certain probability limit theorems.

The theory of information was divided by Shannon into two parts, according to the absence or the presence of noise. Actually this division is an oversimplification, because the *theory of noiseless transmission is not the limit of the theory of transmission in the presence of vanishingly small noise.* For example, a circuit with self-similar errors has a capacity equal to one, so that there seems to be no need for error correction of any kind. Actually, as the word length increases, the limit of the best error-precorrecting code is not identical to the best code corresponding to the limit capacity of one. This type of limit behavior is frequently observed in engineering, and is referred to as being a "singular perturbation." Its classical prototype arises in the comparison of nonviscous fluids with those of very small viscosity. Note that the inefficiency of error-correction has already been pointed out in Fontaine 1961.

IV-C. Joint distribution of the first and last error in a box in error

Consider a box from 0 to 1 containing *at least one error,* designate by Y and $1 - X$ the locations of the first and last of these errors, and evaluate the joint distribution of X and Y. The difference $1 - X - Y = Z$ is ruled by the law ΔN^* of Section IV-A, and the marginal distributions of X and Y are ruled by the law of Section III-D. It can be shown that X and Y have the joint density

$$\frac{P(x)P(y)\Delta N^*(1 - x - y)}{\iint P(s)P(r)\Delta N^*(1 - s - r) ds dr},$$

where the double integral is carried over all x and y such that $x > 0$, $y > 0$, and $x + y < 1$. It will be convenient to represent the three quantities x, y and $z = 1 - x - y$ by a point P in what we shall call the "admissible. triangle." This is an equilateral triangle of height 1, in which x, y and z are the distances from P to the three sides of the triangle.

Suppose in particular that U follows the truncated self-similar law $P(u) = (u/\varepsilon)^{-D}$ for u varying from ε to ∞. It is readily seen that, as $\varepsilon \to 0$, the joint density of x and y tends to the limit

$$\frac{x^{-D} y^{-D} (1 - x - y)^{D-1}}{\iint s^{-D} r^{-D} (1 - s - r)^{D-1} ds \, dr},$$

a member of a family sometimes referred to as "Dirichlet distributions."

In order to evaluate the denominator, write it as

$$\int_0^1 s^{-D} ds \int_0^{1-s} r^{-D} (1 - s - r)^{D-1} ds.$$

Setting $r = w(1 - s)$, this becomes

$$\int_0^1 s^{-D} ds \int_0^1 w^{-D} (1 - w)^{D-1} dw$$

$$= \frac{1}{1-D} \frac{\Gamma(1-D) \, \Gamma(D)}{\Gamma(1)} = \frac{\pi}{(1-D) \sin(D\pi)}$$

finally leaving a joint density for x and y equal to

$$\frac{(1-D) \sin(D\pi)}{\pi} x^{-D} y^{-D} (1 - x - y)^{D-1}.$$

It is important to note that this joint density is very strongly concentrated near the three corners of the "admissible triangle" whose sides correspond to $x = 0$, $y = 0$, and $z = 1 - x - y = 0$. The remaining probability is concentrated near the edges of that triangle, leaving very little for the center area.

The interpretation of this fact is simplest if D is close to one or to zero, and runs as follows.

As $D \to 1$, $(1-x-y)^{D-1}$ is close to one, therefore, the bulk of the probability corresponds to very small x and y and X and Y are practically independent.

As $D \to 0$, on the contrary, the terms x^{-D} and y^{-D} are both practically equal to one. Therefore most of the probability is near $x+y=1$. This means that all the errors within the box are concentrated on a small subinterval located anywhere along the length of the box.

Suppose now that D is somewhere between zero and one, say near one-half. The most likely configurations of the first and last errors inside a unit box are "spanning most of the box" and "concentrated near either of the ends of the boxes."

Let us now rederive the joint density of X and Y more directly. Assume therefore that $P(u) = u^{-D}$ holds from 0 to ∞, and express the joint density of X and Y as product of the marginal density of X by the conditional density of Y, knowing X.

1) It has already been established that the marginal density of X is $(1-D)x^{-D}$.

2) It is clear that the joint density of X and Y is a symmetric function of x and y.

3) Finally, because of self-similarity, it is clear that $\Pr\{Y < y \mid X = x\}$ is a function of x and y through the ratio $y/(1-x)$.

Consider now the conditional density of Y, given X. It is the product of $(1-x)^{-1}$ by some function of $y/(1-x)$ and is uniquely determined by the above three conditions. (*Proof:* if there were two distinct solutions, their ratio would be both symmetric in x and y and a function of $y/(1-x)$; it must therefore be a constant, which in turn implies that it equals one. Q.E.D.) This has allowed me to first derive the above joint density of X and Y by simple inspection. That is, I observed that the conditional density of Y, given X, is

$$\sin(\pi D)\pi^{-1} y^{-D}(1-x-y)^{D-1};$$

this function can be written as

$$\sin(\pi D)\pi^{-1}(1-x)^{-1}\left[\frac{y}{(1-x)}\right]^{-1}\left[-1+\frac{(1-x)}{y}\right]^{D-1},$$

which has the form required by condition 3.

It also follows from the preceding argument that, given X, the conditional density of $Y^* = Y/(1-X)$ is

$$\frac{\sin D}{\pi} y^{*-D}(1-y^*)^{D-1},$$

which is the density of a "Beta distribution." The Beta distribution for $D = 1/2$ is classically known as the "arcsine" law of Paul Lévy.

Finally, note that the marginal density of $Z = 1 - X - Y$ is

$$[\Gamma(2-D)\,\Gamma(2-2D)^{-1}\,\Gamma(D)^{-1}]\,z^{D-1}\,(1-z)^{1-2D}.$$

If $D < 1/2$, Z has a single most probable value $Z = 0$. If $D > 1/2$, there are two most probable values, $Z = 0$ and $Z = 1$. The distribution of Z is symmetric with respect to $Z = 1/2$ if $D - 1 = 1 - 2D$, i.e., if $D = 2/3$.

The expectation of Z is

$$\frac{\Gamma(2-D)}{\Gamma(2-2D)\Gamma(D)}\int_0^1 z^D(1-z)^{1-2D}dz$$

$$= \frac{\Gamma(2-D)}{\Gamma(2-2D)\Gamma(D)}\cdot\frac{\Gamma(2-2D)\Gamma(D+1)}{\Gamma(3-D)} = \frac{D}{(2-D)}.$$

Its value is zero if $D = 0$, one-half if $D = 2/3$, and one if $D = 1$.

IV-D. Interdependence between successive inter-boxes

Consider *three* boxes in error and evaluate the probability that the first two are separated by u' or more and that the last two are separated by u'' or more. If the position of the first error in the middle box designated by Y and the position of the last error by $1 - X$, the result of Section III-D shows that

Pr{ first interblock interval $\geq u'$ and second interblock interval $\geq u''$}

$$= \frac{(1-D)\sin(D\pi)}{\pi} \iint \frac{(x+u''-1)^{-D}}{x^{-D}}$$

$$\cdot \frac{(y+u'-1)^{-D}}{y^{-D}} x^{-D} y^{-D} (1-x-y)^{D-1} dx\, dy$$

$$= \frac{(1-D)\sin(D\pi)}{\pi} \iint (x+u''-1)^{-D} (y+u'-1)^{-D}$$

$$\cdot (1-x-y)^{D-1} dx\, dy.$$

Whenever such a double integration is encountered again in Sections IV-D and IV-E, it will be understood that it is carried over all the values of x and y such that $x > 0$, $y > 0$ and $x + y < 1$.

Assume that u' and u'' are both large. The probability that the first interval equals u' and that the second equals u'' is approximately

$$\left[(1-D)Du'^{-D-1}(1-D)Du'^{-D-1}\right]$$

$$\cdot \left[\frac{1}{\pi(1-D)} \sin(D\pi) \iint (1-x-y)^{D-1} dx\, dy\right].$$

The locations of the brackets were selected in such a way that the first is the product of the marginal probabilities and the second represents the correction due to dependence.

$$\tilde{Q} = \frac{\sin(D\pi)}{\pi D(1-D)(1+D)}.$$

$\tilde{Q}$ is always contained between one-half and one; as D varies from zero to one, it decreases from one to one-half. This means that, if the previous correct interblock interval was large, the probability that the next interblock interval be large is *smaller* than the marginal probability. This must naturally be compensated for. Therefore the probability that U'' be large when U' is small must be *greater* than the marginal probability; the probability of any given small value of U'' must *increase as u' increases*. The intervals between boxes in error exhibit a kind of "anti-clustering tendency."

N7 ◇ ◇ SELF-SIMILARITY AND CONDITIONAL STATIONARITY

The result just obtained will now be rederived somewhat differently in order to further specify the dependence between successive intervals U' and U''. Since $U' \geq u'$, the probability that $U'' \geq u''$ is

$$\frac{(1-D)\sin(D\pi)}{\pi[u'^{1-D} - (u'-1)^{1-D}]}$$

$$\bullet \iint (x+u''-1)^{-D}(u'-1+y)^{-D}(1-x-y)^{D-1} dx\, dy.$$

A convenient numerical measure of the extent of dependence for large u'' is provided by the ratio between the above probability and the unconditioned probability $(1-D)u''^{-D}$. This ratio can be written as

$$\frac{\sin(D\pi)}{\pi\left[u'^{1-D} - (u'-1)^{1-D}\right]} \int_0^1 (u'-1+y)^{-D} dy \int_0^{1-y} (1-x-y)^{D-1} dx$$

$$= \frac{\sin(D\pi)}{D\pi[u'^{1-D} - (u'-1)^{1-D}]} \int_0^1 (u'-1+y)^{-D}(1-y)dy$$

$$= \frac{\sin(D\pi)}{\pi D(1-D)} \frac{\int_0^1 (u'-1+y)^{-D}(1-y)^D dy}{\int_0^1 (u'-1+y)^{-D} dy}$$

$$= \frac{\sin(D\pi)}{\pi D(1-D)} \frac{\int_0^1 (u'-r)^{-D} r^D dr}{\int_0^1 (u'-r)^{-D} dr}$$

$$= \frac{\sin(D\pi)}{\pi D(1-D)} \int_0^1 \frac{(u'-r)^{-D}}{\int_0^1 (u'-s)^{-D} ds} r^D dr.$$

This last expression is a weighted average of r^D, using the weight that becomes increasingly concentrated on small values of r as u' decreases. It follows that the conditional density of U'' decreases with u'. In particular, to know that $u' = 1$ is to know nothing about U', and the expression that measures the extent of dependence indeed equals one for $u' = 1$. As u' becomes very large, on the contrary, the extent of dependence becomes again equal to $\tilde{Q}$, since

$$\frac{\sin(D\pi)}{D\pi(1-D)} \int_0^1 dy (1-y)^D = \frac{\sin(D\pi)}{D(1-D)(1+D)\pi} = \tilde{Q}.$$

IV-E. Distribution of the sum of successive inter-boxes

Another way of assessing the interdependence is to examine

$$\Pr\{U' + U'' \geq u\} = \sum_{u'=1}^{u} \Pr\{U' \geq u' \text{ and } U'' = u - u'\}.$$

If U' and U'' were independent, and u were very large, one would have

$$\Pr\{U' + U'' \geq u\} \sim 2\Pr\{U' \geq u\} \sim 2(1-D)u^{-D}.$$

The dependence between U' and U'' is expressed by the more complicated formula for $\Pr\{U' + U'' \geq u\}$, namely

$$\frac{(1-D)\sin(D\pi)}{\pi} \iint dx\, dy \left\{ \sum_{u'=1}^{u} \left[(x+u-u'-1)^{-D} - (x+u-u')^{-D} \right] \right\}$$
$$(y+u'-1)^{-D}(1-x-y)^{D-1}.$$

If u is very large, this formula becomes

$$\left\{ \sum_{u'=1}^{u} [(x+u-u'-1)^{-D} - (x+u-u')^{-D}] \right\} (y+u'-1)^{-D} \sim 2u^{-D},$$

which reintroduces the $\tilde{Q}$ of Section IV-D, and yields

$$\Pr\{U' + U'' \geq u\} \sim \left[\left(\frac{1}{\pi}\right) \sin(D\pi) \iint (1-x-y)^{D-1} dx\, dy \right] [2(1-D)u^{-D}]$$
$$\sim \tilde{Q}[2(1-D)u^{-D}].$$

For small D, the result simplifies: $\tilde{Q} \sim 1$ and $\Pr\{U' + U'' \geq u\}$ behaves as if U' and U'' were independent.

For sums of more than two successive inter-boxes, the formulas become more involved, but the corrections for dependence remain small when $D \sim 0$.

IV-F. Empirical verification

The preceding results can be summed up by saying that successive intervals between boxes in error are practically independent when D is very close to zero, but become very dependent as D increases to one. The natural test of the independence of $T_{k+1} - T_k$ with respect to the past will consist in plotting $\log \Pr\{T_{k+1} - T_k \geq u\}$ for every pattern of past errors. Unfortunately, the need for very large samples has limited us to first-order transitions and a few of high order. Therefore, we vary u' from 1 to 100,000, and examine the following distributions.

- First, a distribution concerned with *three* successive errors:

$$\log \Pr\{T_{k+1} - T_k \geq u'' \mid T_k - T_{k-1} = u'\}.$$

- Then, a distribution concerned with five successive errors:

$$\log \Pr\{T_{k+1} - T_k \geq u'', \text{ when one knows that}$$

either $T_k - T_{k-1} = u', T_{k-1} - T_{k-2} = 1$ and $T_{k-2} - T_{k-3} = 1$;

or $T_k - T_{k-1} = u', T_{k-1} - T_{k-2} = 1$ and $T_{k+2} - T_{k-3} = 2$;

or $T_k - T_{k-1} = u', T_{k-1} - T_{k-2} = 2$, and $T_{k-2} - T_{k-3} = 1\}$.

- Finally, the distribution of the sum of three successive inter-error intervals, $\log \Pr\{T_{k+1} - T_{k-2} \geq u''\}$.

This last function is compared with the distribution of the sum of three independent random variables having the same distribution as $T_{k+1} - T_k$.

The plots of the first function $\Pr\{T_{k+1} - T_k \geq u'', \text{ if } T_k - T_{k-1} = u'\}$ are given in Figures N5-3 through N5-6 for various values of u', considered as a parameter, and u''. The sample size is a decreasing function of u'. It is of the order of 5000 events for $u' = 1$; 2000 events for $u' = 2$; 1000 events for $u' = 3$; 500 events for $u' = 4$; 500 events for u' between 10 and 20; 100 events for u' between 50 and 60, and 300 events for u' between 100 and 200.

If the inter-error intervals were independent, we would expect to find the same distribution for all values of u'. The conditional frequency distributions are indeed very similar but certainly not identical. This disagrees with the Berger & M model. *But it is fully accounted for* by the present self-similar model. Note in particular that the order of the curves for $u' = 1, 2, 3, 4$ is precisely *as predicted*.

The plots of the second functions are similar to those of the first, with one exception. They single out some specific previous patterns which correspond to typical "bursts" of errors as discussed in the literature. The plots given in Figures N5-6 and N5-7 are necessarily based on small sample sizes of the order of a hundred events. Since they exhibit the same characteristics as the other curves, they further establish our thesis that clusters or "bursts," per se, do not have a separate intrinsic meaning and identity.

The third function, $\Pr\{T_{k+3} - T_k \geq u''\}$, which is plotted in Figure N5-3, is based on a sample of several thousand events. It has the same shape as the marginal law $\Pr\{T_{k+1} - T_k \geq u''\}$, but is translated by a quantity a little less than log 3, *as predicted*.

Alternatively, u' can be considered as the variable and u'' as the parameter. In particular, let us plot the variation of the function $\Pr\{T_{k+1} - T_k = 1$ knowing that $T_k - T_{k-1} = u'\}$, when its argument u' varies from 1 to 10,000. It is clear in Figure N5-13, based upon a sample of several thousand events, that this function is practically independent of u' as long as u' is large. But the value of this function is much smaller for $u' = 2$ and for $u' = 1$. Again, this means – *as predicted* – that sequences of three successive boxes in error are markedly less frequent than they would be if the interblock intervals were independent. That is, the model of independent intervals predicts even more clustering than is actually observed (despite the fact that clustering was not willfully built-in). It also predicts too large a number of "clusters" of the form error-correct-error-error.

The curve for $T_{k+1} - T_k = 2$, also shown in Figure N5-13, is based on a smaller sample. One observes deviations from independence of u', but they are not systematic. This is again in conformity with the self-similar model.

Taken together, these results are in quite definite agreement with the main features of the proposed model. The geometric distribution, which is plotted in Figure N5-5 for several values of its parameter, is certainly inapplicable.

V. PROSPECT

The now classical technique of spectral analysis is inapplicable to the processes examined in this paper but it is sometimes unavoidable. M1967i{N9} will examine what happens when the scientist applies the algorithms of spectral analysis without testing whether or not they have the usual meaning. This investigation will lead to fresh concepts that appear most promising indeed in the context of a statistical study of turbulence, excess noise, and other phenomena where interesting events are intermittent and bunched together.

Acknowledgment. Special thanks are due to my colleague and friend J. M. Berger, with whom I had the pleasure of collaborating in Berger & M 1963{N5}. To the acknowledgment of the Berger-M paper, I wish to add an expression of appreciation for numerous comments made by C. E. Shannon, J. R. Pierce, J. W. Tukey, and the students who took my Harvard University course Ap. Math 210 in Spring, 1964.

PART III: INTERMITTENT 1/f NOISES AND CONDITIONED RANDOM PROCESSES

The "panorama of grid-bound self-affine variability," Chapter N1, shows that self-affinity and an $1/f$ spectrum can reveal themselves in several quite distinct fashions. One is featured in this part, a second in Part IV, and a third will be the central topic of M 1998 H.

This part concerns forms of $1/f$ behavior that are predominantly due to the fact that a process does not vary in "clock time" but in an "intrinsic time" that is fractal. Those $1/f$ noises are called "sporadic" or "absolutely intermittent," and can also be said to be "dustborne" and "acting in fractal time."

There is a sharp contrast between a highly anomalous ("non-white") noise that proceeds in ordinary clock time and a noise whose principal anomaly is that it is restricted to fractal time. The former can be Gaussian (as will be the case in most of M1998H). The latter are necessarily very far from being Gaussian. Nevertheless, it is easy to arrange for them to be "marginally Gaussian" insofar as the quantity $X(t)$ is Gaussian for a given t. This is the case for the "co-indicator" function introduced in M 1977i{N9}, and in a related class of processes discussed in an excerpt from M 1969i that is added to M1967i{N9} as a Post-publication Appendix.

This paper provides formal estimates of Wiener-Khinchin covariance and spectral density for various constructions. However, those formal estimates demand a very specific new interpretation. Starting with the notion of conditional stationarity introduced in Chapter N7, the notion of random process is generalized into the notion of sporadic process. In addition, the spirit but not the letter of the Wiener-Khinchin theory is preserved by introducing a more broadly valid new technique, conditional spectral analysis.

Chapter N8 is an introduction to later chapters, with which it overlaps.

$1/f$ noises and the infrared catastrophe

1. Statement of the problem

Many physical fluctuations have empirical (sample) spectral densities $S^*(f)$ that are nearly proportional to f^{-B}, with B close to 1.

This behavior holds up to wavelengths of the order of the sample size, and it is therefore commonly inferred that the population spectral density $S(f)$ is proportional to f^{-B} But this is inadmissible when $B \geq 1$, since it implies that the low frequency energy $\int_0^1 S(s)ds$ is divergent. This form of divergence, being contrary to a major assumption of spectral theory, is sometimes called "infrared catastrophe." One way to avoid it is to truncate and assume that the proportionality to f^{-B} only holds in the range $0 < f_{min} \leq f \leq f_{max} < \infty$. This approach assigns a central role to the lowest attainable frequency f_{min}, which in fact has never been observed.

The purpose of my current investigations, which the present text summarises, is on the contrary to show that the same evidence can be reinterpreted in such a way that the infrared catastrophe is made to disappear, or "exorcised," without having to resort to any unverifiable assumption. This approach is intimately related to my model of self-similar error-clusters, and to my study of turbulence in fluids. It is familiar fact that one can exorcise ultraviolet catastrophe by time quantization.

2. Some processes with "modulated" variance

The simplest nonstationary processes are *sums* of an oscillatory term and a "slowly varying trend," which may itself be generated by a random process. I don't believe that the infrared catastrophe can be eliminated by following this approach.

I can, on the contrary, show that it can be eliminated with the help of a family of stochastic processes of which the simplest example is a random function $X(t)$ that alternates between "active" and "quiescent" periods. Each active period being divided into equal "ε-intervals" of length ε, $X(t)$ constant over any such ε-interval; its value is, with equal probabilities, equal to either 1 or -1; its values over different ε-intervals are statistically independent. This active behavior is an approximation to a white Gaussian process of unit variance. During the quiescent periods, $X(t)$ is constant.

Most instruments will measure the difference between $X(t)$ and some moving average. A quiescent X will therefore register as equaling zero; but it is better for the theory to assume X to be equal to 1 or -1 with equal probabilities. {P.S.1998: Speaking somewhat loosely, one can say that the variance of $X(t)$ is modulated to be 1 in active periods, and 0 in quiescent ones.}

Gilbert's theory of error-clustering, would suggest, as a first step, that transitions between active and quiescent periods are ruled by a Markovian "master process." However, this suggestion can yield neither the observed structure of error-clusters (see M 1965c and Berger & M 1963{N5}) nor the f^{-B}.

The Berger & M approach to error-clusters suggests a model in which the "runs," or durations of active and quiescent periods, are mutually independent, but are such that the probabilities

$$\Pr\{\text{active run} \geq u\} = P_a(u) \text{ and } \Pr\{\text{quiescent run} \geq u\} = P_q(u),$$

are non-exponential functions. In this spirit, f^{-B} noises are discussed in Chapter 9. The present text concentrates upon the case $B = 1$: it possesses original features, and understanding of these features should make it easy for the reader to translate my earlier results into terms of electrical fluctuations. This text will, moreover, be restricted to the simplest variance modulator.

3. Spectra equivalent to f^{-1}

A spectral density $S(f)$ will be called "equivalent to f^{-1}" if it is of the form $S(f) = f^{-1} R(f)$, where $R(f)$ has the following property: for every $k > 1$, $R(kf)/R(f) \to \infty$ as $f \to 0$. Such functions $R(f)$ are called "slowly varying" in the sense of Karamata. In the range of very small f's, this family of functions $f^{-1}R(f)$ includes improper spectra with infrared catastrophe, such as f^{-1} itself, $f^{-1}(|\log f|)$, and $f^{-1}(|\log f|)^{-1}$. But it also includes proper spectra with a finite low frequency energy, such as $f^{-1}(\log f)^{-2}$, and $(|\log f|)^{-1}(\log|\log f|)^{-2}$. I doubt however that one will ever be able to distinguish experimentally between these various functions. My theory therefore avoids any "non-operational" distinction between different functions $S(f)$ equivalent to f^{-1}.

In order to exorcise the infrared catastrophe, I require a generalization of the usual Wiener-Khinchin form of spectral analysis. The new form is also applicable to cases from which the catastrophe is absent, and it is best to introduce it through those cases.

4. Wienerian spectra equivalent to f^{-1}

Recall that $\int_0^\infty P_q(s)ds = Q$, and $\int_0^\infty P_a(s)ds = A$ are the expected durations of quiescent and active runs. This being granted, consider a process with modulated variance, as defined in Section 2, that satisfies three conditions:

a) $P_q(u)u$ varies slowly as $u \to \infty$

b) $Q < \infty$ and

c) $A < \infty$.

Let t be chosen at random and consider the expression

$$1 - C(\tau) = E\{½ [X(t) - X(t + \tau)]^2\}.$$

With the probability $A(A + Q)^{-1}$, the instant t falls within an active run. For $|\tau| < \varepsilon$ but $\tau \neq 0$, $C(\tau)$ is not needed. For other τ, we have

$$1 - C(\tau) = 1 \text{ if } |\tau| > \varepsilon, \text{ and } 1 - C(0) = 0.$$

With the probability $Q(A + Q)^{-1}$, the instant t falls within a quiescent run. If $t + \tau$ is in the same quiescent run, $X(t) = X(t + \tau)$; if $t + \tau$ is in a different run $|X(t + \tau) - X(t)|$ equals either 0 or 2 with equal probabilities, and $½[X(t) - X(t + \tau)]^2$ equals 1 on the average. Therefore, $1 - C(\tau)$ is equal to the probability that the quiescent run containing t be finished before time $t + \tau$. This probability equals

$$\frac{\int_0^{|\tau|} P_q(s)ds}{\int_0^\infty P_q(s)ds}.$$

Averaging over ts located within active and quiescent runs, we obtain for $|\tau| > \varepsilon$

$$C(\tau) = (A + Q)^{-1}[Q - \int_0^{|\tau|} P_q(s)ds]$$

By a well-known result called Tauberian theorem, the conditions imposed upon $P_q(u)$ are both necessary and sufficient in order for $S(f)$ to be a spectral density that is equivalent to f^{-1}, and is a proper Wiener spectral density, i.e., devoid of infrared catastrophe.

This seems to have solved our problem. But in practice it has not, for the following two reasons.

On the one hand, the value of the population mean Q is altogether modified by practically unsignificant changes in the asymptotic behavior of $P_q(u)$, such as the replacement of

$$P_q(u) = u^{-1}(\log u)^{-1.1} \text{ by } P_q(u) = u^{-1}(\log u)^{-1.5}.$$

On the other hand, the practical estimates of $C(\tau)$ depend upon estimates of Q. In the present case, the theoretical Q is given by a barely convergent integral. {P.S.1998: I probably meant an integral that would become divergent if $P_q(u)$ were replaced by a function that tends to 0 only a little more slowly.} Therefore, the estimated Q is highly dependent upon sample size and sampling variation. That is, longer and longer samples of $X(t)$ will require constant revisions in the estimate of Q. The estimates of $C(\tau)$ or $S(f)$ will also have to be constantly modified. As a result, the problem of determining empirically the C or S must be considered as being ill-posed. This claim is related to amply documented practical difficulties in the measurement of f^{-1} spectra.

It is therefore advisable to postpone any attempt to explain why $fS(f)$ should be slowly varying as $f \to 0$, or why $uP_q(u)$ should be slowly varying as $u \to \infty$. The present study is devoted to a more realistic task of determining which questions concerning f^{-1} noises are well-posed, in the sense that their answers are independent of uncontrollable sampling fluctuations and/or of irrelevant details. Examples of uncontrollable data are the precise value of $\int_0^1 S(s)ds$ and the precise rate at which $\int_f^1 S(s)ds$ increases as $f \to 0$.

5. Conditioned Wienerian spectra related to f^{-1}

We propose to answer the dilemma encountered in Section 4 by observing that the unconditioned expectation of $X(t)X(t + \tau)$ makes little sense when the variance of X is modulated according to the rules of Section 2. The reason is that, although rules b) and c) were so chosen that our X is statistically stationary, the *finite* samples of X are very poorly described by the *asymptotic* results of the Wiener-Khinchin theory.

Consider a long sample of fixed duration T_{max}. As the initial point t^0 moves along, the "qualitative" behavior of $X(t)$ can be shown to exhibit one of three different "regimes:"

1) The sample is entirely contained in a quiescent run, therefore it has the appearance of a direct current. This happens with the probability

$$\int_\infty^{T_{max}} P_q(s)ds \, [A + Q]^{-1}.$$

2) The sample is entirely in an active run, therefore it has the appearance of white noise. This happens with the probability

$$\int_\infty^{T_{max}} P_a(s)ds \, [A + Q]^{-1}.$$

3) With the remaining probability, the sample is mixed, that is contains portions of both active and passive runs.

If both t and $t + \tau$ are between t^0 and $t^0 + T_{max}$, the conditioned covariance in the mixed regime is

$$E[\,X(t)\,X(t+\tau)\,|\,t^0 < t, t+\tau < t^0 + T_{max}\,]$$

$$= C(\tau, T_{max}, F) = \frac{\int_{|\tau|}^{T_{max}} P_q(s)ds}{\int_0^{T_{max}} P_q(s)ds + \int_0^{T_{max}} P_a(s)ds}.$$

This expression depends on T_{max} but is independent of t.

Up to small corrective terms, the usual Wiener-Khinchin unconditioned expectation $E[X(t)X(t+\tau)]$ is a weighted average of the conditioned expectations relative to the above three regimes. For all the usual processes, averaging presents no difficulty. In the present case, however, averaging leads to ill-posed problems, because it involves weights that depend upon sampling fluctuations and are generally ill-determined.

Fortunately, ill-posed problems are avoided if one keeps away from averaging between different regimes. Let us indeed return to the well-determined spectrum of the mixed regime. By simple inspection, this spectrum is independent of the behavior of $P_q(u)$ and $P_a(u)$ for $u > T_{max}$. Thus, given T_{max}, the conditioned spectrum of the mixed regime is not affected by the value of $\int_0^\infty P_q(s)ds$. Moreover, once one knows that a sample of duration T_{max} is mixed, it does not matter which was the a priori probability for a mixed regime to be observed in this sample.

6. Not necessarily Wienerian conditioned spectra related to f^{-1}

The assumption that $Q = \int_0^\infty P_q(s)ds$ is finite cannot be operationally verified. But this assumption is not necessary to define a conditioned spectrum for the mixed regime. In this fashion, any spectrum $S(f)$, such that $fS(f)$ is slowly varying as $f \to 0$, can be related without infrared catastrophe to a function $P_q(u)$ such that $uP_q(u)$ is slowly varying as $u \to \infty$.

To simplify, I shall carry out the argument in the case where it is precisely true that $S(f) \sim f^{-1}$. Let us, moreover, make the following assumptions:

$$T_{max} \gg \varepsilon; \quad \int_0^{T_{max}} P_a(s)ds \ll \int_0^{T_{max}} P_q(s)ds;$$

and $P_q(u) = \varepsilon/U$ for $\varepsilon < u < \infty$, while $P_q(u) = 1$ for $u < \infty$.

If $\varepsilon < \tau < T_{max}$, we obtain

$$C(\tau, T_{max}, F) \sim 1 - \frac{\log(\tau/\varepsilon)}{\log(T_{max}/\varepsilon)}.$$

Consider the Fourier transform $S(f, T_{max}, F)$ of this conditioned covariance. The very definition of $X(t)$ implies that $|X| = 1$ for all t; there-

fore $\int_0^\infty S(s, T_{max}, F)ds = 1$, independently of the value of T_{max}. As to the form of S, we only need the following approximations:
- for small τ, S is near $S(0) \sim T_{max} [\log(T_{max}/\varepsilon)]^{-1}$,
- for $1/T_{max} \ll f \ll 1/\varepsilon$, one has $S \sim f^{-1}[\log(T_{max}/\varepsilon)]^{-1}$.

Test that the approximate $S(0)$ equals the value of the second approximation at $\tau = 1/T_{max}$. Thus, a continuous approximating curve is obtained if the first approximation is used up to $\tau = 1/T_{max}$ and the second is made to apply beyond $\tau = 1/T_{max}$.

As $T_{max} \to \infty$, the lowest meaningful frequency $1/T_{max}$ decreases. But so does the expected amount of energy in any frequency band. The total energy remains independent of T_{max}. *This means that the conditioned spectrum exhibits no infrared catastrophe.*

The prefactor $[\log(T_{max}/\varepsilon)]^{-1}$ is a slowly decreasing function of T_{max} and its effect is likely to be small and hidden by sampling fluctuations. The latter are nonnegligible, as will be shown elsewhere.

7. Conclusion

Analogous arguments can be carried out for other "models" of $1/f$ noise. The general idea remains as follows. As "common sense" tells us, the appearance of infrared catastrophe cannot correspond to an actual divergence, and it may be an appearance due to a natural confusion between ordinary Wiener-Khinchin spectra and conditioned spectra. For the usual processes, the latter are simply conditioned forms of the former; but this is not so for the processes studied in the present text. Sampling spectra of either kind are meaningful for $f > 1/T_{max}$. Extrapolation of the same analytic behavior from this range of frequencies to the range of frequencies $f < 1/T_{max}$ is safe in the case of samples from Wiener spectra. On the other hand, the same kind of extrapolation applied to samples from conditioned spectra may well yield the appearance of infrared catastrophe.

Such is for example the case when $S(f)$ seems proportional to f^{-1}. Those spectra thus appear to characterize a class of random processes that flip between being an uneventful direct current and being in an eventful regime. Some are not stationary in the usual sense but satisfies a weaker form of "conditional" stationarity, described in M 1965c{N7}, and Berger & M 1963{N5}.

I examine elsewhere the addition of large numbers of mutually independent processes following the pattern described in this paper. Addition leaves the conditional covariance unchanged.

Co-indicator functions and related 1/f noises

• **Chapter foreword.** The paper was my first presentation of the philosophy that led to the nonlinear filters described in Chapter N4. It does not offer ready-made answers to any old questions, but it raises new questions that reflect the complexity of nature and open new fields to study. It requires three separate introductions: one concerned with mathematical method, and the other two with substance. The title was rephrased to be forward-looking.

The notion of random co-indicator, or coin function of a set S. The main innovation concerning mathematical method is a very revealing *random* function. Given a set S, the classical *indicator function* $I(P)$ is defined by $I(P) = 1$ if $P \in S$ and $I(P) = 0$ otherwise. To investigate closed sets S whose open complement includes more than one component, I found it useful to start with the indicator function of the *complement* of S, instead of S itself, and to replace it by a random function defined as follows: a) $W(P) = 0$ if $P \in S$; b) $W(P) = $ constant in each open complement of S and c) the W in different open complements are independent.

This paper examines the function $W(P)$ on the line, that is, a function of the form $W(t)$, when S is a discrete time-set that may either be a Lévy dust of dimension $D \in [0, 1]$, or a generalization characterized by a parameter D that satisfies $D \in [1, 2]$ and is not a dimension.

Relevance of W(t) for the resolution of the "infrared catastrophe." Concerning substance, the main innovation is to show that this catastrophe may well be an illusion. Indeed, the coin function with $W > 0$ constitutes an explicit example showing that a function may fail to diffuse and remains more or less bounded (in constrast, for example, to Brownian motion, which diffuses away), despite the fact that its spectral density is of the form f^{-B} with $B > 1$. Without further elaboration, this density would suggest that the "observed" spectral energy in low frequencies is infinite.

The key out of paradox is that the function $W(t)$ is not an ordinary random function endowed with a Wiener-Khinchin spectrum. It is a generalized function introduced for this purpose, called sporadic function in M 1967b{N10}. Therefore, the measurements meant to estimate the ordinary spectrum are, in fact, estimates of a *conditional spectrum*.

This paper's calculations are complicated and tedious, simply to insure that they remain elementary. The calculations in Chapter N10 involve very similar topics; they are more elegant but use advanced mathematics.

The notion of random indicator function of a set S. The usual indicator function of a set S is 0 outside of S, i.e., at points where $W(t)$ does not change, and 1 elsewhere. It is useful to randomize it by defining a function $V(t)$ whose values at the points of S are independent; random and > O fine points of the passage to the continuous time limit, disregarding $V(t)$ is more or less a "rectified" positive transform of the "derivative" of $W(t)$.

The notion of "white function." of a set S. This paper also mentions yet another function of a set S. Its value at different points of S are independent random values of zero expectation and finite variance. It is called "white function," because it has a constant spectral density. The distribution of its values at the points of S may be Gaussian. If so, its distribution is a mixture of a Gaussian and of 0, that is, nonGaussian. The conclusion is far-reaching: while its values are orthogonal, they are very far from independent.

A white function is obtained by taking the "derivative" of $W(t)$ and changing its sign at random. $V(t)$ is a rectified form of the white function.

The white function is not stressed in this paper but acquired great importance in my later work in finance, and is featured prominently in M 1997E. It exemplifies an extremely important phenomenon, namely the blindness of spectral analysis with respect to extreme forms of dependence. As this paper predicted, those forms of dependence are revealed by simply taking absolute values before performing spectral analysis.

"Shorthand" notation. This reprint uses an idiosyncratic notation I am experimenting with. For the sake of mnemonics and conciseness, every sum of the form $\sum_{n=1}^{N} Z(n)$ will be denoted as $Z_\Sigma(N)$.

Editorial changes. In the original, D is denoted by θ, W_Σ by W^0, $\tilde{T}$ by T^0, V_Σ by V_0. The original denoted the *coin function* as *core function*. •

✦ **Abstract.** In thin metallic films, semiconductors, nerve tissues, and many other media, the measured spectral density of noise is proportional to f^{D-2}, where f is the frequency and D a constant $0 \leq D < 2$. The energy of these "f^{D-2} noises" behaves more "erratically" in time than is expected from functions subject to the Wiener-Khinchin spectral theory. Moreover, when $D \leq 1$, blind extrapolation of the expression f^{D-2} to $f=0$ suggests that the total energy is infinite. This divergence, called "infrared catastrophe" raised problems of great theoretical interest and of great practical importance in the design of electronic devices.

This paper interprets these spectral measurements without paradox, by introducing a new concept, "conditional spectrum." Examples are given of random functions that have both the "erratic" behavior and the conditional spectral density observed for f^{D-2} noise.

A conditional spectrum is obtained when a procedure that was meant to measure a sample Wiener-Khinchin spectrum is used beyond its domain of applicability. The conditional spectrum is defined for nonconstant samples from all random functions of the Wiener-Khinchin theory and in addition for nonconstant samples from certain nonstationary random functions, and for nonconstant samples from a new generalization of random functions, called "sporadic functions."

The simplest sporadic functions has a f^{-2} conditional spectral density. It is a direct current broken by a single discontinuity uniformly distributed over the whole time axis. White noise is the $D=2$ limit of f^{D-2} noise. The other f^{D-2} noises to be described partake both of direct current and of white noise, and continuously span the gap between these limits. In many cases, their noise energy can be said to be proportional to the square of their "direct current" component.

Empirical studies are suggested, and the descriptive value of the concepts of direct current component and of spectrum are discussed. ✦

INTRODUCTION

This paper addresses important and puzzling mathematical problems raised by the low-frequency behavior of the f^{D-2} noises. Time is assumed integer-valued, to avoid extraneous difficulties. We are not concerned with the unfinished task of deducing f^{D-2} noises from more fundamental physical laws.

Section I concludes that, if $D \leq 1$, one must go beyond the Wiener-Khinchin spectral theory to account for f^{D-2} noises. If $D > 1$, some parts of the Wiener-Khinchin theory are applicable, but other parts fail.

Section II studies a current that is constant except for one discontinuity at a point whose probability distribution is uniform over $-\infty < t < \infty$. This is not an ordinary random function, but an example of a more general new concept to be called "sporadic generalized random function," and it is shown to be a f^{-2} noise in the sense of the new concept of conditional spectrum.

Section III uses random renewal sequences to construct several classes of functions that include both ordinary random functions and sporadic functions.

Section IV shows in which sense the functions of Sections III are f^{D-2} noises.

Section V shows that for some functions of Section III, the noise energy is proportional to the square of the "dc component."

Finally, Section VI shows that, while the marginal distribution of $X(t)$ may be Gaussian for the functions of Section III, their long-run averages are far from being Gaussian, their extreme values having a high probability of occurrence. Such noises therefore appear as "time-varying" or "noisy."

Other f^{D-2} noises can be constructed; an example covering the range $1 < D < 2$ is studied in M & Van Ness 1968.

This paper is intimately related to the study of "self-similar error clusters" in M 1965c{N7}. Space lacks for any stress upon self-similarity, but it will underline much of our development.

I. BOUNDED f^{D-2} NOISES IN INTEGER TIME

I-A. Measured spectra, and their interpretation by the Wiener-Khinchin theory

Let time be integer-valued. When the finite function $X(t)$ is known for $1 \le t \le T$, the "sample spectrum" is defined for every frequency f such that $0 \le f \le 1/2$, as equal to the function

$$S(f, T, \omega) = \int_0^f \frac{1}{T} \left| \sum_{t=1}^{T} f \exp(-2\pi i t s) X(t) \right|^2 ds.$$

The notation is the one customary in probability theory, namely, the letter ω stands for the whole set of values of the sample function $X(t)$, not only for $1 \leq t \leq T$, but also for any other t for which X may be defined.

The function S is usually evaluated when $X(t)$ is believed to be defined for $-\infty < t < \infty$, and is *either* a periodic function of period T, *or* a "Wiener-Khinchin function," also called "stationary random function of second order," or "covariance-stationary random function." In the latter case, the Wiener-Khinchin theory states that $\lim_{T \to \infty} S(f, T, \omega)$ is almost surely equal to a function $S(f)$ called spectrum that characterizes the process generating $X(t)$. This is the justification of calling $S(f, T, \omega)$ the "sample spectrum." One can estimate $S(f)$ through the value of $S(f, T, \omega)$ for large but finite T. If the following conditions hold: $S(f)$ has a finite derivative $S'(f)$, T is large, and f and Δf are not small with respect to $1/T$, one can similarly estimate $S'(f)$ through

$$\Delta S(f, T, \omega)/\Delta f = (1/\Delta f)[S(f + \Delta f, T, \omega) - S(f, T, \omega)].$$

More refined estimation procedures are known, but we need not be concerned with them.

Knowing only a finite sample for $1 \leq t \leq T$, it is impossible to determine whether $X(t)$ is a Wiener-Khinchin function for $-\infty < t < \infty$. Even from a pragmatic viewpoint, it is difficult to ascertain how useful it is to consider $X(t)(1 \leq t \leq T)$ as a sample of Wiener-Khinchin function. We shall presently see, however, that the dependence of $S(f, T, \omega)$ on its variables f, T and ω tells a great deal about the process generating $X(t)$.

I-B. The f^{D-2} noises

Using informal physical terms, the noises to be studied are defined as satisfying the condition $\Delta S(f, T, \omega)/\Delta f \sim f^{D-2} Q(T, \omega)$, with $0 \leq D < 2$. The sign $\sim$ will mean that the relation holds for all T, and for all meaningful values of f, say for $f > 5/T$.

If $1 < D < 2$, the above form of $\Delta S/\Delta f$ is compatible with the assumptions that $Q = \lim_{T \to \infty} Q(T, \omega)$ satisfies $0 < Q < \infty$, and that $X(t)$ is a Wiener-Khinchin function of spectrum equal to $S(f) = (D-1)^{-1} f^{D-1} Q$. However, this does not mean that all the textbook results can be applied. Indeed, this spectrum satisfies $S'(0) = \infty$, while many specialized results of the Wiener-Khinchin theory require $S' < \infty$ (see Sections VI and VII). Hence, use of the textbook results requires extreme caution. Moreover, $\lim_{T \to \infty} Q(T, \omega) = 0$ is not excluded, as we shall see.

If $0 \leq D \leq 1$ and $0 < Q < \infty$, one has $\int_0^{1/2} Qf^{D-2} df = \infty$. This means that Qf^{D-2} *cannot* be a Wiener-Khinchin spectral density for all $f > 0$. This divergence is called "infrared catastrophe." It implies that it is impossible to postulate that the portion of $X(t)$ for $1 \leq t \leq T$ is an arbitrarily chosen sample of a Wiener-Khinchin function.

In practice, T is bounded by some finite $T_{max} < \infty$. Therefore, one way out of the quandary is to note that there exist Wiener-Khinchin functions such that $S'(f)$ is proportional to f^{D-2} for large f, say for $f > 5/T_{max}$, but the proportionality only holds for $T < T_{max}$. The double truncation $T < T_{max}$ and $f > 5/T_{max}$ would also eliminate the milder difficulty that $S'(0) = \infty$, which is encountered for $1 < D < 2$. However, T_{max} is usually impossible to measure. As a result, physically interesting quantities cannot possibly depend upon T_{max}, and their study is not helped by invoking the asymptotic Wiener-Khinchin theory. In other words, the existence of f^{D-2} noises challenges the mathematician to reinterpret spectral measurements otherwise than in "Wiener-Khinchin" terms.

The thesis of the present paper is that operations meant to measure the Wiener-Khinchin spectrum may unintentionally measure something else, to be called the "conditional spectrum" of a "conditionally covariance stationary" random function. The basic step in the definition of this concept consists, for every $T > 0$, in classifying the functions $X(t)$ into the following two categories.

- The category $B(T)$ of functions that are *not* constant for $1 \leq t \leq T$.
- The category $A(T)$ of functions that *are* constant.

This being done, the function $X(t)$ is then said to be "conditionally stationary" for $1 \leq t \leq T$, if the set of its values over this finite interval can be imbedded in an ordinary stationary process. The idea is that, for every finite set of K instants t_k, with $1 \leq k \leq K < \infty$ and $1 \leq t_k \leq T$, the conditional probability distribution of the random vector

$$X(t_1), \ldots, X(t_k), \ldots, X(t_K) \quad (\text{knowing that } X(t) \in B(T))$$

must be identical to the conditional probability distribution of the imbedding stationary random function. Now define the conditional covariance by the conditional expectation

$$E[X(t)X(t+n) \mid 1 \leq t < t+n \leq T, \quad X(t) \in B(T)].$$

The weaker property of "conditional covariance stationarity" will be said to hold if the conditional covariance is independent of t. Where this condition holds, the "conditional spectrum" of $X(t)$ (under condition $B(T)$) is defined as equal to $S(f, T) = E[S(f, T, \omega) \mid \omega \in B(T)]$.

Clearly, all ordinary stationary random functions are also conditionally stationary and have a conditional spectrum. But the latter concept is of more general validity, its "natural" realm being constituted by "sporadically varying" random functions (see M 1967b{N10}). Unfortunately, this new concept cannot be fully described in this paper, but an attempt will be made to present its flavor.

The conditional spectrum will hopefully make it possible to relate various observed properties of f^{D-2} noises that seem mutually incompatible. However, what conditional spectra tell us about f^{D-2} noises is less than Wiener-Khinchin spectra tell about those Wiener-Khinchin functions, for which $S'(f)$ is defined, and finite, for all $f \geq 0$.

I-C. The $f^{D-2} L(f)$ noises

For f^{D-2} noises, $(\Delta S/\Delta f)f^{2-D}$ is by definition independent of f. A less dogmatic description of the facts is to say that $(\Delta S/\Delta f)f^{2-D}$ "varies slowly when f is small."

The concept of "slow variation" is not uniquely determined, however. In the body of the paper we shall say that a function varies slowly at the origin if it has a positive finite limit for $f \to 0$. A noise will be called "$f^{D-2}L(f)$ noise," if either $(\Delta S/\Delta f)f^{2-D}$ or $S(f)^{1-D}$ varies slowly near $f = 0$. The appendix will refer to a less demanding concept of slow variation, due to Karamata: it is a bit complicated but "mathematically natural."

I-D. Continuous-time f^{D-2} noises and high-frequency difficulties

In addition to low-frequency problems discussed in this paper, continuous-time f^{D-2} noises present high-frequency difficulties that were first encountered for white noise. Physics masters these difficulties by assuming that time is quantized, and mathematics masters them by introducing generalized random processes called "random Schwartz distributions." Spectral densities proportional to f^{D-2} for $f \to \infty$ play an important role in the modern theory of trigonometric series. In fact, diverse physical consideration discussed in the present paper have led to progress in pure mathematics (see Gnedenko & Kolmogorov 1954).

II. A DIRECT CURRENT WITH A SINGLE RANDOM DISCONTINUITY CAN BE CONSIDERED A f^{-2} NOISE

This section treats in detail an example of a function ruled by chance, whose conditional "spectrum" is proportional to f^{-2}. The Wiener-Khinchin spectrum is not defined for this function.

II-A. The "direct current" coin function, or co-indicator function

Let W' and W'' be two independent and identically distributed random variables with zero mean, unit variance, and continuous distribution. Thus, it is almost sure that $W' \neq W''$. A "half-integer" will designate a number of the form $1/2+$ an integer, and $\tilde{T}$ will be a half-integer-valued random variable, independent of W' and W'', and defined by $F(t) = \Pr\{\tilde{T} < t\}$. $\tilde{T} \in (a, b)$, with a and b integers, will mean that the half-integer $\tilde{T}$ satisfies $a < \tilde{T} < b$, or $b < \tilde{T} < a$.

The coin-function $W(t, \tilde{T})$ will then be defined as follows (see Figure 1):

$$\begin{cases} W(t, \tilde{T}) = W' & \text{when } \tilde{T} < t \\ W(t, \tilde{T}) = W'' & \text{when } \tilde{T} > t. \end{cases}$$

The covariance-like function $C_w(t, n)$ is defined as

$$C_W(t, n) = E[W(t, \tilde{T})W(t+n, \tilde{T})]$$
$$= 1 - (1/2)E[W(t, \tilde{T}) - W(t+n, \tilde{T})]^2.$$

If $\tilde{T} \notin (t, t+n)$, $1 - C_W$ vanishes. If $\tilde{T} \in (t, t+n)$, $1 - C_W = 1$. Therefore,

$$C_W(t, n) = \Pr\{\tilde{T} \notin (t, t+n)\} = 1 + |F(t) - F(t+n)|.$$

It is impossible to choose F so that $W(t, \tilde{T})$ is a stationary random function. However, the usual concept of stationarity postulates properties of W over the whole time axis, while only finite samples are ever available to the physicist. When $W(t, \tilde{T})$ is only known for $1 \leq t \leq T < \infty$, it is reasonable to restrict t and $t+n$ by the condition $(t, t+n)$ [] $(1, T)$, and to consider separately the following covariances, relative, respectively, to $\tilde{T} \in (1, T)$ and to $\tilde{T} \notin (1, T)$:

$$E[W(t, \tilde{T})W(t+n, \tilde{T}) \mid \tilde{T} \notin (1, T)] = 1$$

$$E[W(t, \tilde{T})W(t+n, \tilde{T}) \mid \tilde{T} \in (1, T)] = \Pr\{\tilde{T} \notin (t, t+n) \mid \tilde{t} \in (1, T)\}$$

$$= \frac{1 + |F(t) - F(t+n)|}{F(T) - F(1)}.$$

The first expression having the same value in all cases, it need not be given a special name. The second expectation can, therefore, be called, for short, a *conditional covariance*. One can say that $W(t, \tilde{T})$ is *conditionally covariance-stationary over the span* $(1, T)$, if its conditional covariance is a function of n alone, as long as $0 \le t \le T - n$. This requires $\tilde{T} \in (1, T)$ to have a uniform conditional distribution over the half-integers within $(1, T)$. In that case, one has:

$$E[W(t, \tilde{T})W(t+n, \tilde{T}) \mid \tilde{t} \in (1, T)] = 1 - \frac{|n|}{T-1}.$$

One can imbed the $(1, T)$ sample of $[W(t, \tilde{T}) \mid \tilde{T} \in (1, T)]$ into a variety of stationary processes. The simplest, denoted by $W_\Sigma(t, \tilde{T}, T)$, has jumps at almost all time instants of the form $\tilde{T} + kT$ (integer k, $-\infty < k < \infty$), and is such that $W_\Sigma(t)$ and $W_\Sigma(t + kT)$ are independent for all t and k. "Imbedding" now means that $[W(t, \tilde{T}) \mid \tilde{t} \in (1, T)]$ and $W_\Sigma(t, \tilde{T}, T)$ have identical finite probability distributions over $(1, T)$. It is, therefore, tempting to define a *conditional spectral density* for $[W(t, \tilde{T}) \mid \tilde{T}E(1,t)]$ as equal to the Wiener-Khinchin spectral density of $W_\Sigma(t, \tilde{T}, T)$, namely, the Fourier transform of $1 - |n|/(T-1)$. For large T, this density takes the form $(2/\pi)\sin^2(Tf/2)\tilde{T}^{-1}f^{-2}$. If one tries to estimate this density from a

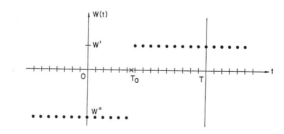

FIGURE N9-1.

sample by using, $\Delta f \sim 5/T$, the oscillations of the function $\sin^2$ will be smoothed away. Thus, one will find that $E[\Delta S(f, T, \omega)/\Delta f]$ is proportional to $\tilde{T} - 1f^{-2}$. (The parameter ω is absent because we average over all the values of ω that are compatible with $\tilde{T} \in (1, T)$.)

As already noted in Section I, the sample spectral density and the sample conditional spectral density concern different functions, but are given by the same algorithm. Therefore, the concept of *conditional spectrum may be needed to interpret measurements that were meant to measure the Wiener-Khinchin spectrum.*

A weakness of the above argument is that it requires a finite upper bound T_{max} to the values of T such that the conditioned variable $[\tilde{T} \mid \tilde{T} \in (1, T)]$ is uniformly distributed over the half-integers within $(1, T)$. If $T > T_{max}$, the conditional covariance depends upon t. One could evade this difficulty by choosing $\Pr\{\tilde{T} = t + 1/2\}$ uniformly among half-integers over an immensely long period of time, say, between, 1000 A.D. and 3000 A.D.. However, as mentioned in Section I, we see little to be gained from invoking the Wiener-Khinchin theory in a case where it is overwhelmingly affected by quantities that cannot be measured. No quantity of physical interest can possibly depend upon the span of linearity of the distribution of $\tilde{T}$. Moreover, limiting this span eliminates the possibility of proving limit theorems. But such theorems could be used in practice to approximate the finite behavior of physically interesting quantities.

An alternative that I prefer, is to make $\tilde{T}$ into "a generalized random variable" distributed with uniform measure over *all* half-integer times. $W(t, \tilde{T})$ then becomes a "generalized random process" that is "sporadically varying" in the sense of M 1967b{N10}. However, the same conditional spectrum is obtained, whether one prefers to work with a nonobservable cutoff $T_{max} < \infty$ or an unbounded measure.

II-B. The function $W(t, \tilde{T})$ is not ergodic

A function $X(t)$ is called ergodic if $E(X)$ is defined and if it is almost sure that

$$\lim_{|t'' - t'| \to \infty} \frac{1}{t'' - t'} \sum_{s = t'+1}^{t''} X(s) = E(X).$$

When $\tilde{T}$ is a proper random variable, the function $W(t, \tilde{T})$ is nonergodic, being highly nonstationary. When $\tilde{T}$ is uniformly distributed over $(-\infty, \infty)$, $W(t, \tilde{T})$ is also nonergodic: it is "almost surely" constant over any prescribed finite time span, and its sample mean equals $W(t')$, which is random and independent of T.

The expression

$$\lim_{T \to \infty} \frac{1}{T} \sum_{t=1}^{T} [W(t, \tilde{T}) \mid \tilde{T} \in (1, T)]$$

is also random, namely, of the form $W'Z + (1 - Z)W''$, where the variable Z is uniformly distributed over $(0, 1)$.

II-C. Sampling distribution of the conditional covariance of $W(t, \tilde{T})$

The expression

$$R = \left[\frac{1}{T} \sum_{t=1}^{T} W(t, \tilde{T}) W(t+n, \tilde{T}) \mid \tilde{T} \in (1, T) \right]$$

satisfies $ER = 1 - |n|/T$. However, it is a corollary of nonergodicity is that, if $T \to \infty$ while n/T remains constant, R does *not* tend to its expectation. For example, when $0 < n < T/2$, the span of variation of $H = \tilde{T}/T$ is to be divided in the following three parts.

From 0 to n/T, $R = (1 - h - n/T)W''^2 + hW'W''$.

From n/T to $1 - n/T$, $R = (1 - n/T)W'^2 + (1 - h - n/T)W''^2 + hW'W''$.

From $1 - n/T$ to 1, $R = (h - n/T)W'^2 + (1 - h)W'W''$.

Thus, for every fixed couple (W', W''), R is the mixture of three uniformly distributed random variables.

II-D. The "indicator function" of $\tilde{T}$

Let $V(t, \tilde{T})$ be defined as follows: If $t + 1/2 \neq \tilde{T}$, $V(t, \tilde{T}) = 0$; if $t + 1/2 = \tilde{T}$, $V(t, \tilde{T})$ is the value of some random variable, with positive values, continuous distribution, unit mean, and finite variance. Except for scale, $V(t, \tilde{T})$ is the absolute value of the finite differential $W(t + 1, \tilde{T}) - W(t, \tilde{T})$.

Again, it is only worth considering the conditional function $[V(t, \tilde{T}) \mid \tilde{T} \in (1, T)]$. If the distribution of $[\tilde{T} \mid \tilde{T} \in (1, T)]$ is uniform, the conditional mean of $V(t, \tilde{T})$ is $E[V(t, \tilde{T}) | \tilde{T} \in (1, T)] = (T-1)^{-1}$ and the covariance of $[V(t, \tilde{T}) | \tilde{t} \in (1, T)]$ satisfies

$$C_V(0) = \frac{1}{T-1} - \frac{1}{(T-1)^2} \quad \text{while} \quad C_V(n) = -\frac{1}{(T-1)^2} \quad \text{when} \quad n \neq 0.$$

Thus, $V(t, \tilde{T})$ tends to become spectrally white as $T \to \infty$, and its conditional spectral density takes the form $f^{-0}/(T-1)$. This example shows, as mentioned in Section I-B that, even if $1 < D < 2$, the $f^{D-2}Q(T, \omega)$ spectrum is compatible with a function $Q(T, \omega)$ such that $Q(T, \omega) \to 0$ as $T \to \infty$.

III. THE INDICATOR FUNCTION $V(t)$ AND THE CO-INDICATOR (OR COIN) FUNCTION $W(t)$ FOR GENERAL RENEWAL PROCESSES: DEFINITIONS

Section II demonstrated that the concentration of energy in low frequencies may be greater for conditional spectra than for any Wiener-Khinchin spectrum. Less extreme examples of this phenomenon will now be found for other functions $X(t)$ of integer valued time, also obtained by "time-modulating" white noise. The main purpose of this paper is to study the functions to be defined in Sections III A,B,C, and to examine conditions under which their conditional spectral density is of the form f^{-B}. One must first make the renewal sequence stationary by selecting its starting point appropriately, as will be done in Section III D. A point of discontinuity of $X(t)$ will be defined as a half-integer such that $X(t - 1/2) \neq X(t + 1/2)$. Thus, for white noise with continuous marginal distribution in integer-valued time, almost *every* half-integer is a point of discontinuity. For the example in Section II, on the contrary, every sample of T values contained at most *one* discontinuity. In the remainder of this paper, we shall see that any conditional spectral density proportional to f^{D-2}, with $0 < D < 2$, can be obtained by selecting properly the points of discontinuity of a properly constructed step function. To simplify, these points will be assumed to constitute a renewal process.

III-A. Definition of the random indicator function $V(t)$ of a renewal process

A renewal process in half-integer-valued time (or process of recurrent events) is a random sequence of half-integer instants T_k, which contains a specified (nonrandom or random) "starting point," and is such that the distances $V_K = T_K - T_{K-1}$ are independent positive random variables with a common distribution $\Pr\{U < u\} = 1 - P(u)$. It is well-known that

$$E(U) = \sum_{n=1}^{\infty} P(n) = P_\Sigma(\infty).$$

Given the renewal times t_k, consider a second sequence of positive, identically distributed and independent random variables V_k, with continuous distribution, unit mean and finite variance. Then define the random indicator function $V(t)$ as follows:

$$\begin{cases} \text{If } t + 1/2 = t_k \text{ for some } k, & V(t) = V_k \\ \text{Otherwise,} & V(t) = 0. \end{cases}$$

The discontinuities of $V(t)$ are almost surely all the instants of the form t_k or t_{k-1}.

III-B. Definition of the "coin function" W(t) of a renewal process

Given the renewal times t_k, consider a third sequence of identically distributed independent random variables W_k, with continuous distribution, zero mean, and unit variance. Then define the coin function by letting $W(t) = W_{k+1}$ when $t_k < t < t_{k+1}$. The discontinuities of $W(t)$ are almost surely all the instants of the form t_k. Except for scale, V is a rectified form of the "derivative" of W (See Figure 2.)

Remark: For many purposes, the independence of the variables W_k could be replaced by the weaker assumption that they are uncorrelated.

III-C. Definition of the "white function" X(t) of a renewal process

Given the renewal times t_k, consider a fourth sequence of identically distributed independent random variables X_k, with continuous distribution, zero mean and unit variance. Then define the white function as follows:

$$\begin{cases} \text{If } t + 1/2 = t_k \text{ for some } k, & X(t) = X_k \\ \text{Otherwise,} & X(t) = 0. \end{cases}$$

III-D. Construction of the stationary sequence $\{T_k\}$, or $\{T_k | U\}$, when $E(U) < \infty$

It is shown in many books on renewal process that the following construction yields a stationary renewal process, to be called $\{T_k\}$ or $\{T_k | U\}$.

The *first step* is to arrange the indices k in such a way that $\tilde{T} = \min\{T_k \mid T_k > 0\}$, and to select $\tilde{T}$ at random with the distribution

$$\Pr\{\tilde{T} = 1/2 + n\} = \frac{P(n+1)}{E(U)}.$$

The *next step* is to choose T_{-1} so that

$$\Pr\{T_{-1} \leq 1/2 - m \mid \tilde{T} = 1/2 + n\} = \frac{P(n+m)}{P(n+1)}.$$

Finally, the other renewal points are chosen as follows: for $k > 0$, $T_k - \tilde{T}$ will be the sum of k independent random variables U_s with the common distribution $P(u)$; for $k < -1$, $\tilde{T} - T_k$ will be the sum of $|k-1|$ independent random variables U_s with the common distribution $P(u)$.

If $E(U) = \infty$, the preceding construction is impossible (unless one replaces the ordinary random processes by some generalization, such as the sporadic processes defined in M 1967b{N10}). However, "nonempty samples of $\{T_k\}$" can be defined *without* constructing $\{T_k\}$ itself (that is, without referring to sporadic processes). For this definition, two additional preliminary definitions will be needed.

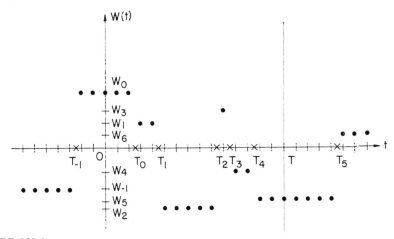

FIGURE N9-2.

III-E. Definition of the truncated random variable U_T

U_T is defined by

$$P_T(u) = \Pr\{U_T \geq u\} = \begin{cases} P(u) & \text{when } u < T \\ 0 & \text{when } u \geq T. \end{cases}$$

Note that, U_T being bounded, $E(U_T)$ is finite for all $T < \infty$.

III-F. Definition of $\{T_k | U_T\}$

This is the stationary renewal sequence built upon U_T. Thus, every half integer time instant has the same probability $1/E(U_T)$ of belonging to $\{T_k | U_T\}$, and every sample of duration T, drawn from $\{T_K | U_T\}$, contains at least one renewal point.

Clearly, if $E(U) < \infty$, a nonempty sample of $\{T_k\}$ of length T can also be obtained by taking an arbitrary segment of $\{T_K | U_T\}$ of length T. This idea is generalized as follows:

III-G. Definition of a conditioned nonempty sample of $\{T_k\}$

A "conditioned" nonempty sample of $\{T_K\}$, of length T, is obtained by arbitrary translation in time of the $(1, T)$ sample of the stationary renewal sequence $\{T_K | U_T\}$.

III-H. The special case of Section II

Suppose that $P(u) \equiv 1$ for all finite values of u, $U_T \equiv T - 1$, and $\{T_k | U_T\}$ is a sequence of uniformly spaced points. Then $(1, T)$ contains exactly one point of $\{T_K | U_T\}$, namely $\tilde{T}$, which is uniformly distributed over $(1, T)$. Thus $P(u) \equiv 1$ corresponds to the construction of Section II.

III-I. Definitions

The spectra of $V(t)$, $W(t)$, or $X(t)$ conditioned by $B(T)$, are defined as the Wiener-Khinchin spectra of the functions $V(t)$, $W(t)$, or $X(t)$ built on the stationary renewal sequence $\{T_K | U_T\}$.

IV. EXPECTED WIENER-KHINCHIN AND CONDITIONAL SPECTRA OF THE FUNCTIONS V(t), W(t), AND X(t)

IV-A. Summary

The low-frequency behavior of the ordinary and conditional spectra of $V(t)$, $W(t)$, or $X(t)$ is related to the behavior of $P(u)$ for $u \to \infty$. This section will evaluate those spectra without making any assumptions on $P(u)$, save for the last step of each subsection. The general treatment of this last step will be postponed to the Appendix, and the present section will relate to very special $P(u)$, to be denoted as $P(u) = P_D(u)$. The characteristic function $\varphi(z) = -\int_0^\infty \exp(izu) dP(u)$, then becomes $\varphi_D(z)$, defined as follows:

If $0 < D < 1$, one has $\varphi_D(z) = 1 - C(1 - e^{iz})^D$, with $0 < C < 1$;

If $1 < D < 2$, one has $\varphi_D(z) = 1 - C'(1 - e^{iz}) + C(1 - e^{iz})^D$.

In the second case, we assume $0 < DC \leq C' \leq C + 1$, so that $C \leq 1/(D-1) < 1$. One readily finds that

for small z, one has $2 - \varphi_D(z) - \varphi_D(-z) \sim 2C |\cos(D\pi/2)| |z|^D$

for large n, one has $P_D(n) \sim \dfrac{n^{-D}}{|\Gamma(1-D)|}$.

The properties of the corresponding functions $V(t)$, $W(t)$, and $X(t)$ can be summarized as follows:

Wiener-Khinchin spectra. They are defined only in the case $1 < D \leq 2$. If so, both $V(t)$ or $W(t)$ have Wiener-Khinchin spectra of the form $S(f) = f^{D-1} L(f)$, where $L(f)$ varies slowly near $f = 0$.

Conditional spectra. They are defined for all D. When $1 < D < 2$, they differ little from the Wiener-Khinchin spectra. But they have the virtue of also being defined when $0 < D < 1$.

• $W(t)$ has a conditional spectrum of the form $S(f, T) = f^{D-1} L(f) Q(T)$, where $5/T \ll f \ll 1/2$. If $1 < D < 2$, $Q(T)$ is unimportant, because $0 < \lim_{T \to \infty} Q(T) < \infty$. If $0 \leq D < 1$, $Q(T)$ is essential and $Q(T)\tilde{T}^{1-D}$ varies slowly for $T \to \infty$.

• $V(t)$ has a conditional spectrum of the form $S(f, T) = f^{-B} L(f) Q(T)$, where B is a function of D satisfying $0 \leq B \leq 1$. If $1 < D < 2$, then $B = 2 - D$, as for the Wiener-Khinchin $S(f)$, and $0 < \lim_{T \to \infty} Q(T) < \infty$. If $0 < D < 1$, then

$B = D$; in that case, $\lim_{T \to \infty} Q(T) = 0$, and $Q(T)\tilde{T}^{1-B}$ varies slowly near $T = \infty$.

• One can readily prove (but we shall not do so) that $X(t)$ has a white Wiener-Khinchin spectrum whenever $E(U) < \infty$, and a white conditional spectrum in all cases.

Another very interesting function, whose spectral properties are identical to those of $W(t)$, is $W_\Sigma(M, t) = M^{-1/2} \sum_1^M W(m, t)$, where the $W(m, t)$ are independent coin functions built on independent renewal processes with the same law $P(u)$. If M is very large, $W_\Sigma(M, t)$ is "locally Gaussian," meaning that the joint distribution of $W_\Sigma(M, t')$ and $W_\Sigma(M, t'')$ is Gaussian if $|t' - t''|$ is small. Space lacks here for a more detailed study of $W_\Sigma(M, t)$.

IV-B. The indicator function $V(t)$ when $E(U) < \infty$

The covariance function of $C_V(n)$ of $V(t)$ satisfies

$$C_V(n) = \frac{[E(V^2) - 1/E(U)]}{E(U)},$$

while for $n < 0$, one has

$$C_V(n) = E[V(t)V(t+n) \mid V(t) \neq 0]E[V(t)] - [E(V)]^2$$
$$= \frac{d\overline{N}_0(n) - 1/E(U)}{E(U)},$$

where $d\overline{N}_0(n)$ is defined as

$$\Pr\{t + n + 1/2 \in \{T_k\} \mid t + 1/2 \in \{T_k\}\}.$$

Since the spectral density of $V(t)$ is

$$S'_v(f) = 2C_V(0) + 4\sum_{n=1}^{\infty} C_V(n) \cos(2\pi nf),$$

it follows that

$$E(U)S'_V(f) = 2E(V^2) - \frac{2}{E(U)} + 2\left\{\sum_{n=1}^{\infty} + \sum_{n=-\infty}^{-1}\right\} \exp(2\pi i nf)\left\{d\overline{N}_0(n) - \frac{1}{E(U)}\right\}.$$

It was shown by Feller 1957 (p. 285) that

$$\sum_{1}^{\infty} \exp(-bn) d\overline{N}_0(n) = \left\{ 1 + \sum_{u=1}^{\infty} \exp(-bu) dP(u) \right\}^{-1}$$

whenever b is such that either side converges. Consider the series

$$\sum_{1}^{\infty} \exp(-bn) \left[d\overline{N}_0(n) - \frac{1}{E(U)} \right]$$

$$= \left\{ 1 + \sum_{n=1}^{\infty} \exp(-bu) dP(u) \right\}^{-1} \frac{-\exp(-b)}{[1 - \exp(-b)] E(U)}.$$

It converges for all values of b, such that $b = 0$, and the real part of b is ≥ 0. Thus, the function $S'_V(f)$ can be written for $f > 0$, as

$$E(U) S'_V(f) = 2E(V^2) - \frac{2}{E(U)} + \frac{2}{1 - \varphi(2\pi f)}$$

$$+ \frac{2}{1 - \varphi(-2\pi f)} - \frac{\exp(2\pi i f)/E(U)}{1 - \exp(2\pi i f)} + \frac{\exp(-2\pi i f)/E(U)}{1 - \exp(-2\pi i f)}.$$

Near $f = 0$, the sum of the fifth and sixth terms is bounded. The behavior of the other terms depends upon the second moment of U:

If $E(U^2) < \infty$, the sum of the third and fourth terms is also bounded, and $S'_V(f)$ is bounded for all $f = 0$. Thus, $V(t)$ is a f^{-0} noise (i.e., $D = 2$) irrespective of the detailed behavior of $P(u)$.

If $E(U^2) = \infty$, while $E(U) < \infty$, one can write $\varphi(z) = 1 - izE(U) + \varphi^0(z)$, where $\varphi^0(z)/z \to 0$ as $z \to 0$. Therefore, $[1 - \varphi(z)][1 - \varphi(-z)]/z^2 E^2(U) \to 1$ as $z \to 0$, and one can write, neglecting bounded terms,

$$E(U) S'_V(f) \sim \frac{2[2 - \varphi(2\pi f) - \varphi(-2\pi f)]}{f^2 E^2(U)}.$$

When $P(u) = P_D(u)$, it follows that

$$0 < \lim_{f \to \infty} S'_V(f) f^{2-D} < \infty,$$

N9 ◇ ◇ CO-INDICATOR FUNCTIONS AND 1/f NOISES

thus, $S'_V(f)f^{2-D}$ varies slowly near $f=0$ and $V(t)$ is a $f^{2-D}L(f)$ noise. Q.E.D.

IV-C. The coin function $W(t)$ when $E(U) < \infty$

Its covariance $C_W(n)$ satisfies

$$C_w(0) - C_w(n) = 1 - C_w(n) = 1/2 E[W(t) - W(t+n)]^2.$$

Since $W(t) - W(t+n)$ vanishes unless $(t, t+n)$ contains at least one event T_k, for which the variance of $W(t) - W(t+n)$ is two, one has

$$\Pr\{\text{for some } k, T_k \in (t, t+n)\} = \Pr\{\text{for some } k, T_k \in (1, n+1)\}$$

$$= \Pr\{\tilde{T} < n+1\} = \frac{P_\Sigma(n)}{E(U)}.$$

Therefore,

$$C_w(n) = 1 - \frac{P_\Sigma(n)}{E(U)}.$$

Hence,

$$S'_W(f) = 2 + 2\left(\sum_{n=1}^{\infty} + \sum_{n=-\infty}^{-1}\right) \frac{\exp(2\pi i n f)[1 - P_\Sigma(n)]}{E(U)}.$$

This is the Fourier transform of the repeated integral of $-dP(u)$, and its behavior again depends upon the second moment of U:

The inequality $E(U^2) < \infty$ implies $\sum_{n=1}^{\infty}[1 - P_\Sigma(n)] < \infty$,

and $S'_W(f)$ is bounded.

The equality $E(U^2) = \infty$ with $E(U) < \infty$ implies that, near $f = 0$,

$$S'_w(f) \sim \frac{2[2 - \varphi(2\pi f) - \varphi(-2\pi f)]}{f^2 E(U)} \sim E^2(U) S'_V(f).$$

Therefore, the discussion relative to $S'_V(f)$ applies without change to S'_W, except for the factor $E^2(U)$. This factor has the dimension of the square of time, and comes up because $E^2(U)E[V^2(t)]$ and $E[W^2(t)]$ are both dimensionless. Q.E.D.

IV-D. The conditioned coin-function W(t, T)

This is the coin-function of a section of length T of the renewal process $\{T_K | U_T\}$. Hence,

$$C_w(n, T) = \begin{cases} 1 - P_\Sigma(n)/P_\Sigma(T-1) & \text{for } n \leq T-1 \\ 0 & \text{for } n \geq T-1. \end{cases}$$

The spectral density of $W(t, T)$ is not monotone: it is better, therefore, to consider the spectrum itself, choosing appropriately the frequency at which $S = 0$.

If $E(U) = \infty$, let $S_W(1/2, T) = 0$. Then

$$-S_w(f, T) = 1 - 2f + \frac{2}{\pi} \sum_{n=1}^{T-1} C_w(n, T) n^{-1}[(-1)^n - \sin(2\pi n f)]$$

$$= 1 - 2f + (\frac{2}{\pi})[P_\Sigma(T-1)]^{-1} \sum_{n=1}^{T-1}[1 - P_\Sigma(n)]^{-1}[(-1)^n - \sin(2\pi n f)].$$

Define $S^*_W(f, T)$ by carrying the sum to ∞ in this last formula. If $P(u) = P_D(u)$, with $0 < D < 1$, one finds that a $S^*_W(f, T)f^{D-1}$ varies slowly near $f = 0$, and that $S^*_W(f, T)$ is the spectrum of an $f^{D-2}L(f)$ noise.

As to the "correction" corresponding to the term $\sum_{n=T}^{n=\infty}$, its absolute value has an upper bound proportional to $T^{-1}f^{-1}$, hence is negligible except in the physically meaningless region of small values of fT. Q.E.D.

If $E(U) < \infty$, one shows similarly that, when $P(u) = P_D(u)$, $W(t, T)$ is an $f^{D-2}L(f)$ noise except for small values of fT. Q.E.D.

IV-E. The conditioned indicator function V(t, T)

Its covariance $C_V(n, T)$ satisfies

$$C_V(0, T) = \frac{E(V^2) - 1/E(U_V)}{E(U_V)}.$$

For $0 > n > T$, one has

$$E[V(t, T)V(t + n, T) \mid V(t, T) \neq 0] = E[V(t)V(t + n) \mid V(t \neq)0] = d\overline{N}_0(n).$$

Thus,

$$C_V(n, T) = \frac{d\overline{N}_0(n) - 1/E(U_T)}{E(U_T)}.$$

Taking $S_V(0, T) = 0$, yields

$$E(U_T)S_V(f, T) = 2f\{E(V^2) - \frac{1}{E(U_T)}\} + \frac{2}{\pi} \%\sum_{n=1}^{T-1} \{\overline{N}(n) - \frac{1}{E(U_T)}\} n^{-1} \sin(2\pi f n).$$

When T and fT are both large, $1/E(U)$ (which may vanish) may replace both occurrences of $1/E(U_T)$ in the right-hand side of the equation and one can carry the sum to $n = \infty$ instead of $n = T - 1$. These changes yield

$$E(U_T)S_V(f, T) \sim E(U_T)S^*_V(f, T),$$

where

$$E(U_T)S^{*\prime}_V(f, T) = 2[E(V^2) - 1/E(U)] + 2\sum_{n=1}^{\infty} \cos(2\pi f n)[\overline{N}(n) - 1/E(U)],$$

$$= 2[E(V^2) - 1/E(U)] + \frac{2}{1 - \varphi(2\pi f)} + \frac{2}{1 - \varphi(-2\pi f)}$$

$$- \frac{\exp(2\pi i f)/E(U)}{1 - \exp(2\pi i f)} + \frac{\exp(-2\pi i f)/E(U)}{1 - \exp(-2\pi i f)}.$$

If $E(U^2) < \infty$, $E(U_T)$ converges rapidly to $E(U)$, and the spectrum reduces to the known spectrum of the unconditioned indicator function. The latter has a bounded density, a characteristic of $f^{-B}L(f)$ noises.

If $E(U^2) = \infty$ but $E(U) < \infty$, one has

$$S_V(f, T) \sim S_V(f)E(U)/E(U_T).$$

This brings us back to the Wiener-Khinchin spectrum of $V(t)$, except for the factor $E(U)/E(U_T)$. When $D \ll 2$, and as $T \to \infty$, $E(U)/E(U_T)$ converges slowly towards its limit 1.

If $E(U) = \infty$, one can write $\varphi(z) = 1 - \varphi^0(z)$, where $\varphi^0(z) \to 0$ with z. For small f, one has $E(U_T)S_V^*(f, T) \sim 2/\varphi(2\pi f) + 2/\varphi^0(-2\pi f)$. In particular, if $P(u) = P_D(u)$, with $0 < D < 1$, one finds

$$\frac{2}{1 - \varphi(2\pi f)} + \frac{2}{1 - \varphi(-2\pi f)} \sim -2C^{-1}[(1 - e^{-2\pi i f})^{-D} + (1 - e^{-2\pi i f})^{-D}]$$

$$\sim |f|^{-D}[4C^{-1}(2\pi)^{-D} \cos(D\pi/2)].$$

Thus, $V(t, T)$ is an $f^{-D}L(f)$ noise. Moreover,

$$\lim_{T \to \infty} [1/E(U_T)] = 0.$$

In the notation of Section I, $\lim_{T \to \infty} Q(T) = 0$, which is thus shown by example to be compatible with $1 < 2 - D < 2$.

V. THE RELATION BETWEEN AVERAGE NOISE ENERGY AND AVERAGE DC COMPONENT

V-A. The mean and variance of the cumulative expressions $V_\Sigma(t)$, $W_\Sigma(t)$, and $X_\Sigma(t)$, when $E(U) < \infty$ and $P(u) = P_D(u)$, with $1 \leq D < 2$

The ergodic theorem applies to $V(t)$, $W(t)$, and $X(t)$. That is, with probability one, we have

$$\lim_{t \to \infty} [t^{-1}W_\Sigma(t)] = E(t^{-1}W_\Sigma(t)) = E[W(t)] = 0$$
$$\lim_{t \to \infty} [t^{-1}V_\Sigma(t)] = E(t^{-1}V_\Sigma(t)) = E[V(t)] > 0$$
$$\lim_{t \to \infty} [t^{-1}X_\Sigma(t)] = E(t^{-1}X_\Sigma(t)) = E[X(t)] = 0$$

Since successive values of $W(t)$, $V(t)$, or $X(t)$ are interdependent, the variance is given by the following well-known formula, due to G.I. Taylor

N9 ◊ ◊ CO-INDICATOR FUNCTIONS AND 1/f NOISES

$$\text{Var}[V_\Sigma(t)] = tC_V(0) + 2(t-1)C_V(1) + \ldots 2C_V(t-1)$$

$$= tC_v(0) + 2\sum_{s=1}^{t-1}\sum_{r=1}^{s} C_V(r).$$

If $P(u) = P_D(u)$ when $1 \leq D < 2$, then $C_V(n)$ equals n^{1-D} multiplied by some slowly varying function, and $\text{Var}[V_\Sigma(t)] - tC_V(0)$ is the product of t^{3-D} and a function that varies slowly as $t \to \infty$. Since $tC_V(0)$ can be neglected in comparison with t^{3-D}, we further see that $\text{Var}[V_\Sigma(t)]t^{-3+D}$ varies slowly as $t \to \infty$.

One can interpret $\text{Var}[V_\Sigma(t)]$ as the energy of noise within a sample of duration t, and $E^2[V_\Sigma(t)]$ as the energy of the "direct current" component. This interpretation implies the existence of means of changing the direct current component, which can, for example, be preformed by speeding up or slowing down the whole process. We see thus that "noise energy" and "(dc intensity)^H" are roughly proportional, where H is the constant $H = (3 - D)/2$.

In the white noise limit $D = 2$, "noise energy" is essentially proportional to "dc intensity," as is well known.

In the case $D = 1$, on the contrary, *"noise energy" is essentially proportional to the square of direct current intensity, which is precisely the nonclassical behavior that is empirically observed for 1/f noises.*

When D barely exceeds 1, the function $Y = X^H$ may be linearly approximated over wide ranges of variation of X. The noise power will, therefore, appear proportional to $(dc)^2$, where the coefficient of proportionality depends upon the dc, and becomes very small as the direct current increases.

An analogous argument shows that $\text{Var}[W_\Sigma(t)]t^{-3+D}$ varies slowly at $t \to \infty$. $\text{Var}[W_\Sigma]$ is the noise energy for every function of the form $W(t) + M$. The "dc" component of $W_\Sigma(t) + Mt$ is Mt, and one can again say that, if $M \neq 0$, the noise energy of $W(t) + M$ is essentially proportional to [direct current intensity of $W(t) + M]^H$.

For $X(t)$, on the contrary, the traditional result

$$\text{Var}[X_\Sigma(t)] = t/E(U) = \text{"dc"}$$

holds for all values of D.

V-B. The ergodic theorem in the case $0 < D < 1$

The ergodic theorem fails to apply to $W(t)$ for reasons similar to those encountered for the function $W(t, \tilde{T})$ of Section II: As $t \to \infty$, $t^{-1}W_\Sigma(t)$ tends in distribution to a *random* limit. The ergodic theorem is trivial for $V(t)$, because the sample mean almost surely tends to zero as $T \to \infty$.

Thus, one cannot speak of the "dc component" of processes such as $W(t)$, with $0 < D < 1$. For processes such as $V(t)$, the direct current component takes the "universal" value 0, which tells us nothing of interest.

One can, however, consider the *conditional* means and variance, and one finds for all $D \in (0, 1)$.

$$\text{Conditional Var}\left[W_\Sigma(t, T)\right] \sim \text{Conditional mean}\left[W_\Sigma(t, T)\right]^2.$$

This is a nonclassical form of the above-mentioned empirical relation which Bernamont discovered for $1/f$ noise: "Noise = (direct current)^2." We shall not dwell here upon arguments needed to interpret it.

VI. SAMPLING FLUCTUATIONS FOR LARGE t, WHEN $E(U) < \infty$ $P(u) = P_D(u)$, AND $1 \leq D < 2$, AND THE NOTION OF WILD RANDOMNESS

The expected values evaluated in Section V tell us little about the functions $V_\Sigma(t)$ and $W_\Sigma(t)$ when $E(U^2) = \infty$. Their behavior turns out to be extremely "erratic." Indeed, the condition $E(U^2) < \infty$ (i.e., $S'(0) < \infty$) is necessary and sufficient for $V_\Sigma(t)$ and $W_\Sigma(t)$ to be asymptotically Gaussian. We shall now examine the case $1 < D < 2$, and show that $V_\Sigma(t)$ and $W_\Sigma(t)$, when properly weighted, tend as $t \to \infty$ to limits having infinite variance. *This means that enormous deviations from average behavior can be expected.* The case $0 < D < 1$ will not be studied here (see M 1967b{N10}). Some statements made in this section require long proofs not available in the literature; these proofs will only be sketched. As a preliminary step, we must study the following auxiliary function $N(t)$.

Definition. $N(t)$ will be the random number of events T_K, such that $1 < T_K < t$.

VI-A. Sampling fluctuations of $N(t)$ for large t, when $\Sigma U < \infty$ and $P(u) = P_D(u)$, with $1 \leq D < 2$

The distribution of $N(t)$ was implicitly derived by Feller 1949, and is sketched in Feller 1966, page 360. Feller assumed $\tilde{T} = 1/2$, which transforms $N(t)$ into the function $N_0(t)$, whose expectation $\overline{N}_0(t)$ was used in Section IV. One can show, however, that $N_0(t)$ and $N(t)$ have identical asymptotic behavior, namely

$$\lim_{t \to \infty} \Pr\{N(t) \geq [t - \lambda(t)x]/E(U)\} = G_D(x),$$

where the function $\lambda(t)$ and G_D are defined as follows: $\lambda(t)$, which is a "scale factor" for $N(t) - E[N(t)] = N(t) - t/E(U)$, can be any function such that

$$\lim_{t \to \infty} tP[\lambda(t)] = (2 - D)D^{-1}E(U).$$

Therefore, one can write $\lambda(t) = L^0(t)t^{1/D}$, where $L^0(t)$ varies slowly for $T \to \infty$. The right-hand side $G_D(x)$ is the distribution function of a random variable Λ_D, called "maximally skew Lévy stable random variable of exponent D." Its density is mostly concentrated within a regular skew "bell"; however, for $x \to -\infty$, $G_D(x) \sim (2-D)D^{-1}|x|^{-D}$, and therefore, $E[\Lambda_D^2] = \infty$.

It may be noted that the upper bound of $N(t)$ is $t - 1$. The standard deviation of $N(t)$ can be seen to be identical to that of $V_\Sigma(t)$, and increases less rapidly than t. But $\lambda(t)$ increases even less rapidly than the standard deviation, which explains why the standard deviation of $N(t)/\lambda(t)$ tends to infinity.

We are now ready to proceed to $V_\Sigma(t)$ and $W_\Sigma(t)$.

VI-B. Sampling fluctuations of $V_\Sigma(t)$

$V_\Sigma(t)$ is the sum of a random number $N(t)$ of random variables V_k, where the $N(t)$ and V_k are all mutually independent. Therefore, the limit $\lim_{N \to \infty} \{[V_\Sigma(t) \mid N(t)] - N(t)\}[N(t)]^{-1/2}$ is Gaussian. Such variations are negligible in comparison with the fluctuations of $N(t)$. It follows that for large t, the distributions of $V_\Sigma(t)$ and $N(t)$ are identical.

VI-C. Sampling fluctuations of $W_\Sigma(t)$

This case is a bit more complicated. Neglecting end-corrections that are negligible in relative value, $W_\Sigma(t)$ becomes

$$\sum_{k=1}^{N(t)} (T_{k+1} - T_k) W_k.$$

This is a sum of a random numbers of nonindependent random variables, each of which has zero mean and infinite variance. When these addends are independent, it is well known in the literature (Gnedenko & Kolmogorov 1954) that one can select a slowly varying function $L^s(N)$, in such a way that

$$\lim_{N \to \infty} L^s(N) N^{-1/D} \sum_{k=1}^{N} (T_{k+1} - T_k) W_k$$

is a Lévy stable random variable of exponent D. As we know, this limit's probability density is concentrated mostly near the origin, but it has *two* very long tails and an infinite variance. When the marginal distribution of W_k is symmetric, the limit has a characteristic function equal to $\exp(-|\xi|^D)$. One extreme case of this limit, corresponding to $D = 2$, is the Gaussian density; the other extreme, $D = 1$, is the Cauchy's density $[\pi(1 + x^2)]^{-1}$. Intermediate cases had to be tabulated numerically; examples are drawn in Figure 3. {P.S. 1998. Those graphs' asymptotic straightness is very conspicuous.}

One can show that the above result also applies to $W_\Sigma(t)$. This implies that the $W_\Sigma(t)$ is asymptotically identical to the sum of $N(t)$ independent random variables of the form $(T_{K+1} - T_k) W_k$.

VII. CONCLUDING DISCUSSION

Section VI showed that *some* $f^{D-2}L(f)$ noises have very erratic sampling behavior. Some *other* f^{D-2} noises, however, are Gaussian, which means that they are perfectly "well-behaved." An example is provided by "fractional white noise" which is the formal differential of the random process described in M & Van Ness 1968. This difference in behavior explains that, as mentioned in Section I, the condition $S'(0) < \infty$ is needed to derive many specific properties of Wiener-Khinchin functions. When $S'(0) = \infty$, the spectrum itself is of less help than when $S'(0) < \infty$. It would not suffice, moreover, to know the spectrum and the marginal distribution: for example, the function $W(t)$ may be chosen to have a Gaussian marginal distribution, but it differs deeply from a Gaussian Wiener-Khinchin function.

These observations suggest two broad programs of research. Theoreticians should construct further illustrative classes of $f^{D-2}L(f)$ noises. Experimentalists should not be content with measurements of spectra, and should use the same long samples of $f^{D-2}L(f)$ noise to measure sampling fluctuations as well.

VIII. APPENDIX

The sole reason for working with the distribution $P_D(u)$ was analytic convenience : simple explicit formulas happen to be available in this case for both $P_D(u)$ and $\varphi_D(z)$. We shall now generalize the results of this paper.

The most widely used definition of "slow variation" is the following (See Gnedenko and Kolmogorov 1954, Sections VIII-8 and VIII-9):

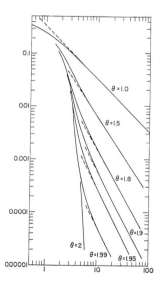

FIGURE N9-3. This is a representation of the positive portions of Paul Lévy's symmetric stable distributions, with $1 \leq D \leq 2$. Abscissa: log u; ordinate: log $\Pr\{U > u\}$. {P.S. 1998: Note that, if u is large enough, one has $\Pr\{U \geq u\} \sim (u/u^*)^{-D}$}

Definition. In the sense of Karamata, one says that $L(f)$ *varies slowly at infinity*, if, for every $h > 0$, $\lim_{f \to \infty} L(hf)/L(f) = 1$.

Condition $K_{d,D}$. Assume $0 \leq D < 2$. The function $P(u)$ will be said to belong to Karamata's class $K_{d,D}$, if $P(u)u^D$ varies slowly at infinity.

Condition $K_{i,D}$. Assume $0 \leq D \leq 2$. The function $P(u)$ will be said to belong to Karamata's class $K_{i,D}$, if $L^D(u) = u^{D-2}\int_0^u s^2 dP(s)$ varies slowly at infinity.

When $0 \leq D < 2$, $K_{i,D}$ and $K_{d,D}$ are equivalent, with $P(u)u^D = (2-D)L^0(u)/D$. However, $K_{i,D}$ remains meaningful when $D=2$ (See Feller 1966, page 544). When $K_{d,D}$ holds and $1 < D \leq 2$, then $E(U) < \infty$.

This definition enters in limit theorems on normed sums of independent and identically distributed random variables, that is, of expressions of the form $a_N \sum_{n=1}^N U_n - b_n$, where $\Pr\{U_n \geq u\} = P(u)$ for every n. If, and only if, condition $K_{i,D}$ is satisfied, one can choose the sequences a_N and b_N in such a way that $\lim_{N \to \infty} \Pr\{a_N \Sigma U_n - b_n < x\}$ exists. Moreover, this limit is a stable probability distribution (see Feller 1966, Gnedenko & Kolmogorov 1954). Hence, $K_{i,D}$ is the necessary and sufficient condition for the existence of a limit distribution for $N(t)$ (see Feller 1949).

It would have been nice if $K_{d,D}$ insured that $[2 - \varphi(z) - \varphi(-z)]|z|^{-D}$ varies slowly at the origin. Such is, however, not the case. Zygmund 1959 vol. I, Section V-2, and the Note on p.379): Condition $K_{i,D}$ on $P(u)$ is equivalent to the following weaker condition on $\varphi(z)$ (Feller 1966, Section XVII.5, especially Lemma 1).

Condition $K_{c,D}$. In the case $0 \leq D \leq 2$, the quantity $\varphi(z)$ will be said to belong to class $K_{c,D}$, if the sequence a_N can be chosen in such a way that, for every z,

$$\lim_{N \to \infty} \{N[2 - \varphi(z/a_N) - \varphi(-z/a_N)]\} = c|z|^D.$$

One readily verifies that the statements in the body of this paper remain correct if the case "slow variation" of $P(u)$ is interpreted by $K_{i,D}$, while $K_{c,D}$ interprets "slow variation" for $\varphi(z)$ and for various spectra.

In particular, $\varphi(z)$ satisfies condition $K_{c,D}$ if it satisfies the condition $\overline{K}_{c,D}$, which expresses that $[2 - \varphi(z) - \varphi(-z)]|z|^{-D}$ varies slowly at $z=0$ in the sense of Karamata. Condition $\overline{K}_{c,D}$ would interpret the rough notion of slow variation for $(\Delta S/\Delta f)f^{D-2}$. Unfortunately, there is no simple way of characterizing the corresponding functions $P(u)$.

&&&&& POST-PUBLICATION APPENDICES &&&&&

A. "LOCALLY-GAUSSIAN" SUMS OF A LARGE BUT FINITE NUMBER OF CO-INDICATOR FUNCTIONS (DISCUSSION OF M1969e)

An important innovation found in the body of this chapter, namely, the coindicator function $W(t)$, promises to have a wealth of applications. In addition, it lends itself to many interesting variants, two of which will be sketched in these appendices.

An early variant of the coin function is described in M1969e, a paper that is far too long and complicated to be reproduced in this book, but sufficiently interesting to be described and commented upon. The input of the construction is a time span T, a large but finite integer M, and a random variable W satisfying $EW = 0$ and $EW^2 = 1$. *Step A.* Construct M independent Lévy dusts on $[0,T]$. *Step B.* Construct the coindicator functions $W_m(t)$ of those M Lévy dusts. *Step C.* Construct a suitably normalized sum $S_M(t)$ of the $W_m(t)$.

With little conceptual change, one may define $S_M(t)$ for all times, instead of restricting it to the interval $[0,T]$. One way is to replace the Lévy dusts on $[0,T]$ by Lévy dusts in which every gap larger than T has been shortened to T. This construction yields a class of random functions that seem to possess mutually contradictory properties. They are statistically stationary, that is, translation invariant in distribution; nevertheless the eye near-irresistibly splits the actual samples into alternating pieces that seem to belong to distinct processes. Illustrations are found in M1969e.

We already know that such splitting is also characteristic of other functions of a Lévy dust. Indeed, that is why Chapters N5 and N7 succeeded in using Lévy dusts to model clustering. The novelty contributed by the function $S_m(t)$ is that its structure is richer; it becomes increasingly rich as M moves from 1 to 2 and on, but one obtains a less interesting Gaussian process as M tends to infinity.

Using a terminology introduced much later, in M1995k, M is called a "lateral parameter", and the most interesting facts occur when M is "large but finite" rather than infinite. Such facts can be called "pre-asymptotic". The concern with pre-asymptotics might have been motivated by a search for mathematical "pathology," but was in this instance motivated by a practical need for ways of accounting for apparent self-contradictions in certain empirical data.

The richness of this structure is traceable to the non-Gaussian character of $S_m(t)$, and the term "locally Gaussian" in the titles of M1969e and of this appendix deserves elaboration. It must be recalled that an essential part of Gaussianity is that the "marginal distribution" defined by Pr $\{X < x\}$ must be Gaussian. Experimental tests of approximate marginal Gaussianity are widely carried out. Whenever they are affirmative, there is a strong temptation to consider that the rich mathematical and statistical theory of Gaussian processes can be utilized with no hesitation. Actually, the full definition of Gaussian processes goes well beyond the marginal distribution; the requirement is that, for every set of times $t_1, t_2, ...t_k$, the vector of coordinates $X(t_1), ...X(t_2), X(t_k)$, must be a Gaussian vector. This full definition is very demanding and practically impossible to test empirically. A contribution of M1969e is to investigate a few things that can go wrong.

Specifically, it provides an example where those k- dimensional vectors are approximately Gaussian for small values of k but not for large values. A consequence is that seemingly safe inferences from Gaussianity fail to hold. For example, consider the version of $S_m(t)$ that is defined on the whole line. Its increment $S_M(t+d) - S_M(t)$ has the pre-asymptotic property of being near-Gaussian for small (and large) values of d, but non-Gaussian on a significant intermediate range of values of d. This significant range disappears when M tends to infinity with a fixed T.

A last feature deserves explanation. M1969e was published in an economics journal because it seemed to me at the time that the pre-asymptotic quandary it addressed was especially acute in finance. As shown in M1997E, my early model of price variability tackled two major "syndromes", namely, the "Noah and Joseph Effects," but had to approach them separately. The near-discontinuities ("Noah Effect") were addressed first, in M1963{E14}, by injecting the L-stable distributions. The cyclic but non-periodic behavior ("Joseph Effect") was addressed soon afterwards, being interpreted as a symptom of long-term dependence; in particular, a puzzle called "Hurst phenomenon" was addressed in M1965h{H} by injecting the fractional Brownian motion. The challenge was that the observed prices combine both characteristics. Today, the best and most elegant combination is provided by the fractional Brownian motion of multifractal time. This construction proceeds as in Chapter N1 and is developed in M1997E. In this context, M1969e, written before I turned to multifractals, was an early and not very successful effort.

B. AN ALGORITHM THAT GENERATES MULTIFRACTALS AS PRODUCTS OF COINDICATOR FUNCTIONS

This appendix is a preview of this book's Part IV. The functions in Appendix A are constructed *additively*. To the contrary, the multifractal measures described in Part IV (and additional thoroughly understood multifractals) are constructed as the *products* of contributions of increasing high frequency. This appendix sketches the role that coindicator functions can assume in a construction of multifractals that can be viewed as intermediate between the pioneering examples described in Chapters 14 and 15.

The examples in M19744f{N15} are constrained to be gridbound, in the sense that their construction implies a recursive subdivision of the unit interval [0, 1] into $b \geq 2$ subintervals, then b^2 sub-sub-intervals, etc... This regular grid is totally antiphysical and was introduced solely because of its mathematical and computational simplicity and convenience. More specifically, an earlier construction given in M1972j{N14} is grid-free. But it was not used much because it was (and remains) computationally tedious and was mathematically intractable until special tools were introduced by J.P. Kahane.

A logical path that proceeds from the M1974 to the M1972 constructions step-by-step is described in M1982F{FGN}, page 379, section 5. Several of the intermediate steps along this path implicitly define additional scale-free multifractal measures and suggest a number of further scale-free multifractals. Those variants will be described elsewhere, but the description of one variant nicely fits in this appendix.

It involves an infinity of multiplicands indexed by an interger k. The first input, a density $\lambda > 0$, makes it possible to define the $k-$ th Poisson set P_k on the time axis, as having the density λ^k. The second input, a tandom variable W satisfying $EW = 0$ and $EW^2 = 1$ makes it possible to define the co-indicator function $W_k(t)$ of P_k. The third input is a scale factor $\sigma > 0$; this σ makes it possible to define the $k-$ th multiplicand as equal to $\exp[\sigma W_k(t)]/E \exp[\sigma W]$.

Increasing λ from 0 to 1 brings each $W_k(t)$ closer in distribution to $W_{k+1}(t)$. Decreasing σ to 0 decreases the effect of each contribution. One can combine $\sigma \to 1-$ and $\sigma \to 0$ to insure that the second moment of the $k-th$ multiplicand remains unchanged. It can be shown that if W is Gaussian, the limit of the product of the multiplcands from $k+0$ to $k = \infty$ converges to the scale-free multifractal described in M1972j{N14}.

Lognomal multiplicands. The preceding construction can be compared to the corresponding step described in M1982F{FGN}, page 379, Section 5. There, the multiplicands are of the form $\exp[\sigma Z_k(t)]$, where Z_k is a Gauss-Markov ("Ornstein-Uhlenbeck") process of time scale equal to λ^k.

The "broken-line process". M1972w{H} investigates yet another simple construction that can replace the coindicator function for many purposes. It has the advantage that the contributions are continuous. Once again, one begins with a Poisson set of points. The difference is that the values of $W(t)$ are not prescribed on the gaps between the points of P_k, but on the points themselves. Then $W_k(t)$ is interpolated linearly between those points, hence the term, "broken-line process."

To be applied to Lévy dusts, which are non-denumerable, the broken-line construction must be modified; it no longer defines $W(t)$ for all t, but for all t except a set of zero mueasure. The idea is to define $W(t)$ as taking independent values for all the end-points of the gaps and then to interpolate linearly.

Sporadic random functions and conditional spectra; self-similar examples and limits

● *Chapter foreword.* Once again, this is a paper that requires two separate introductions: one is concerned with method and the other with substance. Fortunately, the introductions to Chapter N9 largely apply to this chapter, and the reader is referred to them. However, the style of the two chapters exhibit a major difference.

This paper is mathematically complicated and, for reasons described in the *Annotations*, completely devoid of concrete motivation and of graphic illustration. Fortunately, a suitable motivation is found in M 1964w, a report from which Section N5-1 was extracted. Therefore, a few additional excerpts from M 1964w are reproduced immediately following this Foreword, and relevant illustrations are found in Section N3-3.

Furthermore, this paper's original had no abstract; as a substitute, this reprint incorporates a typed sheet that was slipped into the reprints before they were handed out.

It may be worth pointing out that nowhere does the body of the paper state explicitly that "sporadic process" was a new notion in 1965.

Editorial changes. The term *core function* in the original has been replaced in this reprint by the term *coin function*, which is an easy mnemonic for *"co-indicator function."* The fractal dimension is denoted by D instead of θ, the deltavariance by Δ instead of D, and the tilde replaces the asterisks and other awkward symbols. ●

• *Concrete motivation behind the sporadic processes (an excerpt from M1964w)*. This work [is written] from the distinct but related viewpoints of the theory of turbulence, the calculus of probabilities and the theory of communications.

A first aim is to contribute to the description of turbulent flows at very high Reynolds numbers. It is known that these flows appear to be mixtures of "laminar" and "turbulent" regions; however, these regions are difficult to separate in practice. I claim that – from the theoretical point of view – such separation is impossible and should not even be attempted. The fundamental feature of mature turbulence is that it is "intermittent" to such an extreme degree that the conventional Wiener-Khinchin spectral analysis and even the conventional theory of stochastic processes are of no use, or – at best – are extremely awkward to use.

A second aim [therefore] is to begin constructing a theory of extremely intermittent chance phenomena [for which, the usual covariance] "misbehaves." The concepts of covariance and of spectrum can [indeed] be generalized [to fit. The result involves] prefactors that are not numerical constants but depend very much upon chance and upon the sample length.

To account for certain features of turbulence and also for the sake of simplicity, I stress a family of intermittent random functions to be called "self-similar," which have the property that they are invariant with respect to any change of time scale. In their case, the concept of the spectral density is meaningless, however, whichever $\omega' > 0$ and $\omega'' > 0$, one can speak of the ratio of spectral densities at the frequencies ω' and ω'', and one has

$$\frac{S(\omega')}{S(\omega'')} = \left(\frac{\omega'}{\omega''}\right)^{D-2},$$

where D is a parameter between 0 and 1. The little-known Mittag-Leffler distributions play an important role in this context.

Accent will be put upon motivating some rather strange mathematical phenomena that this study unearthed....

A third aim is [predicated upon the idea] that there is a strong resemblance between turbulence in fluids and the apparently unrelated processes called "$1/f$ noises." Therefore, this paper is a further step beyond Berger & M 1963{N5}, and M 1965c{N7}. It is quite possible that self-similar errors reflect some electromagnetic aspects of atmospheric turbulence. My first and third purposes may therefore eventually merge.

It should be repeated and stressed that we shall *not* study the error-clusters or the turbulent region alone, as extracted from the given overall pattern with the help of some seldom explicit rule or criterion. The main thesis underlying this work is rather that *the same model* can in some cases treat both the "active" regions of a pattern (turbulent zones, error-clusters and the like) and its "quiescent" regions (laminar zones, error-free stretches and the like). ●

✦ **Abstract.** *(This text was omitted in the original but inserted when distributing the original's reprints)*. This paper introduces and investigates the new concept of "sporadically varying generalized random function." A sporadic process attributes to the members of a family of functions a set of *non-normalizable* "probability" weights, in such a fashion that every *non-constant* finite sample exhibits the three properties (to be defined) of "intermittency," "non-ergodicity" and "infrared catastrophe." More generally, the concepts of stationarity and of the spectrum fail to be meaningful for sporadic processes, except after conditioning by appropriate events. The properties in question have been observed for such important phenomena as errors of communication, noises (in semiconductors, contacts and thin films), several forms of turbulence in gases and liquids in very large vessels, and (finally) highly clustered patterns in the plane and the space. This paper is motivated by the needs of physics but written in mathematical style. ✦

1. INTRODUCTION

The concept of sporadic random function, a new notion to be introduced in this paper, arose out of a pessimistic evaluation of the suitability of ordinary random functions as models of certain "turbulence-like" chance phenomena. Now that a number of turbulence spectra have been evaluated, it must be admitted that their sampling behavior is often very much at odds with one of the Wiener-Khinchin second order theory. It is true that, given two frequency intervals (λ_1, λ_2) and (λ_3, λ_4), the ratio between the energies in these intervals rapidly tends to a limit, as expected. But the energies within each of the intervals (λ_1, λ_2) or (λ_3, λ_4) continue to fluctuate wildly, however large the sample may be. Another puzzling fact is that some turbulence-like phenomena appear to have an infinite energy in low frequencies, a syndrome often colorfully referred to as an "infrared catastrophe."

The reason the concept of stationary random function must be generalized may be further explained as follows. Many physical time series X are "intermittent." That is, they alternate between periods of activity and quiescent "intermissions" during which X is constant and may even vanish. Moreover, and this is the crucial point, some series are quiescent *most* of the time. Using the intuitive *physical* meaning of the phrase, "almost sure event," such a series would appear to satisfy the following properties, which will entitle it to be called "sporadic:" X is almost surely constant in any prescribed finite span (t', t''), but X almost surely varies sometime. It would be very convenient if each of the above occurrences of the term "almost sure" could also be interpreted in the usual mathematical terms. Unfortunately, the theory of random functions is rigged in such a way that the above two requirements are incompatible (Doob 1953, pp.51 and 70).

There is, however, a simple way of defining "generalized random functions" (g.r.f.) that accommodate the sporadically varying series $X(t)$. It suffices to amend the classical Kolmogorov's probability space triplet $(\Omega, \mathcal{A}, \mu)$ in two ways: (a) the measure μ is assumed unbounded though sigma-finite; and (b) a family $\mathcal{B}$ of conditioning events B is added, where $0 < \mu(B) < \infty$. It is further agreed that, henceforth, the only well-posed questions will be those relative to conditional probabilities Pr $\{A|B\}$ where $B \in \mathcal{B}$.

The first part of the present paper is devoted to the preliminary task of defining mathematically the concept of "sporadic behavior." Sporadic renewal sets will be stressed.

The second part of this paper is devoted to its main purpose, which is to investigate the strange correlation and spectral properties of sporadic processes. Three types of results can be distinguished here.

The first relates to the definition of conditional correlations, given a fixed conditioning event.

The second relates to the limit behavior of conditional correlations as the conditioning is weakened; a central role will be played by functions of "regular variation."

The third type of result relates to the limit behavior of sample values of correlation; here, a central role will be played by the little known "Mittag-Leffler" distribution, and by some self-similar renewal sets and processes.

It should be noted that the "generalized random functions" due to Gelfand and to Itô (Gelfand & Vilenkin, 1964) take a totally different approach. They use Schwartz distributions to solve problems associated with an excessive amount of high frequency energy. The present generalization is, on the contrary, directed towards low frequency problems. However, it has suggested a natural solution to a long standing problem of hard analysis concerning high frequencies; see Kahane and M 1965.

The very convenient notation for our g.r.f. is $(\Omega, \mathcal{A}, \mathcal{B}, \mu)$. It comes from the general axiomatic of conditional probability found in Rényi 1955. But measure theory as such is not essential to the present enterprise, whose spirit is very different from Rényi's. We shall not strive for generality for its own sake, nor for irreducible axiomatics. Our sole object is to select new mathematical objects adapted to the needs of physics, and to prove a few basic results justifying our choice. Examples of applications can be found in M 1965c and 1967i{N7, N9}, each of which includes a more detailed discussion of the motivating empirical findings. For the intermittency of turbulence in fluids, see also Sandborn 1959 and the later portions of Monin & Yaglom 1963. The author will not be surprised if some details of his work turn out to require refinement and correction.

2. SAMPLE SPACES, EVENTS, AND GENERALIZED RANDOM FUNCTIONS BASED UPON SIGMA-FINITE UNBOUNDED MEASURES

As is well-known, a definition of a random function (r.f.) $X(t, \omega)$ begins with a *measurable space* $(\Omega, \mathcal{A})$, where Ω is the *sample space of elementary events* ω, each of which is a *sample function* $x(t)$ of $t \in R$ (that is, $-\infty < t < \infty$). *Simple events* are $\Omega, \emptyset$, and the finite unions, finite intersections, or complements of the ω sets of the form $\{\omega : x_i' < X(t_i, \omega) \leq x_i''\}$, where the x_i', t_i, and x_i'' are rational. The smallest Borel field containing those simple events will be $\mathcal{A}$.

The next step is to form a *measure space* $(\Omega, \mathcal{A}, \mu)$, where the *measure* is a nonnegative completely additive set function defined for $A \in \mathcal{A}$.

The final step in the definition is to assume that $\mu(\Omega) < \infty$; a probability Pr $\{A\}$ can then be defined for all $A \in \mathcal{A}$ by Pr $\{A\} = \mu(A)/\mu(\Omega)$. This is the point where the usual definition has proven inadequate.

When $\mu(A)$ is unbounded, the triplet $(\Omega, \mathcal{A}, \mu)$ specifies $X(t, \omega)$ through measures that cannot be interpreted as nontrivial absolute probabilities. Therefore, in order to specify $X(t, \omega)$ through some kind of probability, one must complete the triplet $(\Omega, \mathcal{A}, \mu)$ by also specifying an appropriate collection $\mathcal{B}$ of admissible conditioning events B, such that $B \in \mathcal{A}$ and $0 < \mu(B) < \infty$.

A *generalized random function* (g.r.f.) will be defined as a quadruplet $(\Omega, \mathcal{A}, \mathcal{B}, \mu)$ and *well-set probabilistic results* will be defined as those relative to conditional probabilities of the form $\Pr\{A|B\} = \mu(A \cap B)/\mu(B)$.

Alternative approach to some g.r.f.'s. In important cases, the measure μ is "sigma-finite," meaning that there exists a denumerable family of events Ω_i with $\Omega = \bigcup_{i=1}^{\infty} \Omega_i$ and $0 < \mu(\Omega_i) < \infty$. Moreover, every B satisfies B with $I(B) < \infty$. Under these two conditions, the g.r.f. $(\Omega, \mathcal{A}, \mathcal{B}, \mu)$ is equivalent to the indexed family of ordinary r.f.'s $(\bigcup_{i=1}^{I} \Omega_i, \mathcal{A}_I, \mu_I)$ where I ranges from 1 to ∞, the elements of $\mathcal{A}_I$ are the sets of the form $A \cap (\bigcup_{i=1}^{I} \Omega_i)$, and μ_I is probability conditioned by $\bigcup_{i=1}^{I} \Omega_i$. These r.f.'s are related to each other by hosts of conditions of compatibility; $\lim_{I \to \infty} (\bigcup_{i=1}^{I} \Omega_i, \mathcal{A}_I, \mu_I)$ is not an ordinary r.f..

Concrete problems cannot involve infinite values of I; therefore, one can always avoid g.r.f.'s by choosing some large but finite "external scale" J, and working with the r.f. $(\bigcup_{i=1}^{J} \Omega_i, \mathcal{A}_J, \mu_J)$. This J will be nonintrinsic, however, and the solution of concrete problems will require the study of "transient small sample behavior" rather than the simpler study of limit theorems. Thus, even from the viewpoint of concrete problems, it is usually simpler to work with the g.r.f. $(\Omega, \mathcal{A}, \mathcal{B}, \mu)$ directly.

3. DEFINITION OF SPORADICALLY VARYING GENERALIZED RANDOM FUNCTIONS AND RELATED CONCEPTS

3.1. Sporadically varying g.r.f.'s

We must first define some sets of Ω. Let

(3.1)
$$A^* = \{\omega : X(t, \omega) \text{ is constant over } -\infty < t < \infty\}$$
$$= \{\omega : \text{g.l.b. } X(t, \omega) = \text{l.u.b. } X(t, \omega) | -\infty < t < \infty\},$$

and for every open interval (t', t'') such that $-\infty < t' < t'' < \infty$, let

(3.2)
$$B(t', t'') = \{\omega : X(t, \omega) \text{ is not constant over } t \in (t', t'')\}$$
$$= \{\omega : t' < t < t'', \text{ g.l.b. } X(t, \omega) < \text{l.u.b. } X(t, \omega)\}.$$

It is necessary that $A^* \in \mathcal{A}$ and $B(t', t'') \in \mathcal{A}$. For that, it is sufficient that time be restricted to some denumerable subset of R. When time is continuous, problems of "separability" may arise. But a discussion of separability is not needed for the examples examined in this paper, and it will therefore be postponed.

The family of all events $B(t', t'')$ will be designated by $\mathcal{B}^*$.

Definition 3.1. *A sporadically varying g.r.f. is a quadruplet $(\Omega, \mathcal{A}, \mathcal{B}, \mu)$, where $(\Omega, \mathcal{A})$ is a measurable space, and where the following conditions are imposed upon the measure μ and upon the subfamily $\mathcal{B}$ of $\mathcal{A}$:*

The measure μ is assumed to satisfy two conditions:

(3.3)
$$\mu(\Omega) = \infty \text{ and } \mu(\emptyset) = 0;$$
$$\mu(A^*) = 0.$$

Two measures μ_1 and μ_2 will not be distinguished if $\mu_1(A)/\mu_2(A)$ is a constant independent of $A \in \mathcal{A}$. The family $\mathcal{B}$ of conditioning events B is assumed to satisfy two conditions:

(3.4)
$$0 < \mu(B) < \infty \text{ if } B \in \mathcal{B};$$
$$\mathcal{B}^* \subset \mathcal{B}.$$

Proposition 3.1. *Conditions (3.3) and (3.4) imply that μ is sigma-finite.*

Proof. Write $\Omega = A^* \cup [\bigcup_{t=-\infty}^{\infty} B(i, i+1+\varepsilon)]$, where $\varepsilon > 0$.

As intended, it is almost sure that a sporadically varying $X(t, \omega)$ does not vary in any *prescribed* bounded interval (t', t''), but varies *somewhere* along R. These characteristics would be mutually incompatible if $X(t, \omega)$ were an ordinary random function.

3.2. Intermissions and set of variation

Given a left-continuous g.r.f. $X(t, \omega)$ and x, the interior of the set $\{t: X(t, \omega) = x\}$ is either empty or is the denumerable union of open intervals. The latter needs a distinguishing name.

Definition 3.2. *An intermission of the g.r.f. $X(t, \omega)$ is a maximal open interval contained in the interior of a set of the form $\{t: X(t, \omega) = x\}$.*

For each ω, the intermissions of $X(t, \omega)$ are denumerable, and can be designated as $(t_h'(\omega), t_h''(\omega))$, where $1 \leq h < \infty$; the ω will usually be omitted.

Definition 3.3. *The set of variation of the g.r.f. $X(t, W)$ will be the closed set $S(\omega) = R - \bigcup_{h=1}^{\infty}(t_h', t_h'')$.*

If $X(t, \omega)$ is sporadically varying, so that $\mu(\Omega) = \infty$, there is a vanishing absolute probability that $S(\omega)$ and (t', t'') intersect. The set $S(\omega)$ will be said to be "sporadically distributed" or "sporadic."

(Some sporadic random sets can be treated directly, by generalizing the definition of random compact set quoted in Blumenthal & Getoor 1962, p. 309 to infinite $\mu(\Omega)$. In this way, one may define functions that satisfy sporadically a property other than the property of being nonconstant.)

An extreme example of sporadic g.r.f.'s is a function with a single randomly located step: $\Omega = R$, where ω is distributed with Lebesgue measure; for $t > \omega$, $X(t, \omega) = X'$ with $|X'| < \infty$; for $t < \omega$, $X(t, \omega) = X'' \neq X'$ with $|X''| < \infty$. Thus, $S(\omega) = \omega$, and there are two intermissions, both of infinite duration. Well-set probabilistic problems relate to step functions with a step randomly located over (t', t'') with the corresponding (finite) Lebesgue measure.

4. MEASURE PRESERVATION AND CONDITIONAL STATIONARITY

4.1. Shift invariant indecomposable measures and conditions

The shift transformation φ_τ is defined as usual. If ω represents the function $X(t)$, then $\varphi_\tau \omega$ represents $X(t + \tau)$, and one assumes $\varphi_\tau \omega \in \Omega$. Similarly, for every $A \in \mathcal{A}$, one defines $\varphi_\tau A$, and one assumes $\varphi_\tau A \in \mathcal{A}$.

If μ is a measure on $(\Omega, \mathcal{A})$, possibly unbounded, the *shift invariance* of μ (that is, the *measure-preserving* character of φ_τ) is defined by the condition:

(a) for every $A \in \mathcal{A}$, and for every τ, one has $\mu(\varphi_\tau A) = \mu(A)$.

The conditioning events $B(t', t'')$, members of $\mathcal{B}^*$, have the following property: for every $B \in \mathcal{B}^*$ and for every τ, $\varphi_\tau B \in \mathcal{B}^*$. Therefore, $\mathcal{B}^*$ is said to be *shift invariant*. More generally, $\mathcal{B}$ is said to be *shift invariant*, if it fulfills the following condition:

(b) for every $B \in \mathcal{B}$, and for every τ, one has $\varphi_\tau B \in \mathcal{B}$. (If $\mathcal{B} - \mathcal{B}^*$ is nonvoid, this is no longer a necessary consequence of (a).)

A set $A \in \mathcal{A}$ is said to be *shift invariant* if it fulfills the usual condition:

(c) for every τ, $\mu[\varphi_\tau A \cup A - \varphi_\tau A \cap A] = 0$.

Such A form a Borel field, designated as $\mathcal{T}$. The least interesting case occurs when every $A \in \mathcal{T}$ is the union of sets in $\mathcal{T}$ having a finite μ measure. The most interesting case is when every set A in $\mathcal{T}$ satisfies either $\mu(A) = 0$ or $\mu(\Omega - A) = 0$; as usual, such a φ_τ will be called indecomposable.

4.2. Conditional restricted stationarity

The usual concept of stationarity, is relative to absolute probabilities. Therefore, it becomes degenerate in the case of g.r.f.'s for which $\mu(\Omega) = \infty$. Partial conditional equivalents are available, however.

Definition 4.1. *The g.r.f.* $(\Omega, \mathcal{A}, \mathcal{B}, \mu)$ *will be said to be* conditionally stationary *if, for every $B \in \mathcal{B}$, there exists a nonvanishing open interval $(t', t'')_B$ with the following property. Let t_i be I time instants, $I < \infty$, and τ a time span such that $[\bigcup \{t_i\}] \bigcup [\bigcup \{t_i + \tau\}]$ and let $A_R^{(i)}$ be I Borel sets of R. Then*

(4.1) $\qquad \Pr \{\forall i, X(t_i, \omega) \in A_R^{(i)} | B\} = \Pr \{\forall i, X(t_i + \tau, \omega) \in A_R^{(i)} | B\}.$

The expression

(4.2) $\qquad \Pr \{X(t, \omega) \in A_R | B\} = \dfrac{\mu\{\omega : [X(t, \omega) \in A_R] \cap [\omega \in B]\}}{\mu(B)},$

which is independent of t as long as $t \in (t', t'')_B$, is a conditional marginal distribution for $X(t, \omega)$, given B.

Returning to the shift invariance of μ and $\mathcal{B}$, note that it has the following obvious consequences

(4.3) $\qquad\qquad \Pr \{\varphi_\tau A | \varphi_\tau B\} = \Pr \{A | B\};$

(4.4) $\qquad\qquad$ if $A \bigcup \varphi_\tau A \subset B$, then $\Pr \{\varphi_\tau A | B\} = \Pr \{A | B\}.$

This (4.4) follows from

(4.5)
$$\Pr\{\varphi_T A | B\} = \frac{\mu(\varphi_T A \cap B)}{\mu(B)} = \frac{\mu(\varphi_T A)}{\mu(B)} = \frac{\mu(A)}{\mu(B)}$$
$$= \frac{\mu(A \cap B)}{\mu(B)} = \Pr\{A | B\}.$$

First example of conditional stationarity. Now, make the stronger assumptions that $\mathcal{B} = \mathcal{B}^*$ and that $X(t, \omega)$ vanishes during its intermissions. Consider the event $A = \{\omega : X(t, \omega) \in A_R\}$. If $0 \notin A_R$ and $t \in (t', t'')$, one has $\mu(A) < \infty$. In particular, $\mu[\omega : X(t, \omega) \neq 0] < \infty$, so that X may be integrable without vanishing identically. For example, if $\mathcal{B} = \mathcal{B}^*$ and $X(t, \omega)$ is bounded and vanishes during the intermissions, it is integrable.

Now consider two instants t and $t + \tau$, both belonging to (t', t''). If $0 \notin A_R$, then $A \cup \varphi_T A$ so that $\Pr\{\varphi_T A | B(t', t'')\} = \Pr\{A | B(t', t'')\}$ by (4.4). If $0 \in A_R$, then $0 \notin R - A_R$ and

$$\Pr\{\varphi_T(\Omega - A) | B(t', t'')\} = \Pr\{(\Omega - A) | B(t', t'')\}.$$

Since $\Pr\{A | B(t', t'')\} = 1 - \Pr\{(\Omega - A) | B(t', t'')\}$, the relation

$$\Pr\{\varphi_T A | B(t', t'')\} = \Pr\{A | B(t', t'')\}$$

is valid for every A of the form $\{\omega : X(t, \omega) \in A_R\}$. In fact

Proposition 4.1. *If $\mathcal{B} = \mathcal{B}^*$, if μ is shift invariant, and if $X(t, \omega)$ vanishes during its intermissions, then $[X(t, \omega) | B(t', t'')]$ is conditionally stationary over some interval $(t', t'')_B$ satisfying $(t', t'')_B \supseteq (t', t'')$.*

Proposition 4.2 *(Converse of 4.1) If $\mathcal{B} = \mathcal{B}^*$, if X vanishes during its intermissions, and if, for every (t', t''), $[X(t, \omega) | B(t', t'')]$ is conditionally stationary over (t', t''), then μ is shift invariant.*

Second example of conditional stationarity. A second important class of g.r.f.'s is one for which the conditional stationarity is *not* a consequence of the shift invariance of μ, and must be postulated separately. This class is defined as having the following property.

For almost all t (that is, except if t belongs to a set of R of vanishing Lebesgue measure), the conditioned marginal distribution $\Pr\{X(t, \omega) \in A_R | B\}$ is the same for all conditions B such that $t \in (t', t'')_B$. One can, therefore, speak in this case of a *nonconditional* "pseudomarginal" distribution for $X(t, \omega)$. The ordinary r.v. whose distribution is $\Pr\{X(t, \omega) < x | B\}$ will be called the "pseudomargin" of X.

Unless X vanishes identically, it is not integrable. However, its pseudomarginal variable can be integrable.

4.3. The structure of $\mu[B(\tilde{u})]$ when μ is shift invariant

If μ is shift invariant, it will often be unnecessary to specify the value of t' in $B(t', t' + \tilde{u})$, because t' is indifferent or obvious from the context. We shall then write $B(t', t' + \tilde{u})$ as $B(\tilde{u})$.

Proposition 4.3. *If μ is shift invariant, then the function $\mu[B(u)]$, defined for $u > 0$, is concave, continuous and right and left differentiable.*

Proof. This will follow classically after it is proved that for every $\tilde{u} > 0$ and $h > 0$, one has

(4.6) $\quad \mu[B(\tilde{t}, \tilde{t} + \tilde{u} + 2h)] + \mu[B(\tilde{t}, \tilde{t} + \tilde{u})] - 2\mu[B(\tilde{t}, \tilde{t} + \tilde{u} + h)] \leq 0.$

Note that $B(\tilde{t}, \tilde{t} + \tilde{u} + 2h) \supseteq B(\tilde{t}, \tilde{t} + \tilde{u} + h) \supseteq B(\tilde{t}, \tilde{t} + \tilde{u})$. Therefore, (4.6) holds if

(4.7) $\quad \begin{aligned} &\mu[B(\tilde{t}, \tilde{t} + \tilde{u} + 2h) - B(\tilde{t}, \tilde{t} + \tilde{u} + h)] \\ &\leq \mu[B(\tilde{t}, \tilde{t} + \tilde{u} + h) - B(\tilde{t}, \tilde{t} + \tilde{u})]. \end{aligned}$

By stationarity, this holds if

(4.8) $\quad \begin{aligned} &\mu[B(\tilde{t} - h, \tilde{t} + \tilde{u} + h) - B(\tilde{t} - h, \tilde{t} + \tilde{u})] \\ &\leq \mu[B(\tilde{t}, \tilde{t} + \tilde{u} + h) - B(\tilde{t}, \tilde{t} + \tilde{u})] \end{aligned}$

which in turn holds if

(4.9) $\quad \begin{aligned} &B(\tilde{t} - h, \tilde{t} + \tilde{u} + h) - B(\tilde{t} - h, \tilde{t} + \tilde{u}) \\ &\subseteq B(\tilde{t}, \tilde{t} + \tilde{u} + h) - B(\tilde{t}, \tilde{t} + \tilde{u}). \end{aligned}$

This last statement is true because the event on the left of $\subseteq$ differs from that on the right by the set of ω such that X varies for $t \in [\tilde{t} + \tilde{u}, \tilde{t} + \tilde{u} + h)$ and $t \in (\tilde{t} - h, \tilde{t}]$ but not for $t \in (\tilde{t}, \tilde{t} + \tilde{u})$.

Proposition 4.4. μ can be written as

$$(4.10) \qquad \mu[B(\tilde{u})] = Q + \int_0^{\tilde{u}} P(s)ds,$$

where $Q \geq 0$ and $P(u)$ is positive nonincreasing, and such that $\int_0^\varepsilon P(s)ds < \infty$ for all $\varepsilon > 0$. $P(0)$ is defined as $\lim_{s \to 0} P(s)$, and may be infinite.

Corollary 4.1. Let

$$\tilde{T}(\omega) = \text{g.l.b.} \{t : \tilde{t} + t \in S(\omega) \mid S(\omega) \cap (\tilde{t}, \tilde{t} + \tilde{u}) \neq \varphi\}.$$

It follows from the above proposition that, given the interval $(\tilde{t}, \tilde{t} + \tilde{u})$,

$$(4.11) \qquad \Pr\{0 < \tilde{T} \leq t_0 \mid 0 \leq \tilde{T} < \tilde{u}\} = \frac{\int_0^{t_0} P(s)ds}{Q + \int_0^{\tilde{u}} P(s)ds},$$

which has the probability density proportional to $P(t_0)$. Moreover,

$$(4.12) \qquad \Pr\{\tilde{T} = 0 \mid 0 \leq \tilde{T} < \tilde{u}\} = \frac{Q}{Q + \int_0^{\tilde{u}} P(s)ds},$$

which is positive if $Q > 0$ and vanishes if $Q = 0$.

Corollary 4.2. If $Q = 0$, and $S(\omega) \cap (t', t'') \neq \varphi$, $S(\omega) \cap (t', t'')$ has zero Lebesgue measure, except for a set of ω of zero μ measure.

Proof. Given $S(\omega)$, choose $\tilde{T}$ with Lebesgue measure over some bounded (t', t'') such that $S(\omega) \cap (t', t'') \neq \emptyset$. Then $\Pr\{\tilde{T} \in S(\omega) \cap (t', t'') \neq \emptyset\} = 0$ is a consequence of $Q = 0$, and it implies that $S(\omega) \cap (t', t'')$ has zero Lebesgue measure with a conditional probability equal to 1.

If $Q > 0$ and $S(\omega) \cap (t', t'') \neq \varphi$, $\mu[\omega : S(\omega) \cap (t', t'')$ has positive Lebesgue measure$] > 0$. But $\mu[\omega : S(\omega) \cap (t', t'')$ has zero Lebesgue measure$]$ may be either positive or zero; examples of both kinds will be described later.

N10 ◇ ◇ SPORADIC FUNCTIONS AND CONDITIONAL SPECTRA

4.4. Degrees of intermittency

The behavior of $P(u)$ allows an important classification of certain ordinary and generalized random processes.

The *nonintermittent case* will be defined by $P \equiv 0$. One must have $Q > 0$ in order that $\mu[B(u)] > 0$. Then $\mu[B(u)] \equiv \mu(\Omega)$ and $\Pr\{\tilde{T} > 0 \mid 0 \leq \tilde{T} < \tilde{u}\} \equiv 0$. Such functions $X(t, \omega)$ have almost surely no open interval of constancy.

The *intermittent cases* will be defined by $P \neq 0$. The constant Q may be ≥ 0. Then $\Pr\{\tilde{T} > 0 \mid 0 \leq \tilde{T} < \tilde{u}\} > 0$ and $X(t, \omega)$ almost surely possesses open intervals of constancy.

A finer classification of intermittent r.f.'s will now be considered.

Finitely intermittent r.f.'s will be those corresponding to $\int_0^\infty P(s)ds < \infty$. They are ordinary random functions that "flip" between a "quiescent" state where X is constant and an "active" state where X may vary.

Infinitely intermittent r.f.'s will be those corresponding to $\int_0^\infty P(s)ds = \infty$. They *are identical to the sporadically varying generalized random functions*. Note that for these functions, it is not excluded that $P(\infty) > 0$.

5. EXAMPLES OF SPORADICALLY VARYING G.R.F.'S AND SPORADIC SETS CONSTRUCTED THROUGH GENERALIZED RENEWAL PROCESSES

The simplest sporadic g.r.f.'s are those whose structure is wholly determined by the function $\mu[B(\tilde{u})]$ implied in the definition of "sporadic." If $Q = 0$, $P(0) < \infty$, and $P(\infty) = 0$, then the function $P(u)$ fully determines a classical renewal process, as we shall now demonstrate.

5.1. Synchronized classical renewal set

This is a sequence $S^0 = \{\tilde{T}_k\}$, with k integer $-\infty < k < \infty$, such that $\tilde{T} = 0 \in S^0$, and such that the intermissions $U_k = \tilde{T}_{k+1} - \tilde{T}_k$ are independent and identically distributed nonnegative random variables with $\Pr\{U \geq u\} = P(u)/P(0)$. The Ω space of the S^0 is that of all sequences containing the origin and such that $S^0 \cap (t', t'')$ is finite or denumerable. By well-known rules (Cox 1962), a measure is attached to the events of this sample space, and $S^0 \cap (t', t'')$ is almost surely finite (it is surely finite if $P(\varepsilon) = P(0)$ for some $\varepsilon > 0$).

5.2. Stationary classical renewals and generalization to $E(U) = \infty$

Starting from the set S^0 of Section 5.1, we shall construct a generalized random set S as follows. The first step is to replace $\tilde{T} = 0$ by a $\tilde{T}$ distributed over $\mathbb{R}^+(t \geq 0)$, with the measure of *density* $P(u_0)$; this measure may be unbounded. The second step is to translate the set $\{\tilde{T}_k, k > 0\}$ to the right by the amount $\tilde{T}$; thus, the measure of T_k is the convolution of the measures of $\tilde{T}$ and $\tilde{T}_k$. The third step is to choose T_{-1} on $\mathbb{R}^-(t < 0)$, taking $\Pr\{T_{-1} \leq -u \mid \tilde{T} = t_0\} = P(u + t_0)/P(t_0)$ as the conditional probability measure; it is easy to see that the unconditioned distribution of T_{-1} has a measure of *density* $P(-u)$. The final step is to translate the set $\{\tilde{T}_k, k < -1\}$ to the left by the amount $T_{-1} - \tilde{T}_{-1}$; thus, the measure of T_k is the convolution of the measures of $\tilde{T}_k$ and $T_{-1} - \tilde{T}_{-1}$.

Proposition 5.1. *The measure of the generalized random set S is shift invariant and indecomposable.*

For example, $\mu[\omega:S(\omega) \cap (t', t'') \neq \emptyset]$ is shift invariant. This result is well-known if $E(U) < \infty$ (Doob 1948); then, S is an ordinary random set (intermittent) and one can normalize $P(s)$ into the probability density $P(u)/[\int_0^\infty P(s)ds]$. Most of the classical proofs of stationarity can also be extended to the sporadic case $E(U) = $ *infinity*.

Proposition 5.2. *Let $E(U) = \infty$. Over the time span $(0, \tilde{u})$, the following processes are identical in law: (1) the stationary sporadic process S constructed above; (2) the nonstationary ordinary process obtained from S^0 by making $\tilde{T}$ random over $(0, \tilde{u})$ with the density $P(t_0)/[\int_0^{\tilde{u}} P(s)ds]$; (3) the stationary ordinary random process constructed, as in Section 5.2, using the truncated $P^*(u, \tilde{u})$ defined by $P^*(u, \tilde{u}) = P(u)$ if $u < \tilde{u}$, $P^*(u, \tilde{u}) = 0$ if $u > \tilde{u}$.*

This last proposition shows how to replace the generalized random set $S(\omega)$ by an indexed family of ordinary random sets (see end of Section 2).

5.3. Sporadically varying g.r.f.'s having S as set of variation

The function $V(t, \omega)$, defined by $V(t, \omega) = 1$ if $t \in S(\omega)$, and by $V(t, \omega) = 0$ if $t \notin S(\omega)$, is a conditionally stationary g.r.f. with the required property, and belonging to the second class of Section 4.2. The integral $K(t, \omega)$ of $V(t, \omega)$ is a nonstationary g.r.f. with the required property.

5.4. The infinitely divisible process technique and the function &N1.

It is convenient to consider the synchronized renewal set S^0 as the set of values of an auxiliary r.f. $\tilde{T}(n)$ of a variable n, not necessarily an integer

(Smith 1960). One will have $\tilde{T}(n_0) = t_0$; $\tilde{T}$ will be left-continuous, it will have independent infinitely divisible increments. That is, given n' and n'', $-\infty < n' < n'' < \infty$, $\tilde{T}(n') - \tilde{T}(n'')$ will be an infinitely divisible positive r.v. whose Lévy's jump function is $1 - P(u)$; its characteristic function (Feller 1966, Gnedenko & Kolmogorov 1954, Lévy 1954, Loève 1963) is

$$(5.1) \qquad \varphi(\zeta) = \exp\left\{ -(n'' - n')\int_0^\infty (e^{i\zeta s} - 1)[-dP(s)] \right\}.$$

The number of jumps of $\tilde{T}$, located in the n-span (n', n'') and having a size in the range $(s, s+ds)$, is a Poisson random variable of mean $|(n'' - n')dP(s)|$. The intervals between successive values of n on which $\tilde{T}$ varies, are exponential random variables of unit expectation.

The function $N^0(t)$ will conversely be defined as being the largest n such that $\tilde{T}(n) \leq t$.

5.5. The fundamental sporadically varying function $\tilde{N}(t)$

Suppose now that $\tilde{T} \in R^+$ has the density $P(t_0)$ (see Section 5.2 concerning T_{-1}). Thus, by randomly shifting the origin of $N^0(t)$, we construct a function $N(t, \omega)$. If $E(U) = \infty$, this g.r.f. (dependent on n_0) constitutes a basic example of a sporadically varying g.r.f.. Its set of variation is $S(\omega)$, independent of n_0.

5.6. Several generalizations of renewal processes

The possibilities excluded so far, $Q > 0$, $P(0) = \infty$, and $P(\infty) > 0$, are readily introduced by generalizing the renewal processes.

The inequality $P(\infty) > 0$ implies that the process is *not* recurrent with probability one. In the infinitely divisible technique, $T(n)$ is not defined for all values of n, but only over a span whose duration is an exponential random variable of mean $1/P(\infty)$. S is then *almost surely a finite set*.

Suppose, in particular, that $P(u) \equiv P(\infty)$ so that $\mu[B(u)] = uP(\infty)$. In that case, S is *almost surely reduced to a single point*, as in Section 3.2.

The most natural way of making $Q > 0$ is to consider the closure of the set of values of the function $T_Q(n) = T(n) + Qn$. Renewals are no longer instantaneous, but their duration is an exponential random variable of expectation Q. The set S^0 is a union of closed intervals, whose total number is almost surely denumerable if $P(\infty) = 0$, and infinite and finite if

$P(\infty) > 0$. If $S(\omega) \cap (t', t'') \neq \emptyset$, this set has a positive Lebesgue measure, except for a set of ω that has vanishing μ measure. Another way of making $Q > 0$ is to have an "act" of duration $Q[P(s') - P(s'')]^{-1}$ follow every intermission of duration $u \in (s', s'')$ [where $P(s') > P(s'')$]. Then, the ω sets, over which $S(\omega) \cap (t', t'')$ has zero or positive Lebesgue measure, are both of positive μ measure.

The infinitely divisible technique is well known to generalize with no difficulty to the case $P(0) = \infty$ with $\int_0^\varepsilon P(s) ds < \infty$. The expected number of jumps of $T(n)$ satisfying the condition that $n' < n < n''$ is almost surely infinite. Moreover, if $S \cap (t', t'') \neq \emptyset$, the number of points of $S \cap (t', t'')$ is almost surely infinite, the Lebesgue measure of $S \cap (t', t'')$ is almost surely zero; the function $N(t)$ is almost surely a singular continuous function (it is the counterpart of the classical Lebesgue function of Cantor's triadic set).

To combine $P(0) = \infty$ with $Q > 0$, consider again the closure of the values of the r.f. $T_Q(n) = T(n) + Qn$. It is then almost sure that S has a non-vanishing Lebesgue measure but contains no closed interval.

The most interesting examples of sporadic sets (those introduced through limit theorems) will be characterized by $Q = 0$ and $P(\infty) = 0$, but $P(0) = \infty$ with $\int_0^\varepsilon P(s) ds < \infty$. They combine the low frequency problems associated with sporadic processes, with the high frequency problems of which an example appears in Kahane and M 1965.

5.7. A generalization that is not a renewal process

Consider finally a process $\tilde{T}(n, \omega)$, whose increments are stationary but are not independent, and whose jumps have a marginal distribution $P(u)$ such that $\int_0^\varepsilon P(s) ds < \infty$ and $\int_0^\infty P(s) ds = \infty$. Let $\tilde{T}$ have the measure of density $P(s)$ (see Section 5.2 concerning T_{-1}). The closure of the set of values of $T(n, \omega)$ is then a sporadically distributed generalized random sequence, but it is not a renewal sequence.

6. THE CONCEPTS OF CONDITIONED DELTAVARIANCE AND COVARIANCE, WEAK STATIONARITY AND COIN FUNCTION

If the stationary ordinary random function $X(t, \omega)$ is such that $E\{[X(t)]^2\} < \infty$, its *covariance* is defined as

$$C(\tau) = E[X(t)X(t + \tau)]$$

N10 ◇ ◇ SPORADIC FUNCTIONS AND CONDITIONAL SPECTRA

It will be convenient to deduce $C(\tau)$ from the function

(6.1) $$D(\tau) = C(0) - C(\tau) = (1/2)E\{[X(t) - X(t+\tau)]^2\}.$$

This $\Delta(\tau)$ is meaningful for all X for which $C(\tau)$ is defined, and also for many others such as the Brownian motion of Bachelier-Wiener-Lévy. Lacking a generally agreed upon term for this important concept, we propose to call it *delta-variance*.

If X is a sporadically varying g.r.f., $\Delta \equiv 0$.

6.1. The conditioned deltavariance and deltavariance stationarity

Given a conditioning event $B \in \mathcal{B}$, consider the function

(6.2) $$D[\tau, B, t] = \frac{1}{2}\int_B [X(t, \omega) - X(t+\tau, \omega)]^2 \mu(d\omega)/\mu[B].$$

Whenever this expression is finite, it may be used to define the *conditional deltavariance* of X, given B.

Definition 6.1. $(\Omega, \mathcal{A}, \mathcal{B}, \mu)$ *is said to be* conditionally deltavariance stationary, *if, for every $B \in \mathcal{B}$, there exists a time span $(t', t'')_B$ of positive duration such that $\Delta(\tau, B, t)$ is independent of t as long as $(t, t+\tau)$ Hence, one can write $\Delta(\tau, B, t)$ as $\Delta(\tau, B)$.*

Proposition 6.1. *Let $\mathcal{B} = \mathcal{B}^*$ and let μ be shift invariant. Then, $\Delta[\tau, B(t', t''), t]$ is conditionally deltavariance stationary for some $(t', t'')_B \supseteq (t', t'')$.*

Proof. The ω such that $X(t, \omega) - X(t+\tau, \omega) = 0$ contribute nothing to Δ. If A_R is a Borel set of R such that $0 \notin A_R$, the ω set

$$\{\omega : [X(t, \omega) - X(t+\tau, \omega)] \in A_R\}$$

belongs to $B(t', t'')$ and its μ measure is independent of t. If $\tau < \tilde{u}$, the numerator of $\Delta[\tau, B(\tilde{u}), t]$ is an integral carried out over $B(\tilde{u})$. However, if $\omega \in B(\tilde{u}) - B(\tau)$ then ω contributes nothing to the integral. That is, one can write

(6.3) $$D[\tau, B(\tilde{u})] = \frac{\Delta^*(\tau)}{\mu[B(\tilde{u})]},$$

where

(6.4) $$D^*(\tau) = \frac{1}{2} \int_{B(\tau)} [X(0, \omega) - X(\tau, \omega)]^2 \mu(d\omega).$$

This $\Delta^*(\tau)$ will be referred to as the *unweighted deltavariance* of X; it is defined only up to multiplication by an arbitrary finite positive number (as is also the case for $P(s)$ and μ). The function $\Delta^*(\tau)$ is nonnegative definite.

If X is an ordinary process of covariance $C(\tau)$, then μ can be chosen so that $\mu(\Omega) = 1$, and Δ boils down to

(6.5) $$D[\tau, B(\tilde{u})] = \frac{\Delta(\tau)}{\mu[B(\tilde{u})]} = \frac{[C(0) - C(\tau)]}{\mu[B(\tilde{u})]}.$$

Fixing τ and varying $\tilde{u}$, we see that $\Delta[\tau, B(\tilde{u})]$ is a decreasing function of $\tilde{u}$, defined for $\tilde{u} > \tau$.

Varying τ, we see that one can eliminate $\tilde{u}$ by forming the *relative deltavariance*, defined as equal to

(6.6) $$\frac{\Delta[\tau', B(\tilde{u})]}{\Delta[\tau'', B(\tilde{u})]} = \frac{\Delta^*(\tau')}{\Delta^*(\tau'')}, \quad 0 < \tau', \tau'' < \tilde{u}.$$

(By considering this ratio, one circumvents the difficulties mentioned in the introduction that were encountered in the empirical estimation of the covariance. See Section 8, also.)

6.2. The conditional covariance and covariance stationarity

Again, given $B \in \mathcal{B}$, consider the function

(6.7) $$C(\tau, B, t) = \frac{\int_B X(t, \omega) X(t + \tau, \omega) \mu(d\omega)}{\mu(B)}.$$

Whenever this expression is finite, it will define the *conditioned covariance* of X, given B.

Definition 6.2. *X is said to be conditionally covariance stationary, if, for every $B \in \mathcal{B}$, there exists a time span $(t', t'')_B$ of positive duration, such that $C(\tau, B, t)$ is independent of t as long as $(t, t + \tau)$ Then, $C(\tau, B, t)$ is denoted by $C(\tau, B)$.*

Proposition 6.2. Let $\mathcal{B} = \mathcal{B}^*$, let μ be shift invariant, and, for every $B \in \mathcal{B}^*$ and $t \in (t', t'')$, let $C(0, B, t) = C(0, B) < \infty$. Then $C[\tau, B(t', t''), t]$ is conditionally covariance stationary for some interval $(t', t'')_B \supseteq (t', t'')$, and one has $|C(\tau, B)| \leq C(0, B)$.

Let us examine two examples that slightly generalize those singled out in Section 4.2.

In the first special class: $\mathcal{B} = \mathcal{B}^*$, μ is shift invariant, and X is such that, if $t \in (t', t'')$ and $0 \notin A_R$, one has $\{\omega:X(t, \omega) \in A_R\}$ This is the case, for example, if X is a g.r.f. that vanishes during its intermissions. Then the numerator of C becomes independent of B, as long as $(t, t + \tau)$ The existence of C is thus reduced to the usual conditions that X must be square integrable, and C can be factored out in the form

$$(6.8) \qquad C[\tau, B(\widetilde{u})] = \frac{C^*(\tau)}{\mu[B(\widetilde{u})]},$$

which serves to define the *unweighted covariance* $C^*(\tau)$.

In the second special class: $\mathcal{B} = \mathcal{B}^*$, μ is shift invariant, and X has a pseudo-marginal distribution (independent of B). Here, $C[0, B(\widetilde{u})]$ is independent of $\widetilde{u}$; it is finite if and only if the pseudomarginal r.v. of $X(t, \omega)$ is square integrable. In that case, one will designate $C[0, B(\widetilde{u})]$ by the same notation $C(0)$ as an ordinary variance, and $C[\tau, B(\widetilde{u})]$ takes the form

$$(6.9) \qquad C[\tau, B(\widetilde{u})] = C(0) - \frac{\Delta^*(\tau)}{\mu[B(\widetilde{u})]}.$$

6.3. The concept of the coin function of a set of zero Lebesgue measure

The r.f. with independent values is the simplest of all r.f.'s, and the only one to be defined fully by its marginal distribution. However, it is not measurable if the allowed values of t are all the points of R.

No such problem arises if time is restricted to a discrete set S of R such as the integers k, so that X is a sequence of independent r.v.. One can then extend X to a left-continuous r.f. of continuous time, $W(t, \omega)$, constant over the intervals of the form $k < t \leq k + 1$; if the marginal distribution of W is continuous, its set of variation is almost surely S.

Now consider a general set S that has a vanishing Lebesgue measure and is of the form $S = R - \bigcup_{h=1}^{\infty}(t_h', t_h'')$, where the (t_h', t_h'') are nonoverlapping. Consider also a r.v. W_M (where M stands for marginal), continuously distributed, of mean zero and of unit variance, and let W_h be a discrete sequence of independent r.v. having the distribution of W_M.

Definition 6.3. *The independent coin function of S, with margin W_M, is defined as follows. If $t_h' < t \leq t_h''$, then $W(t, \omega) = W(t_h'', \omega) = W_h$. If $\{t_k\}$ are K points ($K < \infty$) of S, the $W(t_k, \omega)$ are independent r.v. with the distribution of W_M.*

6.4. Covariance properties of a coin function

If $\omega \notin B(t, t + \tau)$, then $E\{[W(t) - W(t + \tau)]^2\}/2 = 0$. If $\omega \in B(t, t + \tau)$, then $E\{[W(t) - W(t + \tau)]^2\}/2$ is one half of the variance of the difference $W_{h'} - W_{h''}$, where $W_{h'}$ and $W_{h''}$ are independent r.v.'s of variance one. Therefore,

(6.10) $\qquad D[\tau, B(t', t''), t] = \Pr\{\omega \in B(t, t + \tau) \mid \omega \in B(t', t'')\}.$

If, moreover, μ is shift invariant, $(t', t'') = (\tilde{t}, \tilde{t} + \tilde{u})$ and $(t, t + \tau)$

(6.11) $\qquad D[\tau, B(\tilde{u})] = \Pr\{\omega \in B(t, t + \tau) \mid \omega \in B(\tilde{t}, \tilde{t} + \tilde{u})\}$

$$= \frac{\mu[B(\tau)]}{\mu[B(\tilde{u})]} = \frac{Q + \int_0^\tau P(s)ds}{Q + \int_0^{\tilde{u}} P(s)ds}.$$

The relative deltavariance and the unweighted deltavariance, respectively, take the forms

(6.12) $\qquad \dfrac{\mu[B(\tau')]}{\mu[B(\tau'')]} = \dfrac{Q + \int_0^{\tau'} P(s)ds}{Q + \int_0^{\tau''} P(s)ds}$ and $\mu[B(\tau)] = Q + \int_0^\tau P(s)ds.$

Remark. The conditioned deltavariance properties derived above apply irrespective of whether $\mu(\Omega) = \infty$ or $\mu(\Omega) < \infty$. But they are not very useful in the classical case $\mu(\Omega) < \infty$ where W is not sporadic.

6.5. Generalization; the orthogonal coin function

From the viewpoint of the present study of second order properties, the independence of the r.v. W_h is an unnecessarily strong assumption. If W_M is non-Gaussian, the same covariance is obtained if the W_h are only orthogonal. It is not obvious how the function W should be defined for $t \in S - \bigcup_{h=1}^{\infty} t_h''$, but this set has vanishing Lebesgue measure, so that the corresponding values of W do not matter.

7. CONDITIONAL AND ASYMPTOTIC SELF-SIMILARITY

7.1. Self-similarity in time in the sense of conditional deltavariance

Definition 7.1. The process $X(t, \omega)$ is said to be self-similar in time, in the sense of conditional deltavariance, if one has $\Delta[h\tilde{u}, B(\tilde{u})] = = \Delta_L(h)$, where $\Delta_L(h)$, with $0 \leq h \leq 1$, is finite and is not identically zero.

The following theorem expresses the intimate relation between self-similarity and coin functions.

Theorem 7.1. The definition of self-similarity requires that $X(t, \omega) K^{-1/2}$, where $K > 0$, be an orthogonal coin g.r.f. such that $\mu[B(u)] = u^{1-D}$, with $0 \leq D \leq 1$.

Proof. The proof proceeds in several steps.

(a) Since

$$(7.1) \qquad D_L(1) = K = \frac{\Delta^*(\tilde{u})}{\mu[B(\tilde{u})]} = \frac{\Delta^*(\tau)}{\mu[B(\tau)]},$$

$\Delta[\tau, B(\tilde{u})]$ can be written as $K\mu[B(\tau)]/\mu[B(\tilde{u})]$, where the function $\mu[B(\mu)]$ remains to be specified.

(b) The statement that $\Delta^*(\tau) = K\mu[B(\tau)]$ can be rewritten

$$(7.2) \qquad E\{[X(t)K^{-1/2} - X(t+\tau)K^{-1/2}]^2 \mid \omega \in B(t, t+\tau)\}/2 = \mu[B(\tau)],$$

which means that, when $\omega \in B(t, t+\tau)$, the r.v. $X(t)K^{-1/2}$ and $X(t+\tau)K^{-1/2}$ indeed have unit variance and are orthogonal.

(c) Choose any couple (h', h'') such that $0 \leq h' \leq 1$ and $0 \leq h'' \leq 1$. Self-similarity requires

(7.3)
$$\frac{\mu[B(h'h''\tilde{u})]}{\mu[B(h''\tilde{u})]} = \frac{\mu[B(h'\tilde{u})]}{\mu[B(\tilde{u})]}$$

$$\frac{\mu[B(h'h''\tilde{u})]}{\mu[B(\tilde{u})]} = \frac{\mu[B(h''\tilde{u})]}{\mu[B(\tilde{u})]} \cdot \frac{\mu[B(h'\tilde{u})]}{\mu[B(\tilde{u})]};$$

finally,

(7.4) $$\Delta_L(1)\Delta_L(h'h'') = D_L(h')\Delta_L(h'').$$

The solution of this equation is of the form $\Delta_L(h) = Kh^{1-D}$, where D is a constant. Since $\mu[B(u)]$ is only defined up to a positive multiplier, one can choose

(7.5) $$\mu[B(u)] = u^{1-D}.$$

The requirement that $0 \leq D \leq 1$ follows from the convexity of $\mu[B(u)]$ combined with $\int_0^k P(s)ds < \infty$.

The degenerate case $D = 0$. If $D = 0$, $\mu[B(u)] = u$, the set S is almost surely (a.s.) a single point chosen at random on R, as in Section 3.3.

The classical case $D = 1$, $S \equiv R$. If $D = 1$, then $\mu[B(u)]$ is a constant and one obtains the troublesome process of orthogonal values on R.

The sporadic cases $0 < D < 1$. In the interesting case $0 < D < 1$, the Lebesgue measure of $S \cap (t', t'')$ almost surely vanishes. Moreover, $\int_1^\infty P(s)ds = \infty$, and therefore, the g.r.f. $X(t, \omega)$ is sporadic. Since $P(u)$ is defined up to a positive multiplier, one can assume $P(u) = u^{-D}$.

Corollary 7.1. *If X is a g.r.f. that is conditionally deltavariance self-similar in time, it cannot be finitely intermittent.*

Conditional Taylor's scale. G.I. Taylor (see Friedlander & Topper 1961) has proposed the integral $\int_0^\infty C(s)ds$ to define the temporal scale of a random phenomenon whose covariance is $C(s)$. If $X(t, \omega)$ is sporadic, the best that one can do is replace $C(s)$ by a conditional covariance. Moreover, if $X(t, \omega)$ is self-similar in time, this "conditional Taylor's scale" turns out to be $\tilde{u}(1 - D)/(2 - D)$.

7.2. Uniformly self-similar renewal sets

In the present Section, the assumption that $\mu[B(u)] = u^{1-D}$, is combined with the assumption that S is a generalized renewal set, as defined in Section 5.6. In the renewal case, $S(\omega)$ is fully determined by the requirements of nondegenerate deltavariance self-similarity, namely, $Q = 0$, $P(\infty) = 0$, and $P(u) = u^{-D}$ for $0 < u < \infty$, $0 < D < 1$. Insert $P = u^{-D}$ in Lévy's formula of Section 5.3. We obtain S^0 as closure of the set of values of the function $\tilde{T}(n)$, which is defined by the following conditions: $\tilde{T}(0) = t_0$ and $\tilde{T}(n'') - \tilde{T}(n')$ has the characteristic function

$$\begin{aligned}\varphi(\zeta) &= \exp[-(n''-n')\int_0^\infty Ds^{-(D+1)}(e^{i\zeta s}-1)ds]\\ &= \exp\{-(n''-n')|\zeta|^D \Gamma(1-D)\cos(\frac{D\pi}{2})[1-i|\zeta|\zeta^{-1}\tan(\frac{D\pi}{2})]\}.\end{aligned}$$
(7.6)

Such $\tilde{T}$ are called "stable" and $\tilde{T}(n)$ is a Lévy stable process of independent stable increments (Doob 1953, Gnedenko & Kolmogorov 1954, Lévy 1954). The process $\tilde{T}(n)$ varies only by jumps; the positions of the jumps are mutually independent; the number of jumps, whose size is between u and $u + du$ and whose position is between n and $n + dn$, is a Poisson r.v. of expectation equal to $Du^{-(D+1)}dndu$; the degree of "thinness" of S can be described by the fact that its Hausdorff dimension is almost surely D (Blumenthal & Getoor 1960m, p. 267). Finally, S is almost surely a "set of multiplicity" (Kahane & M 1965{N11}).

Proposition 7.1. *After $\tilde{T}$ has been made random, with the unbounded measure of density $(t_0)^{-D}$, one obtains a set S that is uniformly self-similar under change of scale, in the following sense. Given that $S \cap (t', t'') \neq \emptyset$, let $H(t', t'')$ be the set of values of $h(0 \leq h \leq 1)$ such that $t' + h(t'' - t') \in S \cap (t', t'')$. If $H(t', t'')$ is independent in distribution from t' and t'', then $S(\omega)$ is called uniformly self-similar.*

7.3. Asymptotic self-similarity in time

Definition 7.2. *Let the stationary g.r.f. $X(t, \omega)$ be such that, for $0 \leq h \leq 1$,*

$$\lim_{\tilde{u}\to\infty} \Delta[h\tilde{u}, B(\tilde{u})] = \Delta_L(h)$$

is defined, finite and not identically zero. Then, $X(t, \omega)$ will be said to be asymptotically self-similar in time, in the sense of conditional deltavariance.

Theorem 7.2. In order that $\lim_{\tilde{u} \to \infty} D[h\tilde{u}, B(\tilde{u})] = D_L(h)$ be defined, finite and not identically zero, it is necessary and sufficient that μ and Δ^* satisfy

(a) $$\lim_{u \to \infty} \frac{\mu[B(hu)]}{\mu[B(u)]} = h^{1-D}, \text{ with } 0 \leq h \leq 1 \text{ and } 0 \leq D \leq 1.$$

This property means that μ "varies regularly" at infinity, in the sense of Karamata, and expresses a kind of asymptotic self-similarity of S;

(b) $$\lim_{u \to \infty} \frac{\Delta^*(u)}{\mu[B(u)]} = K > 0,$$

which expresses that the correlation between $X(t', \omega)$ and $X(t'', \omega)$, conditioned by $\omega \in B(t', t'')$, tends to zero with $(t'' - t')^{-1}$.

Proof. Letting $h = 1$, we see that it is necessary that

(7.7) $$\lim_{u \to \infty} \frac{\Delta^*(u)}{\mu[B(u)]} \text{ exists.}$$

This limit will define $K = \Delta_L(1)$.

Now let $0 < h < 1$, and write

(7.8) $$\frac{\Delta^*(h\tilde{u})}{u[B(\tilde{u})]} = \frac{\Delta^*(h\tilde{u})}{\mu[B(h\tilde{u})]} \cdot \frac{\mu[B(h\tilde{u})]}{\mu[B(\tilde{u})]}.$$

Thus, it is necessary that

(7.9) $$\lim_{u \to \infty} \frac{\mu[B(h\tilde{u})]}{\mu[B(\tilde{u})]} \text{ exists.}$$

Well-known arguments (Feller 1966, 1967) then show that there must exist a constant D such that

(7.10) $$\lim_{u \to \infty} \frac{\mu[B(h\tilde{u})]}{\mu[B(\tilde{u})]} = h^{1-D} = (\tau/\tilde{u})^{1-D}.$$

Finally, the condition $0 \leq D \leq 1$ follow from the convexity of $\mu[B(u)]$.

Sufficiency is obvious.

Proposition 7.2. *In order that $\int_0^u P(s)\,ds$ be monotone nondecreasing, convex and of regular variation with exponent $0 \le 1 - D \le 1$, it is necessary and sufficient that $P(u)$ be monotone nonincreasing and of regular variation with exponent D.*

Proof: For the proof of sufficiency, see Feller 1967; to prove necessity, one can adapt the proof of theorem 2 of Lamperti 1958t.

If the limit falls within the degenerate case $D = 1$, then $\Delta_L(h)$ is the same as for the process of orthogonal values on R. The intermissions of X are "few" and/or "short" and effectively made negligible by rescaling.

Finally, the interesting limits $0 < D < 1$ imply that $P(\infty) = 0$, and that $\mu[B(u)]$ is not identically constant. The limit function Δ_L is then unaffected by Q, that is, by the Lebesgue measure of R sets of the form $S \cap (t', t'')$.

7.4. Asymptotically self-similar renewal sets

They are characterized by the condition that $\lim_{u \to \infty} P(hu)/P(u) = h^{-D}$. This property is equivalent to saying that the r.v. $\tilde{T}(n'') - \tilde{T}(n')$ belongs to the domain of attraction" of the stable r.v. of exponent D.

8. LIMIT THEOREMS RELATIVE TO THE SAMPLE DELTAVARIANCE IN THE RENEWAL CASE: UNIFORM SELF-SIMILARITY

8.1. The "ergodic" problem of the relation between population and sample means of $X(t, \omega)$; the conditional ergodic problem

Let $(\Omega, \mathcal{A}, \mathcal{B}, \mu)$ be such that the shift transformation φ_T is indecomposable. Birkhoff's individual ergodic theorem applies when X is integrable. We know that such is *not* the case for interesting g.r.f.'s, for example, when X has a pseudomarginal distribution. Even when Birkhoff's theorem applies, it only states that the sample mean tends to zero (or that it is identically zero). Under such circumstances, one becomes interested in the rate at which this limit is achieved, a question that Birkhoff's theorem does not attempt to answer.

It is best, therefore, to start anew, and to attack the problem of the behavior of *conditioned* sample moments. We shall see that as $\tilde{u} \to \infty$, ratios such as

$$\text{(8.1)} \quad \frac{\int_0^{\tilde{u}-\tau} ds\{[X(s,\omega) - X(s+\tau,\omega)]^2 \,|\, \omega \in B(\tilde{u})\}}{(\tilde{u}-\tau) E\{[X(t,\omega) - X(t+\tau,\omega)]^2 \,|\, \omega \in B(\tilde{u})\}}$$

may tend in distribution to limit r.v. other than unity. This property is weaker than ergodicity.

8.2. Commentary upon the relations between successive classes of limit theorem concerning $\sum X(t)$ and $\sum X(t)X(t+\tau)$

As is the case with so many generalizations first suggested by hard facts, the theorems to be described might easily have been first introduced solely to fill in a gap between existing mathematical theories.

The Wiener-Khinchin theory and the "laws of large numbers." The "second order stationary random processes" of Wiener and of Khinchin can be defined through the requirement that the limit in distribution

$$\text{(8.2)} \quad \lim_{\tilde{u}\to\infty} \frac{1}{\tilde{u}} \sum_{t=\tilde{t}}^{\tilde{t}+\tilde{u}} X(t)X(t+\tau)$$

be equal to $E[X(t)X(t+\tau)]$ and be a nonrandom function $C(\tau)$ of τ. This theory is thus parallel in scope to the laws of large numbers that are relative to the case where the first order expression $\tilde{u}^{-1}\sum_{t=\tilde{t}}^{\tilde{t}+\tilde{u}} X(t)$ tends to the nonrandom limit $E(X)$ as $\tilde{u} \to \infty$.

Central limit theorems for normed sums and their counterpart in spectral theory. The next stage of the theory of $\sum X(t)$ was constituted by the central limit theorems relative to the possible behavior of the normed sums of the form

$$\alpha(\tilde{u}) \sum_{t=\tilde{t}}^{\tilde{t}+\tilde{u}} X(t) - \beta(\tilde{u}).$$

Bochner 1959 (p. 295) had introduced normed sums of the form $\alpha(\tilde{u})\int X(t)X(t+\tau)dt$ in his generalization of Wiener's generalized harmonic analysis of nonrandom functions $X(t)$. Stochastic analogs of these theorems do exist, but we do not have room to study them here.

Limit theorem of a form analogous to that of Poisson. For sums of r.v.'s, a third stage of the theory studies expressions of the general form

$$\alpha(\tilde{u}) \sum_0^{K(\tilde{u})} X(t, \tilde{u}) - \beta(\tilde{u}),$$

where the distribution of the $X(t, \tilde{u})$ is allowed to change as $\tilde{u}$ increases.

The theorems of the rest of Section 8 and those of Section 10 can be cast in this "Poisson" mold, with the restriction that $K(\tilde{u})$ is usually $\tilde{u}$ itself, and with the following changes. First, $X(t, \tilde{u})$ is usually replaced by $X(t, \tilde{u})X(t + \tau, \tilde{u})$ or by $[X(t, \tilde{u}) - X(t + \tau, \tilde{u})]^2$. Second, the dependence of these expressions on $\tilde{u}$ is assumed to be of a very explicit kind: it is induced by the time lag $\tilde{u}$ that is characteristic of the conditioning event $B(\tilde{u})$. Third, the summands are dependent.

8.3. The behavior of the function $\tilde{N}(t)$, when $T(n)$ is a process of independent stable increments: The Mittag-Leffler distribution

It is clear that

(8.3) $$\Pr\{N(t) \geq n \mid T(0) = 0\} = \Pr\{T(n) \leq t \mid T(0) = 0\}$$
$$= \Lambda_D(tn^{-1/D}) = \Lambda_D[(nt^{-D})^{-1/D}],$$

where $\Lambda_D(y) = \Pr\{T(n+1) - T(n) \leq y\}$ is the distribution function of a Lévy stable r.v. Feller 1949 was the first to introduce the r.v. whose distribution function is $\Lambda_D(u^{-1/D})$, and he called it a Mittag-Leffler r.v..

Let t_0 be the first point of variation of $N(t)$ such that $t_0 > 0$. If $T(0) = t_0$ does not vanish but is known to be less than $\tilde{u}$, then $[N(\tilde{u}) - N(0)]/(\tilde{u} - t_0)_D$ is a Mittag-Leffler r.v.. Now let t_0 be made random with the fundamental density $(1-D)t_0^{-D}/\tilde{u}^{1-D}$. It follows that the quantity

(8.4) $$\frac{[N(\tilde{u}) - N(0) \mid B(\tilde{u})]}{E[N(\tilde{u}) - N(0) \mid B(\tilde{u})]}$$

is a weighted mixture of Mittag-Leffler r.v. is of exponent D. It will be designated by M_D and called a "modified Mittag-Leffler" r.v.. Since Feller 1949 showed that

(8.5) $$E[N(\tilde{u}) - N(0) \mid T(0) = 0] = \sin(\pi D)(\pi D)^{-1}(\tilde{u})^D,$$

we have

(8.6) $$E\{[N(\tilde{u}) - N(0)] \mid B(\tilde{u})\} = (1 - D)\tilde{u}^D.$$

8.4. Sample deltavariance of W when S is a uniformly self-similar renewal set

Consider the probability

(8.7) $$\Pr\left\{Y_\Delta = \frac{\int_0^{\tilde{u}-\tau} ds\{[W(s) - W(s+\tau)]^2 \mid B(\tilde{u})\}}{(\tilde{u}-\tau)E\{[W(t) - W(t+\tau)]^2 \mid B(\tilde{u})\}} \leq y\right\}.$$

If follows from self-similarity that this expression will be unchanged if $\tilde{u}$ is replaced by $q\tilde{u}$ ($q > 0$, constant) while $B(\tilde{u})$ is replaced by $B(q\tilde{u})$ and τ by $q\tau$. Thus, to each h, $0 \leq h \leq 1$, there corresponds a function $F_\Delta(y, h)$ such that $\Pr\{Y_\Delta \leq y\} = F_\Delta(y, \tau/\tilde{u} = h)$. The form of F_Δ is readily derived when $h = 1$ and when h is very small.

If $h = 1$, then Y_Δ is half of the squared difference between the independent r.v.'s $W(0, \omega)$ and $W(\tilde{u}, \omega)$.

Now let $h < 1$. Since $W(t) - W(t+\tau)$ vanishes if $(t, t+\tau)$ is part of an intermission of S, it suffices to carry the integral

(8.8) $$\frac{1}{2}\int ds\{[W(s) - W(s+\tau)]^2 \mid B(\tilde{u})\}$$

over the set $\Phi^0(\tilde{u}, \tau) = (0, \tilde{u} - \tau) - \bigcup_\tau [(t_h', t_h'' - \tau) \cap (0, \tilde{u} - \tau)]$, where the union $\bigcup_\tau$ is carried out over the values of h such that $t_h'' - t_h' > \tau$.

We shall presently prove the following results for small h. For two different h', h'', the ratio between the sample values of $\Delta[h'\tilde{u}, B(\tilde{u})]$ and $\Delta[h''\tilde{u}, B(\tilde{u})]$ tends to a nonrandom limit as $h' \to 0$ and $h'' \to 0$, while h'/h'' remains constant. As $h \to 0$, the ratio between the population and the sample value of $\Delta[h\tilde{u}, B(\tilde{u})]$ tends to a r.v. M_D.

To avoid the awkward task of letting $h \to 0$, we shall, instead, let $\tilde{u} \to \infty$ with τ fixed. This approach is also advantageous because it increases the generality of the result. Instead of requiring that $P = u^{-D}$, it will be necessary and sufficient that the limit in distribution of the ratio

N10 ⋄ ⋄ SPORADIC FUNCTIONS AND CONDITIONAL SPECTRA

$[N(\tilde{u}) - N(0)|B(\tilde{u})]/E[N(\tilde{u}) - N(0)|B(\tilde{u})]$ be a nondegenerate r.v. different from unity. This is equivalent to saying that $P(u)$ is regularly varying at infinity, with an exponent D such that $0 < D < 1$.

8.5. Limit of the weighted deltavariance of the coin function W of a renewal set, when $\tilde{u} \to \infty$ with fixed τ

Let ω be understood to be conditioned by $\omega \in B(\tilde{u})$, and let $|\Phi^0(\tilde{u}, \tau)|$ be the Lebesgue measure of the set Φ^0. Define P' and P'' by writing

(8.9)
$$P' = \frac{\int_{s \in \Phi^0(\tilde{u}, \tau)} [W(s) - W(s + \tau)]^2 ds}{2|\Phi^0(\tilde{u}, \tau)|}$$

$$P'' = \frac{2|\Phi^0(\tilde{u}, \tau)|}{(\tilde{u} - \tau)E[W(s) - W(s + \tau)]^2}.$$

Now we can write

$$\frac{\int_0^{\tilde{u} - \tau} [W(s) - W(s + \tau)]^2 ds}{(\tilde{u} - \tau)E[W(s) - W(s + \tau)]^2} = P'P''.$$

The denominator of P' almost surely tends to infinity as $\tilde{u} \to \infty$. Moreover, P' itself is nothing but the deltavariance of lag τ of the coin function of an auxiliary ordinary renewal set, based upon the truncated law P^* such that $P^*(u, \tau) = u^{-D}$ for $u \leq \tau$, $P^*(u, \tau) = 0$ for $u > \tau$. For this process, $X(t')$ and $X(t'')$ are independent when $|t' - t''| > \tau$. Therefore, P' is the sample mean of an expression which satisfies the strong law of large numbers. It almost surely tends to its population mean, which is one. Therefore, Y_Δ has the same limit in distribution as P''.

Adapting the argument that led Feller 1949 to the Mittag-Leffler distribution, one obtains the following result:

Theorem 8.1. *Let W be the coin function of a recurrent S, and let $\tilde{u} \to \infty$. In order that the ratio between the sample and expectation values of $\Delta[\tau, B(\tilde{u})]$ have a proper limit in distribution, it is necessary and sufficient that $P(u)u^D$ be slowly varying for some $D \in (0, 1)$. The limit is then a r.v. M_D.*

Corollary 8.1. $W(t)$ being conditioned by $B(\tilde{u})$, the limit in distribution

(8.10)
$$\frac{\lim_{\tilde{u}\to\infty} \int_0^{\tilde{u}} ds[W(t) - W(t+\tau')]^2}{\int_0^{\tilde{u}} ds[W(t) - W(t+\tau'')]^2}$$

is a nonrandom function of τ' and τ''.

Again, by considering this ratio, one circumvents the difficulties that were encountered in the empirical estimation of the covariance.

9. ALTERNATIVE SPECTRA OR FOURIER TRANSFORMS OF ALTERNATIVE DELTAVARIANCES, AND THE POPULATION FORMS OF THE INFRARED CATASTROPHE

For a sporadically varying $X(t, \omega)$, the concept of spectrum is only meaningful in a conditional sense. Moreover, it is ambiguous even when the condition B is fixed, as we shall now see by examples.

9.1. The expectation of the Schuster periodogram

Schuster's periodogram will be defined in Section 10. Its expectation is the most intrinsic form of spectrum and suggests the following expression for the energy to be found in frequencies above λ:

(9.1)
$$G_s(\lambda, \tilde{u}) = \frac{2}{\pi} \int_0^{\tilde{u}} (1 - s/\tilde{u}) C[s, B(\tilde{u})] \sin(2\pi\lambda s) s^{-1} ds.$$

If X were an ordinary r.f., one would have $\lim_{\tilde{u}\to\infty} C[s, B(\tilde{u})] = C(s)$, and $\lim_{\tilde{u}\to\infty} G_s(\lambda, \tilde{u})$ would be the usual spectrum. In the sporadic case, however, *both terms* of the integrand, $C[s, B(\tilde{u})]$ and $(1 - s/\tilde{u})$, depend on $\tilde{u}$, and neither of them tends to a nondegenerate limit as $\tilde{u} \to \infty$.

Suppose, in particular, that the deltavariance is asymptotically self-similar, and $C(0, B) = 1$, so that

(9.2)
$$\lim_{\tilde{u}\to\infty} C[h\tilde{u}, B(\tilde{u})] = 1 - h^{1-D} = 1 - (\tau/\tilde{u})^{1-D}.$$

Then,

(9.3) $\quad G_S(\lambda, \tilde{u}) \to \dfrac{2}{\pi} \displaystyle\int_0^1 (1-s)(1-s^{1-D}) \sin[2\pi(\lambda \tilde{u})] s^{-1} ds = G_S^*(\lambda \tilde{u}).$

This G_S^* is a bounded function such that $G_S^*(\lambda) \sim \lambda^{D-1}$ for $\lambda \to \infty$.

In particular, given that $0 < \lambda', \lambda'' < \infty$,

(9.4) $\quad \dfrac{|G_S(\lambda', \tilde{u}) - G_S(\lambda'', \tilde{u})|}{|\lambda' - \lambda''|}$

is the average spectral density in the frequency span (λ', λ''); in other words, it is the average of the spectral density defined by

(9.5) $\quad 2\displaystyle\int_0^{\tilde{u}} (1 - s/\tilde{u}) C[s, B(\tilde{u})] \cos(2\pi\lambda s) ds.$

This average tends to zero with $1/\tilde{u}$.

On the other hand, the spectral density at $\lambda = 0$ is given by

(9.6) $\quad 2\displaystyle\int_0^{\tilde{u}} (1 - s/\tilde{u}) C[s, B(\tilde{u})] ds = 2\tilde{u} \int_0^1 (1-s)(1-s^{1-D}) ds,$

which tends to infinity proportionately to $\tilde{u}$.

If one examines the spectrum as a whole, the energy will seem to flow to ever lower frequencies as $\tilde{u} \to \infty$.

9.2. The population infrared catastrophe

Given the deltavariance

$$D[\tau, B(\tilde{u})] = \Delta^*(\tau)/\mu[B(\tilde{u})],$$

define the function $G^*(\lambda)$ by

(9.7) $$G^*(\lambda) = \frac{2}{\pi} \int_0^\infty \sin(2\pi\lambda s) s^{-1} \Delta^*(s) ds.$$

When $X(t, \omega)$ is an ordinary r.f. normalized so that $\mu(\Omega) = \mu[B(\infty)] = 1$, Δ^* is a Wiener-Khinchin deltavariance and $|G^*(\lambda') - G^*(\lambda'')|$ is its energy in the spectral interval (λ', λ''). (We use the convention that the energy of a spectral line at λ_0 is split equally between the intervals (λ', λ_0) and (λ_0, λ''), where $(\lambda' < \lambda_0 < \lambda''$.) Moreover, $G^*(0) < \infty$.

For sporadically varying g.r.f.'s, on the contrary, $\mu[B(u)]$ is unbounded and $G^*(0) = \infty$. In the asymptotically self-similar case,

(9.8) $$G^*(\lambda) \sim \lambda^{1-D}, \quad \text{as } \lambda \to 0.$$

In interpreting empirical spectral measurements, one may be tempted to handle G^* as if it were a Wiener-Khinchin spectrum. But $G^*(0) = \infty$ would then be interpreted as meaning that there is an infinite energy in low frequencies, which is impossible physically and therefore "catastrophic" for the identification of G^* to a Wiener-Khinchin spectrum. To distinguish this difficulty from high frequency divergences, it is called an "infrared catastrophe." As introduced in the theory of sporadically varying g.r.f,'s G^* is not a spectrum and its divergence is not impossible physically and hence not catastrophic for the theory.

More reasonable definitions of the spectrum will be proposed presently. They will show that, in order for $|G^*(\lambda'') - G^*(\lambda')|/\mu[B(\tilde{u})]$ to be a rough estimate of the energy in the frequency band (λ', λ''), one must assume that $1/\tilde{u} \leq \lambda' < \lambda'' \leq \infty$. In particular, the energy in the band $(1/\tilde{u}, \infty)$ is roughly $G^*(1/\tilde{u})/\mu[B(\tilde{u})]$. If $G^*(0) = \infty$, then both numerator and denominator increase as $\mu^* \to \infty$, but their ratio may well tend to a finite limit. The energy will seem to flow into ever lower frequencies, but the total *expected* energy will remain fixed.

9.3. An example where it is possible to construct a stationary ordinary r.f. that coincides over $(0, \tilde{u})$ with the sporadic $X(t, \omega)$

Let us now suppose that the conditioned covariance $C[\tau, B(\tilde{u})]$ satisfies $C(\tilde{u}, B(\tilde{u})) = 0$, and consider the function $C_L(\tau, \tilde{u})$ such that $C_L(\tau, \tilde{u}) = C[\tau, B(\tilde{u})]$ if $|\tau| < \tilde{u}$, $C_L(\tau, \tilde{u}) = 0$ otherwise. This function is continuous and is easily seen to be positive definite. Therefore, there exists a stationary ordinary r.f. $X_L(t, \omega, \tilde{u})$, of which C_L is the Wiener-Khinchin covariance. In frequencies above λ, it has an energy equal to

(9.9)
$$G_L(\lambda, \tilde{u}) = \frac{2}{\pi} \int_0^{\tilde{u}} D[s, B(\tilde{u})] \sin(2\pi\lambda s) s^{-1} ds$$
$$+ \frac{2}{\pi} \Delta[\tilde{u}, B(\tilde{u})] \int_{\tilde{u}}^{\infty} \sin(2\pi\lambda s) s^{-1} ds.$$

This function $G_L(\lambda, \tilde{u})$ is bounded, as it should be and varies little for $\lambda < 1/\tilde{u}$. Its behavior for large λ, however, is mostly determined by the behavior of $\Delta[\tau, B(\tilde{u})]$ for small τ: one has $G_L(\lambda, \tilde{u}) \sim G^*(\lambda)/\mu[B(\tilde{u})]$. Therefore, an infrared catastrophe would be brought about if this approximation, valid only for $\lambda \to \infty$, were applied for $\lambda \to 0$.

9.4. A periodic function which coincides with the conditioned sporadic process $X(t, \omega)$ over $(0, \tilde{u})$

The function

(9.10) $$C_P(\tau, \tilde{u}) = (1 - \tau/\tilde{u})C[\tau, B(\tilde{u})] + (\tau/\tilde{u})C[\tilde{u} - \tau, B(\tilde{u})]$$

is the covariance of a periodic function $X_P(t, \omega, \tilde{u})$. As $\tilde{u}$ increases, so does the number of spectral lines whose frequency is between 0 and some fixed λ_0. Energy seems to flow to low frequencies.

10. LIMIT THEOREMS RELATIVE TO THE SAMPLE PERIODOGRAM AND OTHER SAMPLE ESTIMATORS IN THE RENEWAL CASE

10.1. The Schuster periodogram

The Schuster periodogram is defined as being the r.v.

(10.1) $$|Y(\lambda, \omega, \tilde{u})|^2 = (1/\tilde{u}) \left| \int_0^{\tilde{u}} X(s, \omega) e^{-2\pi i s \lambda} ds \right|^2.$$

It is the squared modulus of the Fourier transform

(10.2) $$Y(\lambda, \omega, \tilde{u}) = \tilde{u}^{-1/2} \int_0^{\tilde{u}} X(s, \omega) e^{-2\pi i s \lambda} ds.$$

Let $X(t, \omega)$ be an ordinary r.f. of zero mean, satisfying appropriate additional conditions. It is a standard result that, as $\tilde{u} \to \infty$, the ratio $|Y|^2/E(|Y|^2)$ tends in the limit to an exponential r.v., whose expectation is (naturally) unity. In the present Section, the distribution of $|Y|^2/E(|Y|^2)$ will be examined under the assumption that X is the coin function of a sporadic renewal set S.

If S is self-similar, the ratio between the sample and expected values of the periodogram has $\lambda \tilde{u}$ as its only parameter. However, it will only be assumed that X is the coin function of an *asymptotically* self-similar renewal sequence. Very low and very high values of the parameter $\lambda \tilde{u}$ will lead to very different limit distributions for $|Y|^2/E(|Y|^2)$.

10.2. Sample fluctuation of $Y(0, \omega, \tilde{u})$ and the sample infrared catastrophe

For $\lambda = 0$, one has

$$|Y(0, \omega, \tilde{u})|^2 = \tilde{u}\left[\tilde{u}^{-1}\int_0^{\tilde{u}} W(s)ds\right]^2. \tag{10.3}$$

As $\tilde{u} \to \infty$, $\tilde{u}^{-1}\int W(s)ds$ tends in distribution to a nondegenerate random variable. (This can be easily proven by continuing the argument in Breiman 1965.) Thus, $|Y(0)|^2$ tends to infinity with $\tilde{u}$, both in distribution and almost surely. This result is the same as if the function $W(s)$ were an ordinary stationary r.f. whose population mean has not been removed.

10.3. Sample fluctuation of $Y(\lambda, \omega, \tilde{u})/(E|Y|^2)^{1/2}$ at very high frequencies

As in the study of $\Delta[h\tilde{u}, B(\tilde{u})]$ for small τ, the idea is to cut out from $(0, \tilde{u})$ some stretches that contribute nothing to $Y(\lambda, \omega, \tilde{u})$, while leaving a remainder one can treat by Wiener-Khinchin methods. Note, therefore, that any portion of an intermission of W, whose duration is an integral multiple of $1/\lambda$, contributes nothing to $Y(\lambda, \omega, \tilde{u})$. One can, therefore, carry out the integration of $We^{-2\pi i \lambda s}$ over a subset $\Phi(\tilde{u}, \lambda)$ of $(0, \tilde{u})$. That is, one can write Y as the following product of two independent r.v.,

$$Y = \{|\Phi(\tilde{u}, \lambda)|/\tilde{u}\}^{1/2} Y_\Phi, \tag{10.4}$$

where

(10.5) $$Y_\Phi = |\Phi(\tilde{u}, \lambda)|^{-1/2} \int_{s \in \Phi(\tilde{u}, \lambda)} W(s) e^{-2\pi i \lambda s} ds.$$

This is the sample Fourier vector of a portion of length $\Phi(\tilde{u}, \lambda)$, cut off from an ordinary r.f. so defined that $X(t')$ and $X(t'')$ are independent r.v. when $|t' - t''| > 1/\lambda$. We know, therefore, that as $|\Phi| \to \infty$, Y_Φ tends towards an isotropic Gaussian vector of zero mean.

For $\Phi(\tilde{u}, \lambda)/\tilde{u}$, division by some weight function of $\tilde{u}$ and λ again yields a r.v. whose limit in distribution, as $\tilde{u} \to \infty$, is a r.v. M_D.

Combining the two terms of Y proves that the high frequency fluctuations are *unaffected* by the marginal distribution of $W(t, \omega)$, and we have the following theorem.

Theorem 10.1. *If $\tilde{u} \to \infty$, while λ is constant, the limit in distribution*

(10.6) $$\lim_{\tilde{u} \to \infty} \{Y(\lambda, \omega, \tilde{u})[E|Y(\lambda, \omega, \tilde{u})|^2]^{-1/2}\}$$

is a compound random variable: an isotropic Gaussian vector whose mean square modulus is an M_D r.v..

Corollary 10.1. *If λ is fixed, the limit in distribution*

(10.7) $$\lim_{\tilde{u} \to \infty} \left\{ \frac{|Y(\lambda, \omega, \tilde{u})|^2}{E|Y(\lambda, \omega, \tilde{u})|^2} \right\}$$

is the product of two independent r.v.'s, E and M_D, where E is exponential.

10.4. Joint fluctuations of $Y(\lambda', \omega, \tilde{u})$ and $Y(\lambda'', \omega, \tilde{u})$

The method used for $Y(\lambda, \omega, \tilde{u})$ remains applicable if there exist two positive *integers* q' and q'' such that $\lambda' q' = \lambda'' q'' = \lambda_0$. For brevity, denote $\Phi(\tilde{u}, 1/\lambda_0)$ by Φ, and consider the four-dimensional vector of coordinates

(10.8)
$$\Phi^{-1/2} \text{ Real part of } \int_{s \in \Phi} W(s) e^{-2\pi i \lambda' s} ds,$$
$$\Phi^{-1/2} \text{ Imaginary part of } \int_{s \in \Phi} W(s) e^{-2\pi i \lambda' s} ds,$$
$$\Phi^{-1/2} \text{ Real part of } \int_{s \in \Phi} W(s) e^{-2\pi i \lambda'' s},$$
$$\Phi^{-1/2} \text{ Imaginary part of } \int_{s \in \Phi} W(s) e^{-2\pi i \lambda'' s} ds.$$

As $|\Phi| \to \infty$, this vector tends to an isotropic four-dimensional Gaussian r.v.; therefore, the limits in distribution

(10.9) $$\lim_{\tilde{u} \to \infty} \frac{|Y(\lambda', \omega, \tilde{u})|^2}{\Phi(\tilde{u}, \lambda_0)} \text{ and } \lim_{\tilde{u} \to \infty} \frac{|Y(\lambda'', \omega, \tilde{u})|^2}{\Phi(\tilde{u}, \lambda_0)}$$

are two independent exponential r.v..

Also, one has the following theorem.

Theorem 10.2. *If λ'/λ'' is rational, the two limits in distribution*

(10.10) $$\lim_{\tilde{u} \to \infty} \frac{|Y(\lambda', \omega, \tilde{u})|^2}{E[|Y(\lambda', \omega, \tilde{u})|^2]} \text{ and } \lim_{\tilde{u} \to \infty} \frac{|Y(\lambda'', \omega, \tilde{u})|^2}{E[|Y(\lambda'', \omega, \tilde{u})|^2]}$$

are, respectively, of the form $M_D E'$ and $M_D E''$, where M_D, E' and E'' are mutually independent, and E' and E'' are exponential with unit expectation.

10.5. Weighted spectral estimators

For functions to which the Wiener-Khinchin spectral analysis applies, reliable estimation of the spectral density is obtained through weighted averages of Schuster's periodograms $E_\lambda M_D$. Given a fixed frequency band (λ', λ'') with $\lambda' > 0$, the factors E_λ are asymptotically eliminated. Thus, the ratio between the sample and the population values of the weighted spectral density tends in distribution to M_D as $\tilde{u} \to \infty$. The energy in the variable frequency band $(0, 1/\tilde{u})$ tends in distribution to a limit not degenerate to zero; this is a sampling form of "infrared catastrophe."

Acknowledgment. In the possibly unfinished task of "debugging" the details of this paper, I was helped by D. Chazan, A. G. Konheim, H. P. McKean Jr. and J. W. Van Ness.

&&&&&&&&&& ANNOTATIONS &&&&&&&&&&

How this paper came to be written. Jerzy Neyman started as a very pure mathematician in the 1920s, moved on to statistics in the 1930s, when his work flowered, and became in the 1950s one of the innovators who established statistics as an autonomous academic discipline, quite separate from both mathematics and every one of the sciences. Every fifth year, starting in 1945, statisticians and probabilists trekked to Berkeley to pay him homage and listen to the speakers he selected.

This autonomous statistics is very peripheral to my interests, and Neyman viewed me as peripheral to his. Therefore, he invited me only once. My lecture was more or less based on M1965c{N8}, and also M1964w, the report already mentioned in Chapter N5 and this chapter's foreword. It was poorly received. A group in the audience started a loud private conversation, trying to "make sense about what this speaker was trying to tell them," but finally conceding that, "after all, it might not be total nonsense."

The *Berkeley Proceedings* editors asked me to write my thoughts in concise mathematical format. I did my best; this is why (as already said in the chapter's foreword) the physical motivation was reduced to a minimum. However, the *Berkeley Proceedings* volume on *Probability* having probably become too long, my paper was transferred into the very miscellaneous volume devoted to *Physical Sciences and Engineering.* Effectively, this hid my work from those for whom I was writing.

Random sets of multiplicity for trigonometric series (Kahane & M 1965)

• *Chapter foreword.* While the preceding chapters show that Lévy dusts play an important role in physics, this chapter involves a purely mathematical issue. To include it in this volume may therefore seem an overindulgence. But it is short and illustrates the first instance of a role I came to play repeatedly in mathematics: someone who asks new questions and suggests new directions, rather than a problem solver.

The main result is that, with probability one, the Lévy dusts S are *sets of multiplicity*. This means that a trigonometric series that converges to zero outside of S may fail to converge to zero on S. All the other sets are called *sets of unicity*. The preceding distinction is due to Cantor; he showed that intervals are sets of multiplicity and that finite sets are sets of unicity, and went to define the celebrated triadic Cantor dust while searching for a set of multiplicity of zero measure. Cantor could not determine whether his triadic dust is of unicity or of multiplicity, therefore moved on to other tasks ... and became the immortal founder of set theory. But the problem lived on, and harmonic analysts are proud that Cantor started his career as one of them.

The search for sets of multiplicity went on very gradually, as seen in Kahane & Salem 1963. The new solution given in this chapter is far simpler than all the preceding ones and in time suggested what has been called the "natural method." Kahane 1971 simplified it further: starting with any fractal dust with $D < 0.5$ – even the original Cantor dust – its image by a Wiener Brownian function is a set of multiplicity.

Recent work on this topic is not familiar to me.

In this translation, section titles were added.   •

N11 ◊ ◊ RANDOM SETS OF MULTIPLICITY (WITH J. P. KAHANE)

✦ **Abstract.** Consider a trigonometric series that converges towards zero outside of a set E. Cantor posed the problem of determining whether it necessarily follows that this series also converges towards zero on E. When the answer is positive, E is called a set of unicity; when it is negative, a set of multiplicity. Basic references are Zygmund 1959, Chapter IX, or Kahane & Salem 1963, Chapter V. Cantor's set theory was born out of this difficult problem. The aim of the present text is to show that a new example of a random set of multiplicity is provided by a set that plays an important role in certain models of statistical physics (M 1967b{N10}). ✦

1. Definitions and a construction

E will always denote a closed set on the real line. It is called a "set of multiplicity in the restricted sense," or a set of type M_0, if it carries a measure $d\mu$ whose Fourier transform $\hat{\mu}(y) = \int e^{iyx} d\mu(x)$ goes to zero at infinity. The set E is called of type M_β ($0 \leq \beta \leq \frac{1}{2}$) if it carries a measure such that $\hat{\mu}(y) = o(|y|^{-\beta})$ when $y \to \infty$. The upper bound of α's for which E is of type $M_{\alpha/2}$ has been called the Fourier dimension of E. When the Lebesgue measure of E is 0, the Fourier dimension is always at most equal to the Hausdorff dimension (= Frostman's capacity dimension),

Using a difficult construction (see Salem 1951, Kahane & Salem 1963), Salem has shown that, for every $\alpha \in {]}0, 1{[}$, there exists a set E such that its Hausdorff and Fourier dimensions are both equal to α. Salem constructs a certain random set $E(\omega)$ ($\omega \in$ probability space Ω) and shows that $E(\omega)$ almost surely has the desired property. It will be shown that it suffices to take for $E(\omega)$ the closure of the set of values taken by a stable increasing random process of index α, as defined by Paul Lévy.

Let us recall that an increasing process with independent increments is a mapping $X: [0, \infty{[} \times \Omega \to [0, \infty{[}$ such that, for almost all ω, $X(t, \omega)$ is an increasing function of t, is continuous to the right, such that $X(0, \omega) = 0$ for all ω, and such that, for $0 \leq t_1 \leq t_2 \leq \ldots \leq t_n$, the random variables $X(t_{j+1}) - X(t_j)$, ($j = 1, 2, \ldots, n-1$) are independent. To every such process corresponds a "Lévy measure" $d\sigma$; it is positive, is carried by $]0, \infty[$, and satisfies $\int u d\sigma(u) < \infty$. Conversely, to every Lévy measure verifying these conditions, corresponds a random process such that

$$(*)\, \mathcal{E}\, e^{i[y_1 X_1 + y_2(X_2 - X_1) + \cdots + y_n(X_n - X_{n-1})]} = e^{-[t_1 \psi_1 + (t_2 - t_1)\psi_2 + \cdots + (t_n - t_{n-1})\psi_n]}$$

holds for every increasing sequence $t_1, t_2, \ldots, t_n$ and for every real sequence $y_1, y_2, \ldots, y_n$. ($\mathcal{E}$ will denote the expectation.) Here, we use the notations

$X_j = X(t_j)$, $\Psi_j = \Psi(y_j)$, and $\psi(y) = \int_0^\infty f(1 - e^{iuy}) d\sigma(u)$.

The stable processes of index α correspond to $d\sigma(u)/du = ku^{-1-\alpha}$, with $0 < \alpha < 1$, therefore to $\psi(y) = cy^\alpha$, where $c = c(\alpha)$ is complex.

More generally, let X be a positive process with independent increments, and denote by $E_0 = E_0(\omega)$ the set of values of $X(t)$ on the interval $[0, 1]$, and by E its closure, and set

$$\mu(x) = \mu(x, \omega) = \sup\{t \mid X(t) \leq x\} \%\% \text{and} \%\% h(y) = \inf_{z \geq y} |\psi(z)|.$$

E is the support of the measure $d\mu$. In the case of a stable process of index α, it it known that the Hausdorff dimension of E is almost surely α (Blumenthal & Getoor 1960m, p. 267.)

2. Theorem, corollary and proof

Theorem. – Almost surely,

$$\hat{\mu}(y) = O\left(\sqrt{\log |y|/h(|y|)}\right) \%\% \text{as } |y| \to \infty.$$

Corollary. For a stable random process of index α, the Fourier dimension of E is α.

Proof. It is inspired by Salem 1951. One writes

$$\hat{\mu}(y) = \int_0^1 e^{iyX(\theta)} d\theta.$$

Hence $|\hat{\mu}(y)|^{2p}$ takes the form

$$\int_0^1 \cdots \int_0^1 \exp\{iy[(X(\theta_1) + X(\theta_2) + \cdots + X(\theta_p) - X(\theta'_1) - X(\theta'_2) - \cdots - X(\theta'_p)]\}$$
$$d\theta_1 d\theta_2 \ldots d\theta_p d\theta'_1 d\theta'_2 \ldots d\theta'_p$$

$$= (p!)^2 \int_{0 \leq \theta_1 \leq \theta_2 \leq \cdots \leq \theta_p \leq 1} \int_{0 \leq \theta'_1 \leq \theta'_2 \leq \cdots \leq \theta'_p \leq 1} \cdots$$

$$= (p!)^2 \sum_{\{\varepsilon_j\}} \int_{0 \leq t_1 \leq t_2 \leq \cdots \leq t_{2p} \leq 1} e^{iy(\varepsilon_1 X(t_1) + \varepsilon_2 X(t_2) + \cdots + \varepsilon_{2p} X(t_{2p}))} dt_1 \, dt_2 \ldots dt_{2p}.$$

In the first expression, the dots ... stand for the integrand, and in the third expression the sum is taken over all the systems $\{\varepsilon_j\}$ of values $\varepsilon_j = \pm 1$ $(j = 1, 2, \ldots, 2p)$ such that $\sum_1^{2p} \varepsilon_j = 0$. Apply the rule (*) with $n = 2p$, and set $y_{2p} = \varepsilon_{2p} y$, $y_{2p-1} = (\varepsilon_{2p} + \varepsilon_{2p-1})y$, ... , $y_1 = (\varepsilon_1 + \varepsilon_2 + \cdots + \varepsilon_{2p})y = 0$. One finds

$$\mathcal{E}(|\hat{\mu}(y)|^{2p}) =$$
$$(p!)^2 \sum_{\{\varepsilon_j\}} \int_{0 \leq t_1 \leq t_2 \leq \cdots \leq t_{2p} \leq 1} e^{-(t_1 \psi_1 + (t_2 - t_1)\psi_2 + \cdots + (t_{2p} - t_{2p-1})\psi_{2p})} dt_1 \, dt_2 \ldots dt_{2p}.$$

If j is even, $|\psi_j| \geq h(y)$, and in all cases $\Re \psi_j \geq 0$. For every $\{\varepsilon_j\}$, $t_1, t_3, \ldots, t_{2p-1}$, the integral with respect to the even indices is dominated in modulus by $2^p (h(y))^{-p}$; for every $\{\varepsilon_j\}$, the integral is dominated in modulus by $2^p (h(y))^{-p} (p!)^{-1}$. Therefore,

$$\mathcal{E}(|\hat{\mu}(y)|^{2p}) \leq \frac{(2p)! 2^p}{p! (h(y))^p}.$$

Let $\{y_n\}$ be a real, increasing sequence containing all the multiples of $2^{-\nu}$ for $n > n(\nu)$, and satisfying $\sum_1^\infty y_n^{-2} < \infty$. Set $p_n = [\log y_n]$. One has

$$\mathcal{E}\left\{ \sum_{n_0}^\infty y_n^{-2} \frac{|\hat{\mu}(y_n)|^{2p_n} p_n! (h(y_n))^{p_n}}{(2p_n)! 2^{p_n}} \right\} < \infty,$$

so that, almost surely,

$$\hat{\mu}(y_n) = o\left\{\left[\frac{(2P_n)! y_n^2}{P_n!}\right]^{\frac{1}{2p_n}} \left[\frac{2}{h(y_n)}\right]^{1/2}\right\} (n \to \infty),$$

that is,

$$\hat{\mu}(y_n) = o\left\{\sqrt{\frac{\log y_n}{h(y_n)}}\right\}.$$

Since $d\mu$ has a compact support, this is equivalent to the theorem.

It would be interesting to know under which conditions E is almost surely a rationally independent set. This condition is certainly not compatible with $y^\varepsilon = O[h(y)]$, ($\varepsilon > 0$, $y \to \infty$), because, in that case, $E + E + \cdots + E$ (p times) contains an interval for large enough p. If this condition is compatible with $\log y = O[h(y)]$, one obtains new rationally independent sets of multiplicity. Such sets had been described by Rudin 1960, by using some random sets due to Salem. See also Kahane & Salem 1963, Chapter VIII.

&&&&&&&&&&& ANNOTATIONS &&&&&&&&&&&

The French original was presented to the Académie of Sciences on November 3, 1965 by Paul Lévy.

How this paper came to be written. In a balanced scientific collaboration, skills and motivations are equally shared among the co-authors. Berger & M 1963{N6} was a shining example. This chapter is another.

Jean-Pierre Kahane, a specialist in harmonic analysis, knew how to finish this paper and develop it into the "natural method," but he would have not started it on his own, before fractal geometry. I knew how to start it, but not to finish it by myself. He and I had taken exams together in 1947, and his doctoral advisor was my uncle. He sent me a complimentary copy of Kahane & Salem 1963, where I read about sets of multiplicity.

Raphael Salem (1898-1963) began as a gentleman-mathematician and a full-time banker, but when he was in his forties, World War II made him move on to an impressive full-time career in very technical mathematics.

PART IV: TURBULENCE AND MULTIFRACTALS

As described abundantly in Chapter N1 and N3, one of the main forms of $1/f$ noise is multifractality. This part is largely restricted to works that bear on both turbulence and multifractals.

The centerpiece is Chapter N15, which reproduces M 1974f. It tackles the intermittence of turbulence, but is of broader impact: it was the first to investigate the concept of random multifractal measure. The original paper appeared in a major periodical and is reasonably well-known. But it is significantly incomplete without its companion, M 1974c, translated here as Chapter N16. This companion paper was the first to state the Legendre transform formalism of multifractals, but the original appeared in French and had few readers.

The remaining chapters in Part IV reprint hard-to-find papers on turbulence and/or multifractals that either preceded or followed Chapters N15 and N16, and are necessary for a full understanding of either the history or the technical points made in those key chapters.

Two related topics not discussed in this book deserve brief mention.

Among my papers from the 1970s, two other concern turbulence but would not fit in this book. Since they are confined to the Gaussian approximations, both are withheld to be reprinted in M 1998H. M 1975f deals with the dimensional and spectral analysis of turbulent iso-surfaces, and M 1975b, deals with Poisson approximations to Gaussian fields.

In addition to multifractal measures, Chapter N1 allows for multifractal functions. Section 4.3 of Chapter N1 and many portions of M 1997E attribute special importance to fractional Brownian motions of a multifractal intrinsic time. However, my old work reprinted in this Part was restricted to measures.

Sporadic turbulence

✦ **Abstract.** Turbulence in the atmosphere and the ocean, and often in the laboratory, is "spotty" or "intermittent;" it does not satisfy the homogeneity assumptions of the 1941 Kolmogorov-Obukhov theory. Moreover, Landau & Lifshitz 1953 has pointed out that ε_r, the mean rate of energy dissipation in a volume of radius r, must be a random variable whose law depends upon the ratio of r to the external scale L_e. In 1962 Obukhov gave a theory of intermittency, later developed in Kolmogorov 1962 and Yaglom 1966, that leads to the lognormal law for ε_r. Sporadic turbulence is an alternative to Obukhov's theory and is akin to the theory of intermittency due to Novikov & Stewart 1964.

Sporadic turbulence is extremely intermittent. Like some approaches to homogeneous turbulence, its theory involves no physical "cascade" argument, and is based exclusively upon axioms of self-similarity, dimensional correctness, and local character. The difference is that the axioms are geared to turbulent-laminar mixtures, by always being stated for conditional (rather than for absolute) probability distributions. "Ensemble expectations" are not unique, but depend upon a "conditioning event," so that the ergodic problem is very complex. For example, consider the conditional probability

$$\Pr\{u(x+r) - u(x) \neq 0, \text{ given that } u(0) - u(L) \neq 0 \text{ and } 0 < x < x + r < L\}.$$

By self-similarity, this expression must take the form $(r/L)^{1-D}$, where $0 < D \leq 1$. The exponent D is a measure of "degree of spottiness:" $D = 1$ in the homogeneous case, and $D = 0$ when all turbulent energy is concentrated in a single "puff." Kolmogorov's rule, namely

$$\left\langle [u(x+r) - u(x)]^2 \right\rangle = Cr^{2/3},$$

only applies to the expectation taken under the condition that "$u(y)$ is not constant for $x < y < x+r$". The alternative "laminar" solution, for which

$$\left\langle [u(x+r) - u(x)]^2 \right\rangle = 0,$$

also satisfies Kolmogorov's axioms. It applies if $u(y) \equiv$ constant and plays a central role in sporadic turbulence.

For example, whenever $0 < y < x + r < L$, one has

$$\left\langle [u(x+r) - u(x)]^2 \,|\, u(y) \text{ is non-constant for } 0 < y < L \right\rangle$$

$$= C\, \varepsilon_L^{2/3}\, L^{(D-1)/3}\, r^{1 - D/3}. \qquad \blacklozenge$$

&&&&&&&&& ANNOTATIONS &&&&&&&&&&

How this abstract came to be written. Having become involved with turbulence as described in Chapter N2, I checked immediately what could be learned from the sporadic processes described in M 1967b{N10}.

My "sponsor" at this Kyoto meeting was Robert W. Stewart. The banquet speaker was Joseph Kampé de Fériet (1893-1982), who had written on the mathematics of turbulence with Garrett Birkhoff. He brought my uncle to the University of Lille as a professor in 1928, and later helped me come there in 1957, but we both left Lille after one year! In his speech, Kampé de Fériet observed that the elusive secrets of turbulence may require drastically new tools. Then he singled out my short talk and fractal dimension for special comment.

A much fuller but unfinished draft in my files lists M 1967i{N7} and M 1967b {N10} as being "in the press". I could not think of an audience for any text that could be produced easily and missed the *Proceedings* deadline. Printing this draft now would be unneccesary and inappropriate.

Symposium on Turbulence of Fluids and Plasmas, April 1969, Brooklyn

Intermittent free turbulence

✦ **Abstract.** The distribution of the rate of dissipation of free turbulence, as observed in the ocean and the atmosphere, seldom satisfies the homogeneity assumption of the classic Kolmogorov-Obukhov theory. Analogous "intermittency" (with "clustering" of the "active regions") is also observed for the energy of various "$1/f$ noises," for the spatial distribution of rare resources such as metals, and elsewhere.

By putting together the existing discussions of the various forms of intermittency, one finds two broad approaches. The first was followed by de Wijs (a geophysicist not known outside of his field), and later by Obukhov, Kolmogorov and Yaglom. The second approach was followed by Mandelbrot (who started in the context of $1/f$ noises) and by Novikov & Stewart (see M 1967k{N12}).

The purpose of the present work is to construct further variants of both broad approaches to intermittency, to unify all the approaches, and to develop them.

Of particular interest is that all approaches suggest – in agreement with experience – that any interval containing turbulent energy, when examined more closely, will be found to include inserts that are effectively devoid of turbulence. Mathematically, this may be expressed by saying that turbulence is not carried by intervals, but by "thin" sets with many gaps. Such sets can be characterized as having a "dimension" D less than unity (for a discussion of fractional dimension, see M 1967s). At one extreme, there is no insert, and $D=1$; at the other extreme, turbulence is concentrated in a single "puff," and $D=0$.

Another property of intermittency is the exponent Ω that enters in the expression $k^{-\Omega}$, which gives (up to a factor of proportionality) the spectral density of the rate of dissipation ε. Different approaches to turbulence yield different relations between D and Ω. This suggests either that further assumptions are needed to identify the properties of "the" actual turbulence, or that different models of intermittency are to be used in different contexts.

To conclude the random variable $\varepsilon(r)$, the average dissipation in a sphere of radius r, is never lognormally distributed (which refutes a hasty claim of de Wijs, Kolmogorov, Obukhov and Yaglom). Moments of $\varepsilon(r)$ are obtained. In some cases $\varepsilon(r)$ is found to have infinite population variance. This implies – again in agreement with experience – that observed values of $\varepsilon(r)$ can be very widely scattered. ✦

&&&&&&&&&&& ANNOTATIONS &&&&&&&&&&&

How this abstract came to be written. Who nominated me to give an invited talk to this down-to earth Symposium at this extraordinarily early stage of my involvement with turbulence? It must have been François N. Frenkiel, a student and continuing associate of Joseph Kampé de Fériet, whom I mention in the annotation of Chapter 12. Both had known me for a while, and had heard me speak at the 1983 Kyoto Symposium.

This abstract suffices to show that my multifractal model of turbulence had already reached an advanced stage of completion in 1968. The whole text was never finished; not being able to think of an audience to write for, I missed the deadline.

A much fuller but unfinished draft in my files carries no date and no bibliography that could help date it. Section 2 is clearly an abstract of a text titled *The conditional cosmographic principle and the fractional dimension of the Universe* (M 1971n), which I tried unsuccessfully to get published in the astronomical literature around 1971. Section 3 is an abstract of M 1972c, which the journal received on January 16, 1968. Section 4 (typed later than the earlier Sections and not mentioned in the original Section 1) is an advance abstract of M 1972j. The original draft of M 1972j was read during the summer of 1971. All this suggests that the draft in my files was written before 1971. It makes a number of nice points, but printing it now would be both inappropriate and unnecessary.

Lognormal hypothesis and distribution of energy dissipation in intermittent turbulence

● *Editorial changes.* The type-offset original of this chapter is titled "Possible refinement of the lognormal hypothesis concerning the distribution of energy dissipation in intermittent turbulence." That old text repeatedly refers to Kolmogorov's third hypothesis as "appearing untenable" or as being "probably untenable." After this text had been typeset, and when it was being copy-edited, I shortened the title and took the liberty of skipping the qualifiers "appear" or "probably." Their use was *probably* motivated by the profound admiration for Kolmogorov (expressed in Chapter N2, Section 5), by my consistent reluctance to belabor error, and *probably* also by expected disapproval had I done so.

The exponent q was denoted by p in the original, and the Sections were not numbered. An earlier version of this paper had the more straightforward title, "Note on intermittency obtained through multiplicative perturbations." An excellent "advance abstract" of this chapter is provided by the following excerpts from this earlier version. ●

✦ **Advance abstract.** My very modest aim is to analyze the assertion common to [several authors], that Kolmogorov's ε, and more generally the local intensity of turbulence, should be expected to follow the lognormal distribution ... Unfortunately, except for ... cases of limited interest, [the usual] derivation [of lognormality] will be shown to be seriously in error, and as a consequence the earlier [authors'] assumptions of lognormality are seriously flawed. Indeed, multiplicative perturbations predict a distribution that not only grossly differs from the lognormal, but also depends upon details in the process of multiplicative perturbations that is being used. For those concerned with experimental check, this last aspect, naturally, is disquieting.

My remarks apply equally to [a variety of other] subject matters but I shall phrase them in terms of turbulence.

Multiplicative perturbations in general. [The authors cited], as I read them, argue that when turbulence is intermittent, its local intensity at a point in space-time (Kolmogorov's ε) can be considered the multiplicative resultant of many independent and stationary random functions, each corresponding to a different range of eddy wavelengths. The resultant of the eddies whose frequency λ satisfies $1 < \lambda < f$ is thus an approximate local intensity, which will be designated as $X'(t,f)$. The purpose of models of this kind is to list assumptions from which one can deduce the properties of $X'(t,f)$ and, much more important, the properties of local averages of $X'(t,f)$, as defined in one dimension by the integral $X(t,f) = \int_0^t X'(s,f)ds$....

The reason averages are important is that even the most refined measurements concern zones that are much larger than the range of viscosity in a fluid... The most important issue is the extent to which the properties of $X(t,f)$ depend on f when f is large, and how they behave as $f \to \infty$. Also, given that [all] models are unavoidably artificial and oversimplified, an important issue for such models is the extent to which the properties of $X(t,f)$ continue to depend on the original mechanism postulated for the perturbations.

Roughly, [the previous authors invoke], in succession, a central limit theorem, a mean value theorem and finally a law of large numbers. [But this argument] happens to be grossly incorrect, except that the lognormal law does apply to $X(t,f)$ as a rough approximation for finite f when $|t|$ is not much above $1/f$, and for arbitrary f when the effects of high frequency eddies are very weak – in a sense to be defined below. In all other cases, the correct distribution of $X(t,\infty)$ is far from lognormal; in many ways, it is a more interesting [distribution]. In particular, we have the striking result that if the high frequency eddies are sufficiently strong, – again in a sense to be defined below – [and if f is large, then] $X(t,f)$ [nearly] vanishes, save under circumstances of minute probability, under which it may be enormous. To get farther [away] from ... [asymptotic] averaging would be difficult. ✦

✦ **Abstract.** Obukhov, Kolmogorov and others argued that energy dissipation in intermittent turbulence is lognormally distributed. This hypothesis is shown to be untenable: depending upon the precise formulation chosen, it is either unverifiable or inconsistent. The paper proposes a variant of the generating model leading to the lognormal. This variant is consistent, appears tractable, and for sufficiently small values of its unique parameter μ it yields the lognormal hypothesis as a good approximation. As μ increases, the approximation worsens, and for high enough values of μ, the turbulence ends by concentrating in very few huge "blobs." Still other consistent alternative models of intermittency yield distributions that differ from the lognormal in the opposite direction; these various models in combination suggest several empirical tests. ♦

1. INTRODUCTION

A striking feature of the distributions of turbulent dissipation in the oceans and the high atmosphere is that both are extremely "spotty" or "intermittent" in a hierarchical fashion. In particular, both are very far from being homogeneous in the sense of the 1941 Kolmogorov-Obukhov theory. Nevertheless, many predictions of this classic theory have proved strikingly accurate. Self-similarity and the $k^{-5/3}$ spectrum have not only been observed, but are found to hold beyond their assumed domain of applicability. An unexpected embarrassment of riches, and a puzzle!

For many scientists, studying turbulence is synonymous with attempting to derive its properties, including those listed above, from the Navier-Stokes equations of fluid mechanics. But one can also follow a different tack and view intermittency and self-similar statistical hierarchies as autonomous phenomena.

Early examples of this approach in the literature are few in number, but they go very far back in time, and have involved several disciplines. In the field of cosmology, intermittency had already been faced in the eighteenth century, and its study underwent bursts of activity in the period 1900-1920 and today. Furthermore, concern with intermittency arose, independently and nearly simultaneously, in the fields of turbulence (including work by Obukhov, and later by Gurvich & Kolmogorov, Novikov & Stewart and Yaglom), in the study of geomorphology – especially in the study of the distribution of rare minerals (including work by deWijs and Matheron) – and finally in my own work concerning many non-thermal noises. They go under such names as "burst noise," "impulse noise," "flicker noise," and " $1/f$ noise," and may be considered forms of electromagnetic turbulence. As it happens, despite obvious differences, all the scientists working in these fields have followed the same few generic paths. What has brought these various applications together is not yet clear: it may be either their common underdevelopment or genuine kinship.

We shall be concerned with one of these generic paths, which may be designated as the "method of self-similar random multiplicative perturbations." It had two widely distinct sources, the first was a footnote remark in Landau & Lifshitz 1953 concerning the 1941 Kolmogorov theory of self-similar homogenous turbulence. This remark was taken up by Obukhov 1962, and discussed and developed by Kolmogorov 1962, Yaglom 1966 and Gurvich and Yaglom 1967. The second source lay in works by deWijs 1951, and then Mathéron 1962 and his school, on the distribution of rare minerals.

Using the vocabulary of turbulence, let η and L designate the Kolmogorov micro- and macro-scales, and let $\varepsilon(x, r, \eta, L)$ be the average energy dissipation over the cube of side r and center x. Obukhov and Kolmogorov hypothesize, and de Wijs and Yaglom attempt to derive, the property that $\log[\varepsilon(x, r, \eta, L)]$ is a normal random variable of variance equal to $A(x, t) + \mu \log(L/r)$, where the term $A(x, t)$ depends on the characteristics of the large scale motion and μ is a parameter, possibly a universal constant. The above assertion is usually called "Kolmogorov's third hypothesis."

In addition, the expectation of $\log \varepsilon$ is ordinarily assumed equal to $-(\mu/2) \log(L/r) - A(x, t)/2$. Finally, the averages of $\varepsilon(x, r, \eta, L)$ corresponding to cubes whose scale equals the micro-scale of turbulence, are assumed to have a certain correlation function of the form required by self-similarity;

$$E\left[\varepsilon(x + \Delta x, \eta, \eta, L)\, \varepsilon(x, \eta, \eta, L)\right] \sim (L/|\Delta x|)^{\mu} \text{ for large } \Delta x.$$

I will call this last expression the "Gurvich-Zubkovskii correlation." Observe that neither this correlation nor Kolmogorov's "third" hypothesis involve η explicitly, which expresses that they obey Kolmogorov's "second" hypothesis of 1941, which he had maintained unchanged in 1962.

The purpose of the present paper is, first, to show that the above "third hypothesis" raises serious conceptual difficulties which make it untenable; secondly, to propose an improved alternative. The practical relevance of my criticism has not yet been established. It depends upon the value of μ, and each field of application will have to investigate it specifically.

{P.S.1998 Following the probabilists' custom, "random variable" will be shortened to "r.v.."}

2. CRITIQUES OF VARIOUS FORMS OF LOGNORMALITY

Allow me to make the historical background more precise: Obukhov introduces lognormality as an "approximate hypothesis:" on the ground that the lognormal "represents any essentially positive characteristic." Kolmogorov treats lognormality as a "third hypothesis" to be derived from other assumptions. And deWijs and Yaglom derive lognormality from a "cascade" argument. Each approach requires a separate reexamination and critique.

Obukhov's approximate hypothesis. Because it is approximate it can only be examined on pragmatic grounds. Its weakness is that it cannot support the elaborate calculations of moments which have been built on it, because the population moments of the random variable (r.v.) $\exp(Y)$ are extremely sensitive to small deviations of Y from normality.

For example, consider a normal random variable G, a Poisson r.v., P, and a Bernoulli r.v., B, obtained as the sum of a large number H of binomial r.v.'s B_H. When their respective means and variances are equal and large, those three r.v.'s are indeed considered by probabilists as being "nearly identical." But this concept of "near identity" tells little about higher moments of the same order of G, P, and B. A fortiori, the moments of e^G, e^P and e^B of all orders are so influenced by the tails of the various distributions that their values may be very different. For example, suppose they all have the same mean δ and variance δ, and denoted the possible values and probabilities of the binomials B_h contributing to B by B', B'', π' and π'', with $\pi'B'H \ll \delta$, $\pi''B''B < \delta$, and $\pi'B' > \pi''B''$. We then have:

$$E(e^G)^q = \exp(q\delta + \delta q^2/2) = \exp[\delta(q + q^2/2)]$$

$$E(e^P)^q = \exp(-\delta + \delta e^q) = \exp[\delta(e^q - 1)]$$

$$E(e^B)^q = \exp E(e^{\Sigma B}h)^q = (Ee^{qB}h)^H = (e^{q\pi'B'} + e^{q\pi''B''})^H.$$

For large δ, $E(e^B)^q$ increases like $e^{qH\pi'B'}$, that is, less rapidly than $e^{q\delta}$; $E(e^G)^q$ increases more rapidly, and $E(e^P)^q$ even more rapidly.

Even for $q = 1$, one finds different values, respectively

$E(e^G) = \exp(1.5\delta)$ and $E(e^P) = \exp(1.7\delta)$.

The coefficients of variation are

$$\frac{E[(e^G)^2]}{[E(e^G)]^2} = e^\delta \quad \text{and} \quad \frac{E[(e^P)^2]}{[E(e^P)]^2} = e^{(e-1)^{2\delta}} \sim e^{3\delta}.$$

They differ even more, and higher order moments differ strikingly. In short, as soon as one takes exponentials, B and P *cease* to be good approximations to the normal G. It follows that the significance of moment calculations under Obukhov's approximate hypothesis of lognormality is entirely unclear.

This last finding must be reviewed from the viewpoint of the observation, due to Orszag 1970, that the moments of the lognormal increase too fast to satisfy the so-called Carleman criterion. Consequently, lognormal y intermittent turbulence is *not* determined by its moments. The moments of Poisson intermittent turbulence increase even more rapidly, while those of the binomial do satisfy the criterion. However, we have noted that this property is sensitive to minor deviations from normality, so I hesitate to consider this question solved.

Kolmogorov's suggestion that the lognormal hypothesis be considered as strictly valid. This suggestion encounters a different kind of difficulty. Indeed, let us show that the assumption that the r.v.'s $\log \varepsilon(x, r, \eta, L)$ are normal for every x and every r is incompatible with the assumption that the correlation of ε follows the self-similar (Gurvich-Zubkovskii) form.

Indeed, if $A(x, t)$ is replaced by the constant term $\exp \bar{\varepsilon}$, the lognormal hypothesis yields:

$$E[r^3 \varepsilon(x, r, \eta, L)]^q = \bar{\varepsilon}^q r^{3q - q(q-1)\mu/2} L^{q(q-1)\mu/2}.$$

When r reaches it maximum value, which is $r = L$, all these moments reduce to $\bar{\varepsilon}^q r^{3q}$, as they should. But we must examine more closely *how* they tend to this limit. Suppose $\mu > 3$ and focus on the second moment ($q = 2$). The exponent of r in the above expression takes the value $6 - \mu < 3$. This inequality expresses that, when r is doubled, $E[r^3\varepsilon]^2$ is multiplied by a factor that is *smaller* than 8. On the other hand, the fact that the Gurvich-Zubkovskii correlation is positive implies that the factor in question must be *greater* than 8. This is a contradiction, as previously announced.

When $\mu < 3$, the contradiction moves up to higher moments, namely to moments such that q satisfies

$$3q - q(q-1)\mu/2 < 3, \quad \text{i.e.} \quad q/3 > 2/\mu.$$

This last criterion will be encountered repeatedly in the rest of the paper.

There is another internal contradiction. Consider the variables $\log \varepsilon$ corresponding to 8 neighboring small cubes obtained by subdividing a bigger cube. When they are lognormal, consistency also requires the variables $\log \varepsilon$ corresponding to the big cube to be lognormal. However, sums (and hence averages) of independent lognormal variables are themselves *not* lognormal, which suggests that when the eight small cubes's variables are nearly statistically independent, the above requirement is violated. In particular, when μ is very large, the correlation between variables over neighboring small cubes is very small, which suggests that the dependence is small and that the said requirement is violated.

To sum up, for moderately large values of μ, the lognormal hypothesis could only be consistent with some special rule of dependence for which the correlation function is not positive. I can't imagine any such rule, and circumstantial evidence to be described below makes me doubt such a rule exists. This suggests that *Kolmogorov's strict hypothesis is untenable*.

Lognormality obtained as the conclusion of the deWijs-Yaglom (WY) cascade arguments. One may expect the third form of lognormality flawed. Let us review WY step by step. First step: pave space with a regular grid of eddies: the elementary eddies are cubes of side η, eddies of the next stage are cubes of side 2η, so each contains 8 elementary eddies, etc. Second step: assume that $r = \eta 2^n$ for some integer n while $L = \eta 2^N$ for some integer N, and rewrite $\varepsilon(x, r, \eta, L)$ as the product:

$$\frac{\varepsilon(x, r, \eta, L)}{\varepsilon(x, 2r, \eta, L)} \frac{\varepsilon(x, 2r, \eta, L)}{\varepsilon(x, 4r, \eta, L)} \cdots \frac{\varepsilon(x, 2^{N-1}r, \eta, L)}{\varepsilon(x, 2^N r, \eta, L)} \varepsilon(x, L, \eta, L).$$

Third step: identify the last term as $\bar{\varepsilon}$, and assume the ratios in the above expression to be independent identically distributed r.v.'s. Fourth step: one applies the central limit theorem to the sum of the logarithms of the above rations. Conclusion: when the cube of center x and side r is one of the above eddies, the distribution of the corresponding $\varepsilon(x, r, \eta, L)$ is lognormal.

Our criticisms of Obukhov's and Kolmogorov's approaches extend to the WY generative model. In addition, WY predictions concern the eddies themselves, so that direct verification is impossible. On the other hand, when our cube of center x and radius r is *not* an eddy, $\varepsilon(x, r, \eta, L)$ is *not* lognormal. For example, when r is large and one cube overlaps several big eddies, $\varepsilon(x, r, \eta, L)$ is the average of several independent lognormal variables; as we have seen this implies it is not lognormal. To establish the distribution of ε over an arbitrary cube, one would have to average the distribution corresponding to cubes having the same r and overlapping various numbers of eddies.

3. AN ALTERNATIVE TO LOGNORMALITY: LIMIT LOGNORMAL RANDOM PROCESSES

The basic difficulty with the WY cascade argument is, I think, due to the fact that it imposes *local conservation of dissipation*. This is expressed by the fact that various random ratios of the form $\varepsilon(x, r/2, \eta, L)/\varepsilon(x, r, \eta, L)$ corresponding to different parts of an eddy are required to have an average of one. Especially when μ is large, this requirement implies that such ratios are strongly negatively correlated, a feature which is foreign to the Gurvich-Zubkovskii correlation, but (as we saw) is needed in the Kolmogorov argument.

Conservation on the average. By way of contrast, the variant of the model of multiplicative perturbations proposed in the present paper can be characterized by the feature that *conservation of dissipation is assumed, not on the local, but only on the global level*. That is, this model visualizes the cascade process as being combined with powerful mixing motion, and with exchanges of energy that disperse dissipation and free the above ratios from having to average to one. Moreover, in order to better satisfy self-similarity, the hierarchy of eddy breakdowns is taken as continuous rather than discrete. Under these conditions, one can relate ε to a sequence of random functions (r.f.'s) $F'(x, \lambda, L)$ such that the $\log F'(x, \lambda, L)$ are Gaussian, with the variance $\mu \log(L/\lambda)$, the expectation $-(\mu/2) \log(L/\lambda)$ and a spectral density equal to $\mu/2k$ for $1/L < k < 1/\lambda$, and to 0 elsewhere. Consequently, the covariance $C(s, \lambda)$ of $\log F'(x, \lambda, L)$ will be assumed to satisfy $C(s) = \lim_{\lambda \to \infty} C(s, \lambda) = -\mu \log(2\pi e^\gamma s/L)$. (Here, γ is the Euler constant, whose value is about 0.577.)

For fixed x and L, $F'(x, \lambda, L)$ is clearly a sequence of lognormal r.v.'s whose expectation is identically 1, while their variance, and hence their skewness and their kurtosis, all increase without bound as $\lambda \to \infty$.

$F(x, r, \lambda, L)$ will then be defined as the integral of F' over a cube of center x and side r (which need not be any specific cube designated as "eddy"), and will be viewed as $r^3\varepsilon(x, r, \lambda, L)$, namely as the approximate total dissipation that only takes account of perturbations whose wavelength lies between λ and L.

Note that, in contrast to the WY model, there is no specific grid of eddies in the present model. One resemblance to WY is that when $\eta < \lambda \ll r_1 < r_2 < r_3 \ll L$ and $r_2/r_1 = r_3/r_2$, the ratios

$$\frac{F(x, r_2, \lambda, L)}{F(x, r_3\lambda, L)} \quad \text{and} \quad \frac{F(x, r_3, \lambda, L)}{F(x, r_2, \lambda, L)}$$

have identical distributions. One difference is that those ratios need not be independent. A second difference is that WY assume that randomness in $F(x, L, \lambda, L)$ lies entirely beyond the model, while in the present variant the "A (x, t)" is in part due to eddy action.

Our task is to derive the distribution of $F(x, r, \lambda, L)$. In particular, the smallest value of λ is η, and we must check whether or not the distribution of $F(x, r, \eta, L)$, for $r > \eta$, is independent of η. If it is, then Kolmogorov's second hypothesis, unchanged from 1941 to 1962, is satisfied.

A delicate passage to the limit. Our procedure will be to keep x and r fixed and view $F(x, r, \lambda, L)$ as a r.f. of λ. From the mathematical viewpoint, this r.f. happens to be a "martingale" and, of course, $F \geq 0$. Doob's classical "convergence theorem for positive martingales" (Doob 1953, p. 319) states that $\lim_{\lambda \to 0} F(x, r, \lambda, L) = F(x, r, 0, L)$ exists. This result suggests it may be legitimate for small but positive λ say for $\lambda = \eta$, to view $F(x, r, \lambda, L)$ as differing from $F(x, r, 0, L)$ by a "perturbation term."

However, the convergence theorem allows two possibilities. {P.S. 1998: This is a major but unavoidable complication.} The limit may be either non-degenerate, that is, have a positive probability of being finite and positive, or degenerate, that is, almost surely reduced to 0 {P.S. 1998: In the former case, the expectation of the limit is 1. In the latter case, the expectation of the limit is 0, despite the fact that the limit of the expectations is 1.}

When $F(x, r, 0, L)$ is nondegenerate, then for small but positive values of λ, such as $\lambda = \eta$, a perturbation term dependent on λ is required, but one may consider that Kolmogorov's second hypothesis *holds*.

But when $F(\underline{x}, r, 0, L)$ is degenerate, then for small λ, $F(x, r, \lambda, L)$ either nearly vanishes, with probability nearly 1, or is extraordinarily large, with a very small probability. This last probability tends to zero with λ, but it is finite if $\lambda > 0$, which explains why the normalizing constraint $EF(x, r, \lambda, L) = 1$ could be imposed without contradiction. Nevertheless, the perturbation term is non-negligible, so $F(x, r, 0, L)$ is a bad approximation and Kolmogorov's second hypothesis *fails*. The preceding alternative shows the importance of determining which of the above alternates holds for given μ. More precisely, we shall seek when and to which extent the lognormal approximation to $F(x, r, L)$ is reasonable.

First main result. $F(x, r, L)$ is *degenerate* when $\mu > 6$, and *nondegenerate* when $\mu < 6$.

Second main result. In the nondegenerate case $\mu < 6$, the moment $EF^q(x, r, 0, l)$ is finite when $\mu < 6/q$ and infinite when $\mu > 6/q$.

This behavior suggests that, for large values of u,

$$\Pr\{F(x, r, 0, L) > u\} \sim C(r, L) u^{-6/\mu}.$$

Thus, as $\mu \to 0$, $F(x, r, 0, L)$ acquires an increasing number of finite moments, which are shown to converge towards those of the lognormal. This result constructively establishes that for small μ, the cascade scheme of deWijs and Obukhov can be modified so as to avoid the difficulties that have been listed above without significantly changing the prediction. For large μ, on the other hand, the required changes are significant.

The transition criterion $\mu = 6/q$ was already encountered in the discussion of the inconsistency of Kolmogorov's strict hypothesis. Those same high moments that seemed to behave inconsistently no longer do so here. The reason is that in the present model they are infinite throughout. Proofs of the above assertions will be given in the following two sections; each uses specific mathematical tools appropriate to its goal.

text continues on page 308

FIGURE N14-1.

The graphs found on the next three pages are computer simulated approximations to one-dimensional self-similar limit lognormal r.f.'s $F(x, r, \lambda, L, \mu)$. The are plotted for successive values of x, each a multiple of r.

Method of construction: Define $F'(x, \lambda, L = 10^7, \mu = 2)$ as a Gaussian r.f. for discrete x with $1 \leq x \leq 560{,}000$, with a spectral density that is approximately equal to $1/k$ for $1/L < k < 1/\lambda$. This function was simulated on the IBM System 360/Model 91 for selected values of μ and λ, and $F(x, r, \lambda, L, \mu)$ was computed for x multiple of 1000, using the formula $\Sigma_{u=1}^{1000}[F'(u, \lambda, 10^7, 2)]^{\mu/2}$. The ordinate is the ratio $R(x, \lambda, \mu)$ between F and the median of the values of F along the sample. For each μ, the output of the program is a Calcomp tracing across a broad strip of paper. All the programs were written by Hirsh Lewitan using the fast fractional Gaussian noise algorithm described in M 1971f.

On the next three pages, portions of these graphs are shown for $\mu = 0.5$, $\mu = 1$, and $\mu = 4$, respectively, with λ decreasing down the page.

Analysis of the results. The theory predicts that when μ is small (graphs A, B, and C), the ratio $R(x, \lambda, \mu)$ converges to a limit. The simulations clearly confirm that R soon ceases to vary. The ostensible limit is clearly non-Gaussian, but not extremely so.

As μ increases (graphs D, E, and F), the point of ostensible convergence moves towards decreasing values of λ, and the non-Gaussian character of the ostensible limit of R becomes increasingly apparent. In particular, an increasing proportion of the cumulated F becomes due to a decreasing number of sharp peaks and blobs. (The peaks are truncated at 10 for the sake of legibility, but it was decided *not* to plot log R instead of R.)

Finally, $\mu = 2$ (graphs G, H, and I) is the critical value of μ in one dimension (as $\mu = 6$ was in three dimensions). Fron this value on, R ceases to converge to a limit. This lack of convergence is clearly seen on this simulation.

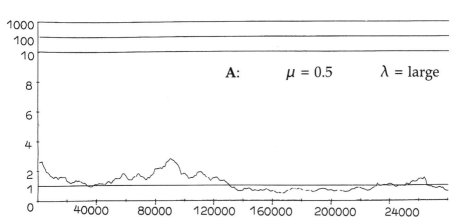

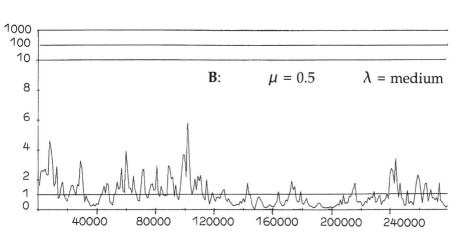

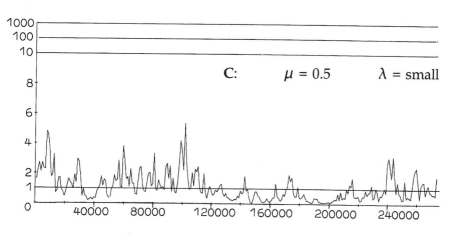

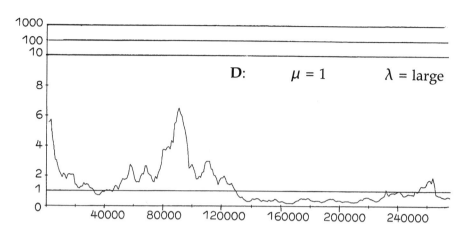
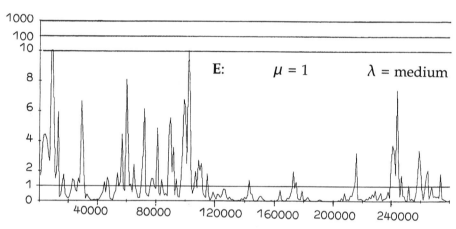
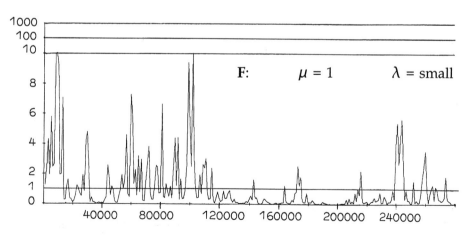

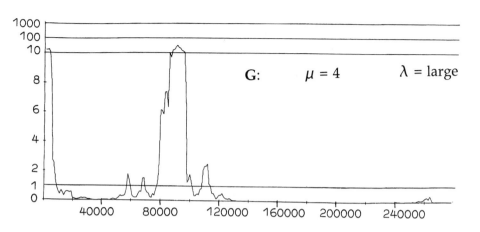

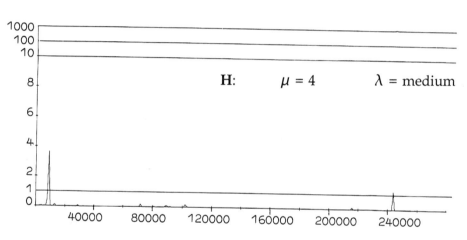

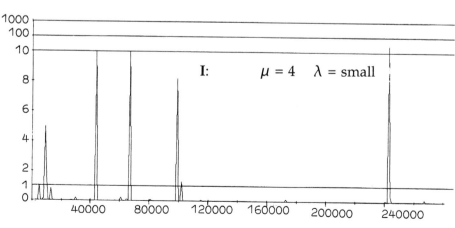

4. THE LIMIT LOGNORMAL MODEL FOR $\mu > 6$: DEGENERATE LIMIT

Our argument will proceed in three steps.

First step: In very skew lognormal distributions, the expectation is overwhelmingly due to occasional large values. Therefore, let $\log F'$ be Gaussian with variance $\mu \log(L/\lambda)$ and expectation $-(\mu/2)\log(L/\lambda)$, implying $F' \equiv 1$, and let $N(\lambda)$ be a function such that

$$\lim_{\lambda \to 0} N(\lambda) = \infty, \text{ while } \lim_{\lambda \to 0} N(\lambda)/\sqrt{\log(L/\lambda)} = 0.$$

Also define three functions, Threshold (λ, L), F'_+, and F'_- as follows:

Threshold $(\lambda, L) = (L/\lambda)^{\mu/2} \exp[-N(\lambda)\sqrt{\mu \log(L/\lambda)}\,]$;

if $F'(x, \lambda, L) \geq$ Threshold (λ, L), then $F'_+ = F'$ and $F'_- = 0$

if $F'(x, \lambda, L) <$ Threshold (λ, L), then $F'_+ = 0$ and $F'_- = F'$.

Finally, let F_+ and F_- be the integrals of F'_+ and F'_-.

The motivation of the above definitions lies in the value of the expectation

$$EF_+(x, r, \lambda, L) = \frac{r^3}{\sqrt{2\pi\mu \log(L/\lambda)}} \int \exp\left\{x - \frac{[x + (\mu/2)\log(L/\lambda)]^2}{4\mu \log(L/\lambda)}\right\} dx,$$

with integration from $\log\{$ Threshold $(\lambda, L)\}$ to infinity. Transforming the integrand into

$$\exp\left\{-\frac{[x - (\mu/2)\log(L/\lambda)]^2}{4\mu \log(L/\lambda)}\right\},$$

and then changing the variable of integration, we obtain

$$EF_+(x, r, \lambda, F) = \frac{r^3}{\sqrt{2\pi}} \int_{-N(\lambda)}^{\infty} f \exp(-z^2/2)\, dz.$$

Conclusion. This form of EF_+ shows that the contribution of F_- to F is asymptotically negligible for $\lambda \to 0$, and that the above choice of $N(\lambda)$ has been appropriate to insure that F is arbitrarily closely approximated by F_+. Moreover, for $\lambda > 0$, the function F' is a.s. continuous, so the variation of F is a.s. concentrated on those intervals where $F' = F'_+$.

Second step: Over any cube of side λ, F' and therefore F'_+, is near constant. Hence

$$F(x, r, \lambda, L) \sim \lambda^3 \sum F'+(x, \lambda, L)$$

with summation carried out over those points of a regular lattice of side λ for which $F' > \text{Threshold}(\lambda, L)$.

Third step: We suspect that there exist circumstances under which $\lim_{\lambda \to 0} F(x, r, \lambda, L) = 0$. At least some of these circumstances may also fulfill the stronger sufficient condition that in a cube of side L, the random number of lattice sites for which $F'_+ > 0$ tends to 0 *almost surely*. A sufficient condition for the latter property is that the number in question should tend to 0 *on the average*. This expected number equals

$$(L/\lambda)^{-3} \Pr\{F'(x, \lambda, L) > \text{Threshold}(\lambda, L)\}.$$

In terms of the r.v.

$$\frac{\log F' + (\mu/2)\log(L/\lambda)}{\sqrt{\mu \log(L/\lambda)}},$$

which is a reduced Gaussian G, the Pr in the equation before last becomes

$$\Pr\{G > \sqrt{\mu \log(L/\lambda)} - N(\lambda)\}.$$

Using the standard tail approximation of G, the expected number in question is about

$$\frac{(L/\lambda)^3 \exp[-(\mu/2)\log(L/\lambda)]}{\sqrt{2\pi\mu\log(L/\lambda)}} = \frac{(L/\lambda)^{3-\mu/2}}{\sqrt{2\pi\mu\log(L/\lambda)}}.$$

Note that this last approximation is independent of $N(\lambda)$ for $\lambda \to 0$.

For this expression to tend to 0 with λ, *a sufficient* condition is $\mu > 6$. (It is also a necessary condition, but this is besides the point; see below.) It follows that, when $\mu > 6$, $\lim_{\lambda \to 0} F(x, r, \lambda, L) = 0$ almost surely. Obviously, the limit is far from being distributed lognormally.

The preceding argument is heuristic, but it is the best I can do in three dimensions. The one dimensional version of the limit lognormal process is easier to study, and the heuristics can be made rigorous by using the Rice formula for the extreme values of a random function. (This is one more reason why it would be desirable to generalize the Rice formula to higher dimensions.)

Extended to the case $\mu < 6$, the preceding argument suggests that the bulk of the variation of F concentrates in approximately $(L/\lambda)^{3-\mu/2}$ cubes of side λ. As $\lambda \to 0$, each cube either is eliminated or becomes subdivided into numerous subcubes. This conclusion is correct, but the above heuristic proof is mathematically very incomplete. The reason is that condition $\lim_{\lambda \to 0} X(\lambda) < \infty$ does *not* exclude the possibility that $\lim_{\lambda \to 0} X(\lambda) = 0$ almost surely. Mathematical concerns of such nature are usually dismissed by physicists, but in the present instance the misbehavior of F for $\mu > 6$ suggests that extreme care is necessary, and different tools are needed to tackle $\mu < 6$.

5. THE LIMIT LOGNORMAL MODEL FOR $\mu<6$: NON-DEGENERATE LIMIT

In this Section, the moments $EF^q(x, r, 0, L)$ will be evaluated for integer q, and then compared with the moments $E[rF'(x, r, L)]$. This last lognormal r.v. provides some kind of link with the WY model. Indeed, it is tempting to reason as follows: $F'(x, r, L)$ varies little over a cube of side r, while the ratio $F'(x, \lambda, L)/F'(x, r, L)$, which equals $F'(x, \lambda, r)$, varies rapidly. Averaged over the cube of side r, this last ratio is bound to be close to its expectation, which is equal to one. This would imply that $F'(x, r, \lambda, L)$ is approximated reasonably by $rF'(x, r, L)$. We shall now check whether or not this is really the case.

The case q = 2. Integrating over the domain where all coordinates of **u** and **v** lie between 0 and r, we have

$$EF^2(x, r, \lambda, L) = E \int \int \int \int \int \int \exp[\log F'(\mathbf{u}, \lambda, L) + \log F'(\mathbf{v}, \lambda, L)] d\mathbf{u} \, d\mathbf{v}$$

$$= \int \int \int \int \int \int E \exp[\log F'(\mathbf{u}, \lambda, L) + \log F'(\mathbf{v}, \lambda, L)] d\mathbf{u} \, d\mathbf{v}.$$

The expression in the exponential is a Gaussian r.v. of expectation $-\mu \log(L/\lambda)$ and variance $2\mu \log(L/\lambda) + 2C(|\mathbf{u} - \mathbf{v}|, \lambda)$. As a result,

$$EF^2(x, r, \lambda, L) = \int \int \int \int \int \int \exp[C(\mathbf{u} - \mathbf{v}, \lambda)] d\mathbf{u} \, d\mathbf{v}.$$

Keep r fixed, with $r \ll L$, and let $\lambda \to 0$. The preceding integral continues to converge if and only if $\mu/2 < 3/2 = 3/q$, in which case its limit for $\lambda \to 0$ equals

$$(2\pi e^\gamma / L)^{-\mu} \int \int \int \int \int \int |\mathbf{u} - \mathbf{v}|^{-\mu} d\mathbf{u} \, d\mathbf{v}.$$

Alternatively, carry the integration over the variables, $\mathbf{u}' = \mathbf{u}/r$ and $\mathbf{v}' = \mathbf{v}/r$, whose values vary from 0 to 1. Then the above second moment converges to

$$r^{6-\mu} L^\mu [(2\pi e^\gamma)^{-\mu} \int \int \int \int \int \int |\mathbf{u}' - \mathbf{v}'|^{-\mu} d\mathbf{u}' \, d\mathbf{v}'].$$

By way of contrast, the would-be approximating lognormal $rF'(x, r, L)$ has a second moment equal to $r^{6-\mu} L^\mu$. The ratio between the limit and the approximate moment is the quantity in brackets. As $\mu \to 0$, its integrand and it prefactor $[2\pi e^\gamma]^{-\mu}$ both tend to 1, and so does the ratio itself.

Suppose that it is true that $\lim_{\lambda \to 0} EF^q(x, r, \lambda, L) = EF^q(x, r, 0, L)$, which is unfortunately not established by the preceding formal calculation. If this were true, it would follow that as $\mu \to 0$, $rF'(x, r, L)$ becomes a good second order approximation to $F(x, r, 0, L)$.

The case $q \geq 3$. By a completely similar calculation, we find that

$$\text{iff } \mu/2 < 3/q, \ \lim_{\lambda \to 0} EF^q(x, r, \lambda, L) < \infty.$$

This suggests that $rF'(x, r, L)$ is a good approximation to $F(x, r, 0, L)$ up to the order $6/\mu$. When μ is very small, F^q has very many finite moments and its low order moments lie near those of the lognormal; one is tempted to describe F itself as being near lognormal.

Now we must tackle the mathematical difficulty concerning the agreement or discrepancy between $\lim_{\lambda \to 0} EF^q(x, r, \lambda, L)$ and $EF^q(x, r, 0, L)$. I am able to give only an incomplete answer to this question. Let $\tilde{p}(\mu)$ be the largest integer satisfying $\mu/2 < 3/\tilde{p}(\mu)$. When $\tilde{p}(\mu) \geq 3$, which implies $\mu < 2$, a standard theorem on martingales (Doob 1953, p.319, Theorem VII, 4.1, clause iii) suffices to establish that $\mu < 2$ is a sufficient condition for $F(x, r, 0, L)$ to be nondegenerate, meaning that $\Pr \{F(x, r, 0, L) > 0\} > 0$. In addition, this theorem establishes that $EF^q(x, r, 0, L) = \lim_{\lambda \to 0} EF^q(x, r, \lambda, L)$. In particular, since $2 < \tilde{p}(\mu)$, the above obtained $\lim_{\lambda \to 0} EF^2(x, r, \lambda, L)$ is indeed the second moment of $F(x, r, 0, L)$.

A bit of additional manipulation establishes that, for all values of q

$$EF^q(x, r, 0, L) = (r/L)^{q(q-1)\mu/2} E(\mu, q),$$

where $0 < E(\mu, q) < \infty$ if $r \ll L$ and $q < \tilde{p}(\mu)$, and $E(\mu, 1) \equiv 1$.

6. MISCELLANEOUS REMARKS

Different forms of correlation. The preceding theory of distribution and of correlation concerns *cubes* of side r or η. Experimental measurements, on the contrary, generally concern averages of ε along thin cylinders of fixed uniform cross section and varying length r. Appropriate changes must be made to extend our results to this case.

Experimental verification of the probability distribution predicted for $F(x, r, 0, L)$. One question must be addressed: are the above results specific to the lognormal model, or do they apply more generally? It has been noted that in the scheme of multiplicative perturbations, the set on which the bulk of variation of $X(t,f)$ occurs is greatly influenced by the tails of the distribution of $\log X'(t,f)$. The central limit theorem gives no information about those tails. More generally, different models of multiplicative perturbations may seem to differ by inconsequential details, yet yield different predictions for the distribution of Kolmogorov's ε. In addition, the alternative models of intermittency belonging to the second broad class mentioned in the introduction, namely the models of Novikov & Stewart 1964 and M 1965c and 1967b{N7, N10}, lead to still different concentration sets, and to probability distributions that are *less scattered than the lognormal*. In other words, the multiplicative model is extremely sensitive to its inputs, and appropriately selected variants could account for distributions that are more scattered or less scattered than the lognormal. In truth, the theory in its present stage offers few predictions that the experimentalist can verify.

Generative models of the scaling law. The interplay we have observed between multiplicative perturbations and the lognormal and scaling distributions has incidental applications in other fields of science where very skew probability distributions are encountered. Notable examples occur in economics, e.g., in the study of the distribution of income. Having mentioned that fact, I leave its elaboration to a more appropriate occasion.

&&&&&&&&&& ANNOTATIONS &&&&&&&&&&

Technical comment on the last paragraph. The last paragraph of M1972j alludes to the following circumstances. M. & Taylor 1967 {E21} had pointed out that the stable processes can be represented (using today's words) as Wiener Brownian motions followed in fractal time. This, my first paper on multifractals, instantly suggested that replacing fractal by multifractal time would yield a new and more general mathematical process showing promise in empirical investigations. In addition, Wiener Brownian motion could be replaced by the fractional Brownian motion introduced in M 1965h{H}.

However, the "more appropriate occasion" called for in the last words of this paragraph did not materialize until after a 25 year delay. Details and references are given in Chapter E6 of M 1997E, the *Selecta* volume devoted to Finance.

Technical comment on multifractals considered as 1/f noises. A side result of this chapter is that the limit lognormal measure has a positive correlation function proportional to t^{-Q}. In loose current terminology, this Q is referred to as a "correlation dimension." In other words, multifractal measures provide an example of f^{-B} noise. This topic is discussed in greater detail in Section N2.2 and the annotations of next chapter.

Roots of lognormality and multiplicative effects in economics. In turbulent dissipation much of the total dissipation is due to deviations that are large, but the very largest peaks are too few to have a significant total contribution. I was prepared to scrutinize the claims for lognormality in Kolmogorov 1962, because I had encountered their counterparts in the totally different context of economics. (See Part III of M 1997E.) That is, I was sensitive to a number of very serious difficulties that were described much later in Chapter E9 of M1997E, unambiguously titled *A case against the lognormal distribution.*

Since most readers of this book are unfamiliar with statistics, it is good to insert at this point some background concerning multiplicative perturbations and lognormality. The standard reference when I was dealing with these matters in the context of economics was Aitchison & Brown 1957. This reference states that the lognormal distribution was first considered in 1879 by a student of Francis Galton. It was rediscovered independently many times.

A great increase in the popularity of multiplicative effects and the lognormal distribution occurred with Gibrat 1931. Robert Gibrat (a French engineer, manager and economist) focussed on economic inequalities such as those in the distribution of personal income. He found that the middle incomes are distributed lognormally, and never faced the fact that the high income tail is definitely *not* lognormal. Pareto's law asserts that this tail follows the scaling distribution. Thus, deviations from lognormality were familiar to experimentalists in fields far removed from the study of turbulence. But those tails were disregarded and in the 1930s Gibrat convinced many statisticians and scientists that the lognormal distribution is, in some way, a basic building block of randomness in nature. That is, many authors feel that no specific justification is needed when randomness is either Gaussian or lognormal, while a specific justification is required for other distributions.

It can be revealed that I have long been dubious about Gibrat's theoretical argument. Contrary to Kolmogorov, Gibrat did not put lognormality as an absolute hypothesis. Instead, very much as Obukhov 1962 was to do for turbulence, Gibrat postulated that log (income) is the sum of many factors, and applied the central limit theorem. No one seemed to be concerned by the fact, that a) one rarely deals with sums of many factors, and b) the central limit theorem says nothing about the tails, while – in economics just like for multifractals – the tails are the interesting portions of the distribution both empirically and theoretically.

Multiplicative effects with a reflecting boundary; critique of their widely advocated use in economics and finance. It is easy to modify Gibrat's argument so that, instead of the lognormal, it leads to the scaling distribution called for by Pareto's law. It suffices to set up a reflecting lower boundary. This is an ancient idea that keeps being resurrected by investigators who wish to use the methods of physics in studies of the social sciences. Strong reservations on those matters are described in Chapter E10 of M 1997E.

The difficult rigorous theory of the limit lognormal multifractal measures. While I had full confidence in the validity of this paper's heuristic results, I was eager to see a rigorous mathematical treatment argument to buttress them. This is why I approached Jean-Pierre Kahane again, sometime in 1972 or 1973, showing him my conjectures. Years before, my heuristics of turbulence inspired Kahane & M 1965{N11}, opening new and interesting "natural" developments in harmonic analysis. That fascinating discipline started with Newton's *spectrum* of light, and with the decomposition of sound into its *harmonics*. With Wiener, it became powerfully affected by the analysis of electrical noises. Authors like Zygmund remained aware of old concrete problems. But others (like my uncle), preferred to "purify" harmonic analysis by forgetting its bright and loud roots. Constant low-key irritation against this attitude makes me welcome every opportunity to demonstrate the continuing power of the "applications" to inspire "pure" mathematics.

In addition to the random multiplicative measures described in this paper, the examples shown to Kahane in 1972 or 1973 included the measures that were later described in M 1974f{N15} and M 1974c{N16}. The latter have a well-defined integer base b, hence can be characterized as as *base-bound*. By contrast, the limit lognormal measures described in this paper can be characterized *base-free*.

As argued in Chapter N1, integer bases are not part of nature, only a mathematical convenience. (I did not know then that physicists use b heavily in renormalization theory; see Chapter N3.) Therefore, I have always strongly favored the base-free measures. Unfortunately, Kahane could not handle them rigorously, as of 1972-3. The base-bound measures, to the contrary, were tackled immediately in Kahane & Peyrière 1976{N17}. It is a well-known fact that when physics and mathematics tackle the same problem, no relation need exist between the levels of technical difficulty that they encounter.

The difficulties Kahane encountered in 1972-3 proved serious, and spurred him to develop delicate new mathematical tools to tackle my construction. He confirmed my base-free conjectures and provided a generalized formal restatement of the limit lognormal multifractals, with new results, as exemplified in Kahane 1987a,b, 1989, 1991a.

Intermittent turbulence in self-similar cascades: divergence of high moments and dimension of the carrier

✦ **Abstract.** Kolmogorov's "third hypothesis" asserts that in intermittent turbulence the average $\bar{\varepsilon}$ of the dissipation ε, taken over any domain $\mathcal{D}$, is ruled by the lognormal probability distribution. This hypothesis will be shown to be logically inconsistent, save under assumptions that are extreme and unlikely. A widely used justification of lognormality due to Yaglom and based on probabilistic argument involving a self-similar cascade, will also be discussed. In this model, lognormality indeed applies strictly when $\mathcal{D}$ is "an eddy," typically a three-dimensional box embedded in a self-similar hierarchy, and may perhaps remain a reasonable approximation when $\mathcal{D}$ consists of a few such eddies. On the other hand, the experimental situation is better described by considering averages taken over essentially one-dimensional domains $\mathcal{D}$.

The first purpose of this paper is to carry out Yaglom's cascade argument, labelled as "microcanonical," for such averaging domains. The second is to replace Yaglom's model by a different, less constrained one, based upon the concept of "canonical cascade." It will be shown, both for one-dimensional domains in a microcanonical cascade, and for all domains in canonical cascades, that in every non-degenerate case the distribution of $\bar{\varepsilon}$ differs from the lognormal distribution. Depending upon various parameters, the discrepancy may be moderate, considerable, or even extreme. In the latter two cases, one finds that the moment $\langle \bar{\varepsilon}^q \rangle$ is infinite if q is high enough. This avoids various paradoxes (to be explored) that are present in Kolmogorov's and Yaglom's approaches.

The paper's third purpose is to note that high-order moments become infinite only when the number of levels of the cascade tends to infinity, as is the case when the internal scale η tends to zero. Granted the usual value of η, this number of levels is actually small, so the limit may not be representative. This issue was investigated through computer simulation. The results bear upon the question of whether Kolmogorov's second hypothesis applies in the face of intermittency.

The paper's fourth purpose is as follows: Yaglom noted that the cascade model predicts that dissipation occurs only in a portion of space of very small total volume. In order to describe the structure of this portion of space, the concept of the "intrinsic fractional dimension" D of the carrier of intermittent turbulence will be introduced

The paper's fifth purpose is to study the relation between the parameters ruling the distribution of $\bar{\varepsilon}$, and those ruling its spectral and dimensional properties. Both conceptually and numerically, these various parameters turn out to be distinct, opening several problems for empirical study. ✦

1. INTRODUCTION AND SYNOPSIS

A striking feature of the distributions of turbulent dissipation in the oceans and the high atmosphere is that both are extremely "spotty" or "intermittent," and that their intermittency is hierarchical. In particular, both are very far from being homogeneous in the sense of the 1941 Kolmogorov-Obukhov theory, in which the rate of dissipation ε was assumed to be uniform in space and constant in time. In intermittent turbulence, ε must be considered a function of time and space. Let $\bar{\varepsilon}\ (\mathcal{D})$ be its spatial average over a domain $\mathcal{D}$. Several approaches to intermittency view $\bar{\varepsilon}$ as lognormally distributed: in Obukhov 1962, lognormality is a pragmatic assumption; in Kolmogorov 1962, it is a basic "third hypothesis" applicable to every domain $\mathcal{D}$; in Yaglom 1966, it is derived from a self-similar cascade model. Yaglom also finds that there is equality between the parameter μ of the lognormal distribution and the exponent in the expressions ruling the correlation and spectral properties of $\bar{\varepsilon}$.

A closely analogous cascade was considered in de Wijs 1951 & 1953, by a geomorphologist concerned with the variability in the distribution of the ores of rare metals. The results in the present paper may therefore be of help outside turbulence theory. A further incidental purpose of this paper is to provide background material to discussion of instances of interplay between multiplicative perturbations and the log-normal and scaling distributions. Such interplay occurs in other fields of science where very skew probability distributions are encountered, notably in economics.

Having mentioned this broader scope of the methods to be described, I shall leave its elaborations to other more appropriate occasions.

While substantial effort is currently being devoted to testing the lognormality experimentally, the purpose of the present paper is to probe the conceptual foundations of lognormality. Like the works of Kolmogorov and Yaglom, this discussion shall be concerned with a phenomenology whose contact with physics remains remote. In particular, the central role of dissipation will not be questioned. On the other hand, greater care will be devoted to matters of internal logical consistency and to details of the assumptions. The relation between theory and experiment will be explored, and will provide a basis for further development of the theory.

Since this paper is somewhat lengthy, the mathematics that has as yet no other application in fluid mechanics will be postponed to Sections 4 and 5. The main results will be stated without proof in this section and in Section 2. Section 3 will elaborate on the important distinction between microcanonical and canonical cascades.

(a) Part of this paper is devoted to a new calculation relative to Yaglom's cascade model for Kolmogorov's hypothesis of lognormality. Let $\bar{\varepsilon}(\mathcal{D})$ be the average of the dissipation ε over a spatial domain $\mathcal{D}$. One form of Yaglom's model assumes that $\mathcal{D}$ is an "eddy," perhaps a three-dimensional cube embedded in a self-similar hierarchy. On the other hand, in all actually observed averages, $\mathcal{D}$ is not a cube, but is more nearly a very thin cylinder. By following up the consequences of Yaglom's model in this case, it will be shown that $\bar{\varepsilon}(\mathcal{D})$ is never lognormal, and that its "qualitative" behavior can fall into any one of three classes:

• In a first class, which is drastically extreme and which will be called "regular," $\bar{\varepsilon}(\mathcal{D})$ is not far from being lognormal.

• In a second class, which is equally extreme and which will be called "degenerate," all dissipation concentrates in a few huge blobs.

• In a third class, which is intermediate between the above two and which will be called "irregular," $\bar{\varepsilon}(\mathcal{D})$ is non-degenerate but is far from lognormal.

The most striking characteristic of the third class is a parameter α_1, satisfying $1 < \alpha_1 < \infty$, which rules the moments (ensemble averages) $\langle \bar{\varepsilon}^q(\mathcal{D}) \rangle$. When $q < \alpha_1$, one has $\langle \bar{\varepsilon}^q(\mathcal{D}) \rangle < \infty$ for all values of the inner scale η, but when $q > \alpha_1$ and one has $\eta = 0$, $\langle \bar{\varepsilon}^q(\mathcal{D}) \rangle = \infty$. Finally, when $q > \alpha_1$ and η is positive but small, $\langle \bar{\varepsilon}^q(\mathcal{D}) \rangle$ is huge, and its precise value is so dependent upon η as to be meaningless. The regular class can be viewed as being the limiting case $\alpha_1 = \infty$, and the degenerate class as corresponding to $\alpha_1 \leq 1$. This eliminates numerous inconsistencies that have been noted in the literature, concerning the behavior of the moments of $\bar{\varepsilon}$ under the lognormal hypothesis.

(b) Yaglom's model involves, although only implicitly, a hypothesis of rigorous local conservation of dissipation within eddies. This feature will be is said to characterize his cascade as being "microcanonical." Another part of this paper will view conservation as holding only on the average; and the resulting cascades, called "canonical," will be investigated. When a cascade is canonical, the behavior of $\bar{\varepsilon}(\mathcal{D})$ will be seen to fall under the same three classes as have been defined above under (a), when $\mathcal{D}$ is cubic eddy, except that the parameter α_1 must be replaced by a new parameter $\alpha_3 > \alpha_1$. In the same cascade, averages taken over cylinders and eddies may fall in different classes; for example, a regular $\bar{\varepsilon}(\mathcal{D})$ when $\mathcal{D}$ is an eddy is compatible with an irregular $\bar{\varepsilon}(\mathcal{D})$ when $\mathcal{D}$ is a cylinder; also, an irregular $\bar{\varepsilon}(\mathcal{D})$ when $\mathcal{D}$ is an eddy is compatible with a degenerate $\bar{\varepsilon}(\mathcal{D})$ when $\mathcal{D}$ is a cylinder.

(c) Another aspect of this paper is purely critical. It concerns Kolmogorov's second hypothesis, which asserts that the value of η does not influence $\bar{\varepsilon}(\mathcal{D})$ in the similarity range. This will indeed be confirmed when $\bar{\varepsilon}(\mathcal{D})$ is in the regular class for every domain $\mathcal{D}$, but will be disproved when all $\bar{\varepsilon}(\mathcal{D})$ are in the degenerate class. In all other cases, the hypothesis is doubtful. Thus, the domains of validity of the second and third hypothesis are related.

(d) This paper introduces, in passing, a new concept, which will be developed fully elsewhere. In the regular and irregular classes, the bulk of intermittent dissipation is shown to occur over a very small portion of space, which will be shown to be best characterized by a parameter D called the "intrinsic fractional dimension" of the carrier. The parameter D is preferable to the relative volume, because the volume is very small and too dependent upon η.

(e) Yaglom's theory introduces yet another parameter, which characterizes the spectrum of $\bar{\varepsilon}$ and is related to a correction factor to the exponent $-5/3$ of the classic Kolmogorov power law. This parameter will be denoted by Q. The parameters α_1, D and Q will be shown to be conceptually distinct. Naturally, the introduction of any additional assumption about the cascade introduces a relation among these parameters. For example, one may, under a special assumption, come close to Kolmogorov-Yaglom theory, and find that α_1, D and Q are functions of a single parameter μ. The question of whether or not the actual parameters are distinct suggests much work to the experimentalist.

(f) For the sake of numerical illustration, a variety of one-dimensional canonical cascades was simulated on a digital computer, IBM System 360/Model 91. The results, unfortunately, cannot be described in this paper. Suffice to say that they confirm the theoretical predictions concerning the limiting behavior, but throw doubt upon the rapidity of convergence to the limit.

2. BACKGROUND AND PRINCIPAL RESULTS

2.1. Background: Yaglom's postulate of independence and lognormality

The purpose of this section is to amplify items (a), (b), (c) and (f) of Section 1. To do so, we shall first describe Yaglom's cascade model in narrower and more specific form. (It is hoped that the spirit of Yaglom's approach is thereby left unaltered.)

To begin with, the skeleton of the cascade process is taken to be made of "eddies" that are prescribed form the outset and which are cubes such that each cubic eddy at a given hierarchical level includes C cubic eddies of the immediately lower level. (C is the initial of "cell number.") This expresses the fact that the grid of eddies is self-similar in the range from η to L. Obviously, $C^{1/3}$ must be assumed to be an integer and is denoted by b. The sides (edge lengths) of the largest and the smallest eddies are equal to the external scale L and internal scale η respectively.

The unit of length will be chosen such that η and L are only dimensionless powers of b. The density of turbulent dissipation at the point $\mathbf{x}$ is denoted by $\varepsilon(\mathbf{x}, L, \eta)$ and the density average over the domain $\mathcal{D}$ will be denoted by $\bar{\varepsilon}(\mathcal{D}, L, \eta)$. Units of dissipation will again be such that $\bar{\varepsilon}$ is dimensionless. When $\mathcal{D}$ is a cubic eddy of side r and center $\mathbf{x}$ (with $-\log_b r$ an integer) we write

$$\bar{\varepsilon}(\mathcal{D}, L, \eta) = \bar{\varepsilon}_r(\mathbf{x}, L, \eta).$$

It is further assumed that the distribution of dissipation over its self-similar grid is itself self-similar. This means that, whenever $\eta \ll r < rb \ll L$, the ratio $\bar{\varepsilon}_{r/b}(\mathbf{x}_s, L, \eta)/\bar{\varepsilon}_r(\mathbf{x}, L, \eta)$ is a random variable, to be denoted by Y_s, having a distribution independent of r. Here, $\{\mathbf{x}_s\}$ is a regular grid of centers of subeddies.

Next (an assumption that goes beyond self-similarity), the successive ratios $\bar{\varepsilon}_{L/b}/\bar{\varepsilon}_L$, $\bar{\varepsilon}_{L/b^2}/\bar{\varepsilon}_{L/b}$, etc. down to $\bar{\varepsilon}_r/\bar{\varepsilon}_{rb}$, are assumed independent. This makes $\log \bar{\varepsilon}_r - \log \bar{\varepsilon}_L$ the sum of $\log_b(L/r)$ independent expressions, each of which is of the form $\log Y$. Finally, assume $\langle (\log Y)^2 \rangle < \infty$, a condition which implies that $\Pr\{Y=0\} = 0$. This means that $\log \bar{\varepsilon}_r - \log \bar{\varepsilon}_L$ is a finite sum from a series that would, if carried out to infinity, satisfy the central limit theorem. One concludes that $\log \bar{\varepsilon}_r$ is approximately Gaussian, meaning that $\bar{\varepsilon}_r$ is approximately lognormal.

At this point, the reader may digress to the appendices A1 and A2, which comment about lognormality.

2.2. Dissipation averaged over thin cylinders

Nevertheless, there are several reasons why, even when all of Yaglom's assumptions are accepted, the argument sketched above does not suffice to justify Kolmogorov's third hypothesis, that $\bar{\varepsilon}$ is lognormal for all $\mathcal{D}$. First (not the basic reason), Yaglom's argument is rigorous only when $\mathcal{D}$ is a cubic eddy. When $\mathcal{D}$ (while three-dimensional) is not an eddy, lognormality is at best approximate. The reason why this argument is not basic is that, for every three-dimensional $\mathcal{D}$ the moments $\langle \bar{\varepsilon}^q \rangle$ are finite for all q. A second argument is more basic and concerns the comparison of theory and experiment. Even though averages taken over three-dimensional domains, $\mathcal{D}$ may be appropriate for a theoretical characterization of turbulence (including the hoped-for linkage between the present phenomenology and actual physics), such averages cannot be measured experimentally. Actual measurements, by necessity, involve averages taken over thin cylinders in time and space. By G.I. Taylor's "frozen turbulence hypothesis," such domains can be replaced by thin cylinders through the spatial flow. When the radius of such a $\mathcal{D}$ is of the same order of magnitude as the inner scale η of the turbulence, $\mathcal{D}$ can be approximated by a one-dimensional straight segment. Thus one must raise the question of whether or not the distribution of $\bar{\varepsilon}(\mathcal{D}, L, \eta)$ remains approximately lognormal when $\mathcal{D}$ is one dimensional. Yaglom does not raise explicitly this question in his works, nor do later writers concerned with the extent to which the observed data fit the lognormal distribution. But all these authors imply that the dimensionality of the averaging domain $\mathcal{D}$ has little effect on the distribution of $\bar{\varepsilon}(\mathcal{D}, L, \eta)$.

This paper will show this implicit belief to be unwarranted. More precisely, whenever $\mathcal{D}$ is not an eddy, the distribution of $\bar{\varepsilon}(\mathcal{D}, L, \eta)$ changes depending which detailed assumptions are made about the cascade process. The assumptions, made in this paper will now be described. We shall evaluate the average of $\bar{\varepsilon}(\mathcal{D}_\eta, L, \eta)$ when $\mathcal{D}_\eta$ is a cylinder of length r and radius η, and we shall find that, except in trivial circumstances, the average is *not* lognormal.

Throughout, the hierarchy of eddies is not viewed as a physical phenomenon, but as a formal device for constructing the fully cascaded state of the medium. Every stage of the division of space will be assumed to preserve the total dissipation. Hence the average over the whole sample of the local averages over eddies. The simplest procedure is to assume nothing else about the corresponding Yaglom ratios Y. The resulting cascade will be called "microcanonical." Consider successive ratios of the form $\bar{\varepsilon}(\mathcal{D}', L, \eta)/\bar{\varepsilon}(\mathcal{D}'', L, \eta)$, where $\mathcal{D}'$ and $\mathcal{D}''$ are cylinders of identical length r but different cross-sections (with $\mathcal{D}'$ embedded in $\mathcal{D}''$). It will be found that these ratios are *not* independent. In order to formalize the limit process of Yaglom, we shall view the internal scale η as a variable tending to zero. In Sections 3 and 4 it will be proved that, except in a trivial case, the distribution of the limit $\bar{\varepsilon}(\mathcal{D}_0, L, 0)$ where $\mathcal{D}_0$ is an infinitely thin cylinder, is never lognormal. In some cases, the difference is small, but in other cases it is great, implying that the influence of the dimension of $\mathcal{D}$ over the distribution of $\bar{\varepsilon}(\mathcal{D}_0, L, 0)$ may be critical. The extent of the divergence of the distribution from the lognormal is expressed to a significant extent by the value of a parameter, denoted as α_1, which is defined as the second zero (the first being $q = 1$) of the expression

$$\tau_1(q) = \log_b \langle Y^q \rangle - (q-1).$$

The definition of α_1 is motivated in Section 4.3, and illustrated on Figure 1. (The latter uses the notation W instead of Y; the relationship between the two will be explained in Section 3.)

• The first class is called "regular," and includes all Y that are bounded by b. It is characterized by $\alpha_1 = \infty$. The resulting $\bar{\varepsilon}(\mathcal{D}_0, L, 0)$ differ little from lognormality. This factor is random but essentially independent of η, and all its moments are finite.

• The second class is called "degenerate," and corresponds to Y's that are extremely scattered. It is characterized by $\alpha_1 \leq 1$. The resulting $\bar{\varepsilon}(\mathcal{D}_0, L, 0)$ vanishes almost surely. In particular, $\langle \bar{\varepsilon}^q(\mathcal{D}_0, L, 0) \rangle = 0$ for every q.

Even though "physical intuition" suggests the opposite, the fact that $\bar{\varepsilon}(\mathcal{D}_0, L, 0) \equiv 0$, hence $\langle \bar{\varepsilon}(\mathcal{D}_0, L, 0) \rangle = 0$, is perfectly compatible with the combination of $\lim_{\eta \to 0} \langle \bar{\varepsilon}(\mathcal{D}_\eta, L, \eta) \rangle = 1$, $\lim_{\eta \to 0} \langle \bar{\varepsilon}^q(\mathcal{D}_\eta, L, \eta) \rangle = 0$ for $q < 1$, and

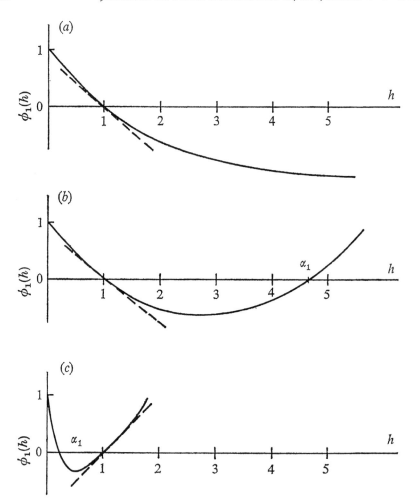

FIGURE N15-1. The distribution of the averaged dissipation $\bar{\varepsilon}$ is determined by that of the random weight W, which is roughly speaking Yaglom's ratio between the average dissipation with a subeddy and an eddy. We plot the function $\bar{T}_1(q) = \log_b \langle W^q \rangle - (q-1)$, which is always cup-convex. Through $\bar{T}_1(q)$, all presently interesting aspects of intermittency are described as follows. The spectral properties of $\bar{\varepsilon}$ (the only one to have been examined before the present study) are determined by the value of $\bar{T}_1(2)$. In addition, the distribution properties (finiteness of moments) depend on the value of α_1, defined as the root (other than $q=1$) of the equation $\bar{T}_1(q) = 0$. Finally, the fractional dimension of the carrier of turbulence depends on the values of $\bar{T}'_1(1)$. Thus, from the viewpoint of properties of $\bar{\varepsilon}$, its distributions falls into the following three classes. (a) Regular class: $\bar{T}'_1(1) < 0$, $\bar{T}_1(2) < 0$ and $\alpha_1 = \infty$. (b) Irregular class: $\bar{T}'_1(1) < 0$, $\bar{T}'_1(2) < 0$ and $1 < \alpha_1 < \infty$. (c) Degenerate class: $\bar{T}'_1(1) > 0$, $\bar{T}_1(2) > 0$ and $0 < \alpha_1 < 1$.

$\lim_{\eta \to 0} \langle \bar{\varepsilon}^q(\mathcal{D}_\eta, L, \eta) \rangle = \infty$ for $q > 1$. Such a drastic discrepancy between the moments of the limits and the limits of the moments is ordinarily achieved by a deliberate effort to create a mathematical pathology. A classical illustration is the sequence for which $\bar{\varepsilon}(\mathcal{D}, L, \eta)$ equals $1/\eta$ with probability η, and equals zero with probability $1 - \eta$. Here, to the contrary, this discrepancy direct practical consequences. (This is a bit reminiscent of the singularity, familiar in fluid mechanics, encountered when the coefficient of viscosity tends to zero.) Thus the degenerate case suggests that, when η is non-zero but small, dissipation concentrates in a few huge blobs.

• The third class is called "irregular," and includes all Y's that are not scattered beyond reason, but can exceed b. It is characterized by $1 < \alpha_1 < \infty$. In this class, $\langle \bar{\varepsilon}(\mathcal{D}, L, \eta) \rangle$ remains identically equal to one, while higher moments $\langle \bar{\varepsilon}^q \rangle$ behave as follows: when $q < \alpha_1$, they remain finite as $\eta \to 0$, but when $q > \alpha_1$, they tend to infinity. This implies that when η is positive but small, their values are extremely large and in practice can be considered infinite.

When a probabilist knows that moments behave as stated above, with the loose additional requirement that the function $\Pr\{\bar{\varepsilon} > x\}$ is "smooth," the simplest distribution he is likely to envisage is the "scaling," defined as follows: $\min \bar{\varepsilon} = x_0 = \alpha_1/(\alpha_1 - 1) > 0$ and $\Pr\{\bar{\varepsilon} > x\} = (x/x_0)^{-\alpha_1}$. The next simplest possibility is $\Pr\{\bar{\varepsilon} > x\} = C(x)x^{-\alpha_1}$, where the prefactor $C(x)$ is a function that varies "smoothly and slowly" as $x \to \infty$. (Examples are functions with a non-trivial limit, and functions that vary like $\log x$ or $1/\log x$). Such random variables $\bar{\varepsilon}$ are called "asymptotically scaling" or "Paretian." To test for their occurrence, it is common practice to plot $\log \Pr\{\bar{\varepsilon} > x\}$ as a function of $\log x$: the tail of the resulting curve should be straight and of slope α_1. However, the more interesting prediction concerns the case when η is very small but positive. In that case, all moments of $\bar{\varepsilon}(\mathcal{D}, L, \eta)$ are very large but finite. If its distribution is again plotted in log-log coordinates, it must end on a tail that plunges down more rapidly than any straight line of finite slope. But, the behavior of the moments of $\bar{\varepsilon}$ as $\eta \to 0$ also yields a definite prediction for small η, namely: *the log-log plot of the distribution is expected to include a long "penultimate" range within which it is straight and of slope α_1. This is one of the principal predictions of the present work.*

This contrast between Yaglom's conclusions and mine turns out to be parallel to the contrast between two classic chapters of probability theory. (a) In the theory of sums of many nearly independent random variables, the asymptotic distribution is, under wide conditions, universal: it is Gaussian. (b) In the theory of the number of offspring in a birth-and-death process, the asymptotic distribution depends upon the distribution of a number of offspring per generation: it is not universal. Using statistical mechanics, the thermodynamics properties of matter had been reduced to theory (a) above, which is why they are largely independent of microscopic mechanical detail. What Yaglom claims in effect is that the same is true of turbulent intermittency. To the contrary, Section 4.3 shows that turbulence is closer to theory (b), the resulting absence of universality probably being intrinsic. More precisely, the theory underlying this paper is an aspect of the "theory of birth, death and random walk."

The ease of verifying this prediction increases as the slope α_1 becomes less steep. In an approximation discussed in Section 4.8, the value of α_1 can be inferred from the spectral exponent $Q = \mu$, as being equal to about $2/\mu \sim 4$. This suggests that moments should misbehave for $q \geq 4$. Further discussion of the validity of this prediction must be postponed until more data are available.

2.3. Validity of the microcanonical assumption and reasons for introducing the canonical cascades

The second purpose of this paper is to probe Yaglom's assumption that the ratios of the form $\bar{\varepsilon}(\mathcal{D}_s, L, \eta)/\bar{\varepsilon}(\mathcal{D}, L, \eta)$, relative to a subeddy $\mathcal{D}_s$ and to an eddy $\mathcal{D}$ containing $\mathcal{D}_s$, are independent. We noted that this is satisfied by the microcanonical model, in which the cascade is merely a way of splitting up space. But less formalistic interpretations are conceivable.

For example, one may keep to the approximation that a cascade divides an eddy exactly into subeddies, but combine splitting with some kind of diffusion, in such a way that conservation of dissipation only holds on the average. (Physically, the "dissipation" invoked in this proposal would correspond to the energy transfer between eddy sizes, rather than to the ultimate conversion of eddy kinetic energy into heat.)

The resulting model is to be called "canonical." It is interesting because (a) when $\mathcal{D}$ is a cylinder the results it yields are essentially the same as in the microcanonical model, and (b) Yaglom's ratios turn out to be so strongly interdependent that $\bar{\varepsilon}(\mathcal{D})$ fails to be lognormal even when $\mathcal{D}$ is an eddy. The theory of the canonical $\bar{\varepsilon}(\mathcal{D})$ with an eddy $\mathcal{D}$ follows the same pattern as the theory of the microcanonical $\bar{\varepsilon}(\mathcal{D})$, with $\mathcal{D}$ a cylinder. Thus, it can fall into any of the three classes noted in Section 2.2, with the change that one must replace α_1 by a new parameter α_3. Ordinarily, $\alpha_3 > \alpha_1$.

Since in some cases the predictions of the canonical and the microcanonical models are very different, the degree of validity of Yaglom's model depends on the solidity of the foundations of the microcanonical assumption. It would be nice if either kind of cascade turned out to have a more precise relationship with the physical breakdown of eddies but, so far, no connection has been established. As a matter of fact, the accepted role dissipation plays in the current phenomenological approach to turbulence should perhaps be downgraded, and the canonical model of a cascade be rephrased in terms of energy transfer between different scale sizes. Nevertheless, attempting this task would go beyond the purpose of the present work, and we shall stick to the logical analysis of the cascades. The relative advantages and disadvantages of the two main models are as follows.

Yaglom's argument. In the case of cylinders, this argument requires amplification that may lead to substantially non-lognormal results; in the case of cubes, it is disputable.

The canonical alternative. In the case of cylinders, this alternative appears to be a nearly inevitable approximation, and in the case of cubes, it may well be an improvement.

2.4. Kolmogorov's "second hypothesis of similarity"

The possibility of peculiar behavior of the moments leads us to probe Kolmogorov's second hypothesis, which was stated originally (1941) for homogeneous turbulence and was generalized in Kolmogorov 1962 to intermittent turbulence. Intuitively, if $\mathcal{D}$ is a domain of characteristic scale $\gg \eta$, the second hypothesis states that the distribution of $\bar{\varepsilon}(\mathcal{D}, L, \eta)$ is nearly independent of η. To restate it rigorously, let us make η into a parameter and let it tend to zero. Kolmogorov's second hypothesis may merely state that $\lim_{\eta \to 0} \bar{\varepsilon}(\mathcal{D}, L, \eta)$ exists. If it does, and if (as is usual in mathematics) the concept of a limit is interpreted through "convergence of probability distributions," then for both the canonical and the microcanonical cascades the second hypothesis will indeed be satisfied. But mathematical convergence need not be intuitively satisfactory, and the second hypothesis ought perhaps to be interpreted in stronger terms.

When $\mathcal{D}$ is an eddy of the microcanonical model, and for other D's models leading to the regular class, we have $\bar{\varepsilon}(\mathcal{D}, L, \eta) \to \bar{\varepsilon}(\mathcal{D}, L, 0)$ mathematically and, for all $q > 0$, $\langle \bar{\varepsilon}^q(\mathcal{D}, L, \eta) \rangle \to \langle \bar{\varepsilon}^q(\mathcal{D}, L, 0) \rangle$. In this case, as long as η is small, the "error term" $\bar{\varepsilon}(\mathcal{D}, L, \eta) - \bar{\varepsilon}(\mathcal{D}, L, 0)$ should be expected to be small. Kolmogorov's intuitive second hypothesis holds uncontroversially. When convergence is regular for both eddies and cylinders, the Kolmogorov-Yaglom lognormal approximation is (up to a fixed correction factor) workable.

In the degenerate convergence class, on the contrary, Pr$\{\bar{\epsilon}(\mathcal{D}, L, 0) = 0\} = 1$. For small but positive η, $\bar{\epsilon}(\mathcal{D}, L, \eta)$ and $\bar{\epsilon}(\mathcal{D}, L, 0)$ may be mathematically close, but are intuitively very different. The actual behavior of $\bar{\epsilon}$ when convergence is degenerate appears to resemble the illustrative example given above. As a result, Kolmogorov's second hypothesis is not really applicable to this class.

When convergence is irregular and η is small, the error term $\bar{\epsilon}(\mathcal{D}, L, \eta) - \bar{\epsilon}(\mathcal{D}, L, 0)$ is extremely likely to be small. But in cases when it is not small, it may be very large, and its own moments of high order may be infinite. In this case, the Kolmogorov second hypothesis is controversial, but its degree of validity improves as α_1 increases.

2.5. Relationship between the canonical cascade and the "limiting lognormal" model in M 1972j{N14}

Both the canonical and the microcanonical variants allow the distribution of dissipation between neighboring subeddies to be highly discontinuous. However, M 1972j{N14}, has investigated yet another alternative cascade model, using a "limiting lognormal process." Its principal characteristic is that it generates its own eddies of different shapes, and that the distribution of dissipation within eddies is continuous. This feature will appear especially attractive when the study of the geometry of the carrier of turbulence is pushed beyond the concept of fractional dimension, to include matters of connectedness. The limiting lognormal model can be viewed, though it was developed first, as an improvement upon a canonical cascade with a lognormal weight W. Section A4 will describe its main characteristics.

2.6. Study by computer simulation of the rapidity of convergence in the canonical cascade process

As always in the application of probability theory, limit cascades (involving infinitely many stages) are of practical interest primarily because the formulae relative to actual cascades (in which the number of stages is large but finite) are unmanageable. The present paper goes a step further, by including "qualitative" arguments about the nature of error terms for finite cascades. In addition, I have arranged numerous computer simulations. The very tentative conclusions are (a) that many of the involved discrepancies from lognormality should manifest themselves only in a relatively small number of exceptionally large observations, and (b) that they depend greatly upon high-order moments of the Yaglom ratio Y, which express comparatively minute characteristics of the cascade model.

If these inference are confirmed, then lognormality may combine the worst of two worlds: it could prove fairly reasonable qualitatively, while its use for any calculation that involves moments could not be trusted. If so, even Obukhov, who did evaluate moments, would prove less pragmatic than he thought. However, having expressed those fears, I hasten to say that I do not share them, and that I believe the study of intermittency to be very enlightening as to the nature of turbulence.

3. INTRODUCTION TO CANONICAL AND MICROCANONICAL CASCADES

3.1. A detailed cascade model

To be able to make a prediction about $\bar{\varepsilon}(\mathcal{D}, L, \eta)$ when $\mathcal{D}$ is a cylinder, one must make assumptions about the local distribution of ε within eddies. We shall build a model by making η smaller and smaller.

Initially, $\eta = L$ and the original dissipation $\bar{\varepsilon}(x, L, L)$ is uniformly distributed in space. At the beginning of each successive stage of the cascade, one assumes that dissipation density is uniform within each eddy of side r. This is also the initial distribution one observes if $\eta = r$; it can therefore be denoted by $\varepsilon(\mathbf{x}, L, r)$. At the end of each stage of the cascade, the dissipation density is uniform in each subeddy of side r/b. When the center of an eddy of side r is denoted by $\mathbf{x}$ the centers of the immediately smaller subeddies will be denoted by $\mathbf{x}_s$, with $0 \leq s \leq C - 1$; they form a regular lattice. The corresponding densities will be denoted by $\varepsilon(\mathbf{x}_s, L, r/b)$. Next, designate the random variable $\varepsilon(\mathbf{x}_s, L, r/b)/\varepsilon(\mathbf{x}, L, r)$ by W_s. The ratio W and Yaglom's ratio Y differ by the fact that W involves local densities, while Y involves averages, but in the microcanonical model the concepts of W and Y will merge. Homogeneity suggests that, at each cascade stage, the s random variables of the form W_s have the same distribution. Self-similarity and Kolmogorov's second hypothesis suggest in addition that the distribution is the same for all values of s, r, L and η. The final stage ends with eddies of side η, and with density $\bar{\varepsilon}(\mathbf{x}, L, \eta)$.

The low- and high-frequency multiplicative factors of $\bar{\varepsilon}(\Omega, L, \eta)$. The random variable $\varepsilon(\mathbf{x}, L, \eta)$ resulting from the above cascade has a single parameter: L/η. Moreover, since the actions of eddies of sides above and below r are quite separate, $\varepsilon_r(\mathbf{x}, L, \eta)$ can be written as the product of two statistically independent factors, which can be studied separately. These factors are, respectively, $\bar{\varepsilon}_r(\mathbf{x}, L, r)$ and $\bar{\varepsilon}_r(\mathbf{x}, r, \eta)$. The former is independent of η and has r/L as its sole parameter; it is a "low frequency factor." The latter is independent of L and has r/η as the sole parameter; it is a "high frequency factor." More generally, when $\mathcal{D}$ is not an eddy but is included in an eddy of side r,

$$\bar{\varepsilon}(\mathcal{D}, L, \eta) = \bar{\varepsilon}_r(\mathbf{x}, L, r)\bar{\varepsilon}(\mathcal{D}, r, \eta).$$

3.2. The approximate lognormality of the low frequency factor $\bar{\varepsilon}_r(\mathbf{x}, L, r)$ and the question of whether or not W can take the value zero

To study the low frequency factor, it suffices to follow Yaglom, as in Section 2.1. One notes that $\log \bar{\varepsilon}_r(\mathbf{x}, L, r)$ is the sum of $\log_b(L/r)$ random factors of the form W. Assuming $\langle (\log W)^2 \rangle < \infty$, the Gaussian central limit theorem seems to suggest that $\log \bar{\varepsilon}_r \mathbf{x}, L, r$ is approximately normally distributed, it would follow that $\bar{\varepsilon}_r(\mathbf{x}, L, r)$ is approximately lognormal.

A finite $\langle (\log W)^2 \rangle$ implies in particular that $W = 0$ has zero probability. On the other hand, there is a model by Novikov & Stewart 1964 which assumes that $W = 0$ has a positive probability. In that case, $\bar{\varepsilon}_r(\mathbf{x}, L, r)$ is usually a mixture: with some positive probability, it vanishes, and with the remaining probability, it is lognormal. In the present paper, to allow $W = 0$ will not cause any complication, and in fact will allow consideration of useful simple examples.

3.3. The high frequency factor; limit behavior for $\eta \to 0$

This limiting behavior is ruled by the following theorem (stated at the intermediate level of generality at which the proof is simplest, a level more general than is required and less general than is possible).

Theorem. Let the domain $\mathcal{D}$ be simple, meaning that $\mathcal{D}$ is the sum of a finite number of eddies when $E = 3$, and the sum of a finite number of eddy edges when $E = 1$. Consider $\bar{\varepsilon}(\mathcal{D}, L, \eta)$ (for fixed $\mathcal{D}$ and L) as a random function of η. Assume $\langle W \rangle = 1$ and let $\eta \to 0$. Then, with probability equal to 1, $\bar{\varepsilon}$ tends to a finite limit random variable.

Proof. This proof is written as a digression addressed to readers having an elementary knowledge of the theory of "martingales." This theory is the next most obvious mathematical generalization of the theory of products of independent random variables of unit expectation, such as Yaglom ratios. In order to conform to the usual presentation of martingales, let us view the actual value of the inner scale η as the "present value," values $\eta' < \eta$ and $\eta' > \eta$ being viewed respectively as "future" and "past." A martingale is a random function such that the expectation of a "future" value, conditioned by knowing the present value and any number of past values, is equal to the present value. Here "time" is discrete, being equal to $-\log_b \eta$. Assume that $\mathcal{D}$ is an eddy of side r; a similar argument applies to other simple $\mathcal{D}$'s. Denote its subeddies of side r/b by $\mathcal{D}_s$. We know that

$$\bar{\varepsilon}(\mathcal{D}, L, \eta/b) = \frac{1}{C} \sum_{s=0}^{C-1} W_s \bar{\varepsilon}(\mathcal{D}_s, L, \eta).$$

Designate by E_C the conditional expectation, given the present and any one of past values of $\bar{\varepsilon}(\mathcal{D}, L, \eta)$. Since $\langle W \rangle = 1$, we have

$$E_C \bar{\varepsilon}(\mathcal{D}, L, \eta/b) = \frac{1}{C} \sum_{s=0}^{C-1} E_C \bar{\varepsilon}(\mathcal{D}_s, L, \eta) = E_C \bar{\varepsilon}(\mathcal{D}, L, \eta) = \bar{\varepsilon}(\mathcal{D}, L, \eta).$$

This proves that $\bar{\varepsilon}(\mathcal{D}, L, \eta)$ is a martingale. Being non-negative, $\bar{\varepsilon}$ obeys a convergence theorem (Doob 1953, p. 319): as $\eta \to 0$, $\bar{\varepsilon}(\mathcal{D}, L, \eta)$ has a limit random variable to be denoted by $\bar{\varepsilon}(\mathcal{D}, L, 0)$.

Corollary. In the case of cubic eddies, $\bar{\varepsilon}_r(\mathbf{x}, r, \eta)$ converges to a limit $\bar{\varepsilon}_r(\mathbf{x}, r, 0)$. By self-similarity, the limit is independent of r, so it can be denoted by $\bar{\varepsilon}_1(\mathbf{x}, 1, 0)$.

Remark. The above theorem means that, when $r/\eta \gg 1$, one knows $\bar{\varepsilon}(\mathcal{D}, L, \eta)$. "approximately" without knowing the exact value of η. However, any more detailed information about the quality of approximation involves the character of the convergence of $\bar{\varepsilon}(\mathcal{D}, L, \eta)$ to $\bar{\varepsilon}(\mathcal{D}, L, 0)$ (regular, irregular or degenerate), and in turn requires more detailed assumptions about the model (e.g., about the set of random variables W).

3.4. The microcanonical cascade

Definition. *A cascade will be called microcanonical if the sum $\sum_{s=0}^{C-1} W_s$ of the weights W_s corresponding to all the subeddies of any eddy is precisely equal to C.* {P.S. My more recent papers use the more self-explanatory term, *conservative..*}

As a corollary, $\langle W \rangle = 1$ and $W < C$.

The microcanonical condition expresses that, at each cascade stage, the total dissipation $r^3\varepsilon(\mathbf{x}, L, r)$ within an original eddy is replaced by an *equal dissipation* distributed among its C subeddies of centers $\mathbf{x}_s$, namely

$$\sum_{s=0}^{C-1} \frac{r^3}{C} \varepsilon(\mathbf{x}_s, L, r/b) = \sum_{s=0}^{C-1} \frac{W_s}{C} [r^3 \varepsilon(\mathbf{x}, L, r)]$$

Hence, as long as $\eta < r$, one has $\bar{\varepsilon}_r(\mathbf{x}, L, \eta) = \varepsilon(\mathbf{x}, L, \eta)$. This result is independent of η, and shows that the high frequency factor $\bar{\varepsilon}_r(\mathbf{x}, L, \eta)$ is identically equal to 1, which makes it independent of η. Consequently, Yaglom's ratio Y_s coincides with W_s and his postulate of independence is satisfied. Thus, *the theory of microcanonical averages taken over three-dimensional eddies is seen to coincide with Yaglom's theory.*

The converse, that Yaglom's theory is identical to the microcanonical theory, is also true, under certain additional constraints, but there is no need to digress for the proof.

Notice that the microcanonical weights W_s are statistically dependent. In particular, if $s \neq t$,

$$\langle W_s W_t \rangle = \langle W_s E_C(W_t | \text{ knowing } W_s) \rangle = \left\langle \frac{W_s(C - W_s)}{(C - 1)} \right\rangle$$

$$= 1 - (\langle W^2 \rangle - 1)/(C - 1) < 1.$$

Therefore, $\langle W_s W_t \rangle < \langle W_s \rangle \langle W_t \rangle$. This inequality expresses that any two weights are negatively correlated (see Section A5). There are analogous results for higher cross-moments; for example, $\langle W_s^2 W_t \rangle < 1$.

3.5. The canonical cascade

Definition. *A cascade will be called canonical if the weights W_s are statistically independent and satisfy $\langle W \rangle = 1$,* meaning that the sum of the weights is equal to C on the average. In order to obtain features that go beyond the microcanonical case, it is critically important to allow W to exceed the ceiling $W = C$.

The canonical variant as an approximation for cylinder averages in a microcanonical cascade. Consider a cylinder of length r constituted by a string of elementary eddies of side η hugging one edge (to be called the "marked edge") of a cubic eddy of side r. The dissipation in this cylinder can be obtained through a sequence of two different subcascades. The first subcascade, applicable until an eddy of side r has been reached, follows the mechanism described in Section 3.3, typically ending up with a lognormal $\bar{\varepsilon}_r(\mathbf{x}, L, r)$. The second subcascade is ruled by a different mechanism. The first difference is that each stage only picks those subeddies placed along the marked edge, but we know their number is not C but $b = C^{1/3}$. The second difference is that the conditions imposed on the corresponding weights are

$$(a): W_s < C, \quad (b) \sum_{s=0}^{b-1} W_s \le C \text{ and } (c) \langle W_s \rangle = 1.$$

By contrast, if the second subcascade had been microcanonical with b subeddies per eddy, the weights would have obeyed the conditions

$$(a'): W_s < b \text{ and } (b'): \sum_{s=0}^{b-1} W_s = b,$$

which are much stronger. As $C \to \infty$ and $b/C \to 0$, conditions (a) and (b) above become increasingly less demanding in comparison with (a') and (b').

This observation gives us a choice between two procedures. The line sections can be studied directly and rigorously. But there is a more attractive alternative: *the second subcascade generating a line average can be approximated by a canonical cascade*. In a canonical cascade, the condition $W < C$ may, in a first approximation, be waived. One may even approximate W by a lognormal random variable, despite the fact that the lognormal is unbounded.

Hence, even if the cascade ruling the cubic eddies is microcanonical, the theory of canonical cascades turns out to be a useful approximation. Incidentally, its most striking result, divergence of high moments, is confirmed by direct argument.

The effect of the condition $W < C$ on the difference between the results of the microcanonical and the canonical models. When three-dimensional canonical eddies are regular, the values of $\bar{\varepsilon}$ given by the canonical and microcanonical theories are identical except for a random prefactor. In this case, the necessary and sufficient condition for regularity is $W < C$. Under the more demanding condition $W < b$, one-dimensional averages are also regular but when $b < \max W < C$, three dimensional and one-dimensional averages belong to different classes and may differ significantly.

4. CLASSIFICATION OF CASCADES ACCORDING TO THE BEHAVIOR OF THE MOMENTS OF $\bar{\varepsilon}$

4.1. A basic recurrence relation for $\bar{\varepsilon}(D, L, \eta)$

Let $\mathcal{D}$ be an eddy of dimension $E = 3$. The definition of Section 3.3 yields, irrespective of the rule of dependence between the W's,

$$\bar{\varepsilon}_{br}(\mathbf{x}, L, \eta) = C^{-1} \sum_{s=0}^{C-1} \bar{\varepsilon}_r(\mathbf{x}_s, L, \eta),$$

where $\{\mathbf{x}_s\}$ is a regular grid of centers of subeddies. Factor the ε on both sides into products of low and high frequency components as follows:

$$\bar{\varepsilon}_{br}(\mathbf{x}, L, br) \bullet \bar{\varepsilon}_{br}(\mathbf{x}, br, \eta) = C^{-1} \sum_{s=0}^{C-1} \bar{\varepsilon}_r(\mathbf{x}_s, L, r) \bullet \bar{\varepsilon}_r(\mathbf{x}_s, r, \eta).$$

Next replace $\bar{\varepsilon}_r(\mathbf{x}_s, L, r)$ by $W_s \bar{\varepsilon}_{br}(\mathbf{x}, L, br)$ and divide both sides by $\bar{\varepsilon}_{br}(\mathbf{x}, L, br)$. We obtain

$$\bar{\varepsilon}_{br}(\mathbf{x}, br, \eta) = C^{-1} \sum_{s=0}^{C-1} W_s \bar{\varepsilon}_r(\mathbf{x}_s, r, \eta).$$

Finally, taking account of self-similarity, we obtain the following basic recurrence relation:

$$\bar{\varepsilon}_1(\mathbf{x}, 1, \eta/br) = C^{-1} \sum_{s=0}^{C-1} W_s \bar{\varepsilon}_1(\mathbf{x}_s, 1, \eta/r).$$

When $E = 1$, so that $\mathcal{D}$ and $\mathcal{D}_s$ are straight intervals of length r, one has the very similar relation

$$\bar{\varepsilon}(\mathcal{D}, 1, \eta/br) = b^{-1} \sum_{s=0}^{b-1} W_s \bar{\varepsilon}(\mathcal{D}_s, 1, \eta/r).$$

Derivation of the moments of eddy averages from the basic recurrence relation. For $q = 1$, it suffices to check that the relation $\langle \bar{\varepsilon}_1(x, 1, \eta) \rangle = 1$ and the above recurrence relation are compatible. For $q > 1$, the recurrence relation for $\bar{\varepsilon}_1$ can be used to deduce a recurrence relation for the sequence of the moments $\langle \bar{\varepsilon}_1^q(x, 1, b^{-k}) \rangle$. The form of the latter depends on the rule of dependence between the W's. Throughout, we shall set $r = 1$, which will simplify the notation.

The microcanonical case. We know that $\bar{\varepsilon}_1(x, 1, \eta) = 1$, but we want to verify that $\langle \bar{\varepsilon}_1^q(x, 1, \eta) \rangle = 1$. Indeed, for $q = 2$, we have

$$\langle \bar{\varepsilon}_1^2(x, 1, \eta/b) \rangle = C \langle (\frac{W}{C})^2 \rangle \langle \bar{\varepsilon}_1^2(x, 1, \eta) \rangle + C(C-1) \langle \frac{W_s}{C} \frac{W_t}{C} \rangle [\langle \bar{\varepsilon}_1(x, 1, \eta) \rangle]^2$$

$$= \frac{\langle W^2 \rangle}{C} \langle \bar{\varepsilon}_1^2(x, 1, \eta) \rangle + \frac{C-1}{C} \left[1 - \frac{\langle W^2 \rangle - 1}{C - 1} \right].$$

Starting from $\langle \bar{\varepsilon}_1^2(x, 1, 1) \rangle = 1$, we obtain

$$\langle \bar{\varepsilon}_1^2(x, 1, 1/b) \rangle = \langle W^2 \rangle / C + 1 - C^{-1} - \langle W^2 \rangle / C + C^{-1} = 1.$$

The recurrence relation reduces to the identity $1 = 1$, as it should. The recurrence relations for $q > 2$ also reduce to identities.

The canonical case. Now, the recurrence relation for the moments takes the form

$$\langle \bar\varepsilon_1^2(x, 1, \eta/b)\rangle = C\langle (W/C)^2\rangle\langle \bar\varepsilon_1^2(x, 1, \eta)\rangle + 2[(1/2)C(C-1)][\langle (W/C)\rangle\langle \varepsilon_1\rangle]^2$$

$$= (\langle W^2\rangle/C)\langle \bar\varepsilon_1^2(x, 1, \eta)\rangle + (C-1)/C.$$

This is no longer an identity, but rather it establishes that the necessary and sufficient condition for $\lim_{\eta\to 0}\langle \bar\varepsilon_1^2(x, 1, \eta)\rangle < \infty$ is $\langle W^2\rangle/C < 1$. Similarly, we have the following important property

$$\lim_{\eta\to 0}\langle \bar\varepsilon_1^q(x, 1, \eta)\rangle < \infty \text{ if and only if } \frac{\langle W^q\rangle}{C^{q-1}} < 1.$$

Conclusion. For eddy averages, the asymptotic behavior of the moments depends on the nature of the cascade.

A necessary and sufficient condition. In order for the inequality $\langle W^q\rangle/C^{q-1} < 1$ to hold for all q, it is necessary and sufficient that $W < C$.

Proof of necessity. The inequality $\langle W^q\rangle/C^{q-1} < 1$, i.e., $\langle (W/C)^q\rangle < 1/C$, implies that

$$\max(W/C) = \lim_{q\to\infty}[\langle (W/C)^q\rangle]^{1/q} < \lim_{q\to\infty} C^{-1/q} = 1.$$

Proof of sufficiency. Knowing that $\langle W\rangle = 1$ and $W < C$, $\langle (W/C)^q\rangle$ is maximized by setting $\Pr\{W = C\} = 1/C$ and $\Pr\{W = 0\} = 1 - VC$. In this extreme case, $\langle (W/C)^q\rangle = 1/C$, so in all other cases $\langle W^q\rangle/C^{q-1} < 1$.

Derivation of the moments of line averages from the basic recurrence: the microcanonical case. The recurrence relation for moments is now replaced by

$$\langle \bar{\varepsilon}^2(\mathcal{D}, 1, \eta/b) \rangle = b \langle (W/C')^2 \rangle \langle \bar{\varepsilon}^2(\mathcal{D}_s, 1, \eta) \rangle + (b-1) b^{-1} \langle W_s W_t \rangle$$

$$= \frac{\langle W^2 \rangle}{b} \langle \bar{\varepsilon}^2(\mathcal{D}_s, 1, \eta) \rangle + \frac{b-1}{b} \left[1 - \frac{\langle W^2 \rangle - 1}{(C-1)} \right].$$

This is no longer an identity: the necessary and sufficient condition for

$$\lim_{\eta \to 0} \langle \bar{\varepsilon}^2(\mathcal{D}, 1, \eta) \rangle < \infty$$

has become $\langle W^2 \rangle / b < 1$. Similarly, we have the following important property:

$$\lim_{\eta \to 0} \langle \bar{\varepsilon}_q^2(\mathcal{D}, 1, \eta) \rangle < \infty \text{ if and only if } \frac{\langle W^q \rangle}{b^{q-1}} < 1.$$

The canonical case. The recurrence relation is unchanged when the dimension changes from $E = 3$ to $E = 1$, except for the replacement of C by b. Therefore, we fall back on the condition $\langle W^q \rangle / b^{q-1} < 1$ of the preceding paragraph.

Conclusion. For line averages, the finiteness of the limiting moments is *not* dependent on the nature of the cascade. On the other hand, the value of the limiting moment, when finite, is smaller when the cascade is microcanonical; for example, for $q = 2$, it is smaller by the factor $1 - (\langle W^2 \rangle - 1)/(C-1)$.

4.2. The determining functions $\Psi(q)$

In order to apply the above results to classify cascades, and in order to carry the theory further, we form the expression

$$\Psi(q) = \log_C \langle W^q \rangle,$$

which will be called the "determining function." More specifically, when $\mathcal{D}$ is E-dimensional, we shall need the quantities

$$\overline{\tau}_E(q) = (3/E)\Psi(q) - (q-1).$$

To define various parameters of dissipation, different features of these functions must be examined. First of all, $\langle W^q \rangle / (C^{E/3})^{q-1} < 1$ is synonymous with $\overline{\tau}_E(q) < 0$, and so the values of the zeros of $\overline{\tau}_E(q)$ are of interest.

For all q, a general theorem of probability theory shows that $\Psi(q)$ is a convex function of q (see Feller 1971, p. 155), and so are all the functions $\overline{\tau}_E$. Hence, $\Psi(1) = \overline{\tau}_E(1) = 0$, and $\overline{\tau}_E(q)$ has, at most, one root other than 1. This root it will be designated by α_E. The conditions $\overline{\tau}_1(q) < 0$ and $\overline{\tau}_2(q) < 0$ are both at least as demanding as $\overline{\tau}_3(q) < 0$, so when $\alpha_1 > 1$, the α_E satisfy $\alpha_1 \leq \alpha_2 \leq \alpha_3$.

A further investigation of the $\overline{\tau}_E$ involves their slopes for $q = 1$, more specifically the expressions

$$D_E = -E\overline{\tau}'_E(1) = -E\langle W \log_{C^{E/3}}(W/C)^{E/3} \rangle$$

$$= -3\langle W \log_C W \rangle + E.$$

Writing $D_3 = D$, we have $D_2 = D - 1$ and $D_1 = D - 2$. The value of D_E will be useful, because an E-dimensional average in a canonical cascade is degenerate when $D_E < 0$ and non-degenerate when $D_E > 0$. In particular, when $W < C^{E/3}$, $D_E > 0$. (The transition case $D_1 = 0$ deserves the attention of the mathematicians, but is too complicated to be tackled in this paper.) More precisely, in the degenerate case $D_1 < 0$, one has $\alpha_E < 1$, and the value of α_E plays no special role. But in the non-degenerate case $D_E > 0$, α_E satisfies $\alpha_E > 1$, and its value serves to determine whether the cascade is regular ($\alpha_E = \infty$) or irregular ($\alpha_E < \infty$).

We know that for all $q > 1$, $\alpha_E = \infty$ is equivalent to $\langle W^q \rangle / (C^{E/3})^{q-1} < 1$. It follows that the necessary and sufficient condition for $\alpha_E = \infty$ is $W < C^{E/3}$, an inequality already featured in Section 2.2.

When $D_E > 0$, the quantity D_E plays an independent role as the intrinsic dimension of the support of $\overline{\varepsilon}$ within an E-dimensional $\mathcal{D}$ (see Section 4.8).

Since $W \log W$ is concave, $\langle W \log W \rangle > \langle W \rangle \log \langle W \rangle = 0$; therefore, the intrinsic dimension D_E never exceeds the embedding dimension E.

Finally, the "second-order" dependence properties of the dissipation, namely its correlation and its spectrum, depend on the value of $\Psi(2)$. Yaglom showed the correlation between the averages taken over small domains $\mathcal{D}$, separated by the distances d, to be proportional to d^{-Q}, with

$$Q = 3 \log_C \langle W^2 \rangle = 3\Psi(2) = 3[\overline{T}_3(2) + 1].$$

To obtain a lower bound on Q, note that from $\langle W \rangle = 1$ it follows that $\langle W^2 \rangle > 1$ and hence that $Q > 0$. More precisely,

$$Q > 3(1 + \overline{T}'_3(1)) > 3 - D.$$

As for the upper bounds, under the constraints $\langle W \rangle = 1$ and $W < C$, we know that the maximum of $\langle W^2 \rangle$ occurs when $\Pr\{W = C\} = 1/C$ and $\Pr\{W = 0\} = 1 - 1/C$, in which case $\langle W^2 \rangle = C$ and so $Q = 3$. More generally, $W < C^{E/3}$ implies $Q < E$. When $W/C^{E/3} > 1$ on the contrary, it may happen that $Q > E$.

Studies involving correlations of higher order q depend similarly on values of f up to the argument q. Since we shall stop at the second order, our classification of canonical cascades will depend solely on the values of Q, D and α_E. These parameters are conceptually distinct, and their numerical values are only related by the conditions of compatibility $Q > 3 - D$ and $(\alpha_E - 1)D_E > 0$. The question of whether or not their actual values are related should be investigated experimentally.

The relationship between W, $\Psi(q)$ and the various other parameters deserves additional mathematical investigations. A knowledge of C and of the distribution of W determines $\Psi(q)$ for all q, and thus determines all the parameters. On the other hand, a knowledge of C and of the values of $\Psi(q)$ for integer values of q an integer need not determine W uniquely. A sufficient condition is that the moments satisfy the Carleman criterion (see Section A2). This technicality is important because this criterion fails in the case of a lognormal W.

4.3. Examples of determining functions

Rectilinear determining functions. Ψ and $\bar{T}_E$ are linear functions of q if W is binomial, i.e., $\Pr\{W = 1/p\} = p$ and $\Pr\{W = 0\} = 1 - 1/p$. If so,

$$\langle W^q \rangle = pp^{-q} = p^{1-q}, \text{ thus } \bar{T}_3(q) = (1-q)\log_C(pC),$$

which is a degenerate form of convex function.

Digression. This example reduces to the classical theory of birth-and-death processes (see Harris 1963). After K "generations," each elementary subeddy either is empty or includes a non-random mass of turbulence equal to p^{-K}. Discarding this last factor, the mass in an eddy and the number of its non-empty elementary subeddies are equal. Their probability distribution is readily determined: between each generation and the next, non-empty elementary eddy can be interpreted as acquiring random "offspring" made of M lower order elementary eddies, with M following a binomial distribution of expectation $C^{E/3}p$. When $M = 0$, the eddy "dies out." When $M > 1$, new eddies are born. Classical results on birth-and-death processes show that the number of offspring after the kth generation is ruled by the following rule. When $p \leq 1/C^{E/3}$, so that $D_E \leq 0$, it is almost certain that the offspring will eventually die out. When $p > 1/C^{E/3}$, so that $D_E > 0$, one forms the ratio of the number of offspring to its expected value $(C^{E/3}p)^K = b^{D_E K}$. One finds that this ratio tends asymptotically towards a non-degenerate limiting random variable that has finite moments of every positive order.

Asymptotically rectilinear determining functions. Now let us suppose only that W is bounded. Designate its greatest attainable value by max W. This means that $\Pr\{W > \max W\} = 0$, but $\Pr\{W > \max W - \theta\} > 0$ for all $\theta > 0$. (A more correct mathematical idiom for max W is "almost sure supremum.") It follows that $\lim_{q \to \infty}(\log_C \langle W^q \rangle / q) = \log_C \max W$, which implies that $\Psi(q)$ has an asymptotic direction of finite slope $\log_C \max W$. Conversely, in order for this asymptotic slope to be finite, it is necessary that $W < \max W < \infty$. Also, $\bar{T}_E(q)$ has an asymptotic direction of slope $-1 + \log \max W / \log C^{E/3}$. When $\max W < C^{E/3}$, this slope is negative and $\bar{T}_E(q) = 0$ has no root other than 1; in other words, $\alpha_E = \infty$. When $\max W > C^{E/3}$, and particularly when $\max W = \infty$, one has $\alpha_E < \infty$. This confirms our assertion that, in general (except for some inequalities), the values of D, Q and the α_E are independent. See the caption of Figure 2.

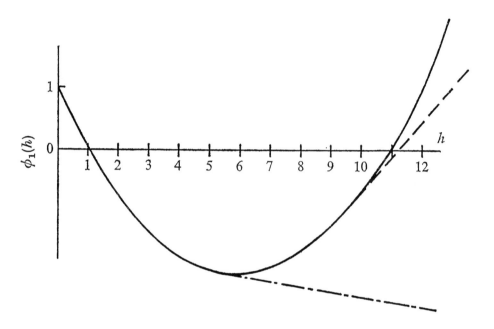

FIGURE N15-2. Characterization of the distribution of $\bar{\varepsilon}(\mathcal{D})$ when (a) log W is normal; $\bar{\tau}_1(q)$ is then a parabola and $\bar{\tau}_1(q) = 0$ has two finite roots (solid line); (b) log W is a sum of sufficiently many terms to be a good approximation to the normal distribution; $\bar{\tau}_1(q)$ is then nearly parabolic for $q < \alpha_1$ (dashed line); (c) log W is the sum of comparatively few terms; even when the quality of approximation to the normal distribution is good by other standards, it may be poor from the viewpoint of $\bar{e}$; in the zone of interest, $\bar{\tau}_1(q)$ is far from parabolic and $\bar{\tau}_1(q) = 0$ may have a single finite root, i.e. $\alpha_1 = \infty$ (dash-dot line). Thus the degree of sensitivity of various properties of $\bar{\varepsilon}$ are very different. On the one hand, the moment properties of the distribution of $\bar{\varepsilon}$ depend greatly upon fine details, namely the tail of the distribution of W: a lognormal W never fails in the regular class, but a "nearly log normal" W may do so. On the other hand, the value of $\bar{\tau}_1(2)$, hence of the spectral properties of $\bar{\varepsilon}$, and even more the value of $\bar{\tau}'_1(1)$, hence of the fractional dimension, will be essentially the same for the three cases as drawn.

Parabolic determining functions. Suppose that log W is Gaussian of mean and variance $\langle \log W \rangle$ and $\sigma^2 \log W = \langle (\log W)^2 \rangle - \langle \log W \rangle^2$. Then W is lognormal, and

$$\langle W^q \rangle = \exp(q \langle \log W \rangle + q^2 2^{-1} \sigma^2 \log W).$$

That is,

$$\Psi(q) = \log_C \langle W^q \rangle = (q \langle \log W \rangle + q^2 2^{-1} \sigma^2 \log W) \log_C e.$$

This function $\Psi(q)$, and the functions $\bar{\tau}_E(q)$ are represented by parabola.

Digression. The lognormal distribution not being fully determined by its moments (see Section 2), other weights W may lead to the same $\Psi(q)$).

To insure that $\langle W \rangle = 1$, we must have $-\langle \log W \rangle = (1/2)\sigma^2 \log W$, a quantity to be denoted (in order to fit Kolmogorov's notation) by $(\mu/6) \log C$. It follows that

$$\Psi(q) = \frac{1}{6}(q-1)q\mu.$$

Hence, $D_E = -3\Psi'(1) + E = E - \mu/2$, $\alpha_E = 2E/\mu$ and $Q = \mu$. Here, the values of D, Q and the α's are all functions of μ, and are strongly interdependent; this is an exceptional circumstance (see Section 4.9).

The determining function when $\log W$ is a sum of many uniform random variables (Figure 2). Suppose that $\log W$ is bounded, for example a sum of many uniformly distributed random variables. Such a sum is near-Gaussian according to the customary definition of nearness. When q is small, the resulting $\bar{\tau}_E(q)$ will nearly coincide with the parabola in the preceding paragraph, but asymptotically it will be a straight line. The graph of $\bar{\tau}_E(q)$ will be a portion of parabola continued by a straight tangent. If we add many uniform components, this tangent will have positive slope, therefore $\log W$ and its Gaussian approximation will yield about the same value for α. If, on the other hand, we add few uniform random variables, the asymptotic tangent will have negative slope and the lognormal approximation will be entirely worthless.

As an illustration, let us describe one of our early computer simulations of a cascade. We thought that W was ostensibly lognormal and we expected $\alpha_E < \infty$. But the actual results were completely at variance with the expectations. They remained mysterious until it was recalled that to generate the Gaussian random variable, our computer added 12 random variables uniformly distributed on $[-0.5, +0.5]$. When the same program was run again, this time adding 48 and then 192 uniform random variables, the results changed to full conformity with the expectations. (See also Sections 4.3 and 4.9.)

4.4. Regular classes

A classification of cascades can be based either on a single value of E, or on two or three values, typically $E = 1$ and $E = 3$.

The regular class for fixed E. Here $\bar{\varepsilon}(\mathcal{D}_E, L, 0)$ is, by definition, a non-degenerate random variable with all moments finite. In a canonical cascade, the necessary and sufficient condition for $\lim_{\eta \to 0} \langle \bar{\varepsilon}_1^q(x, L, \eta) \rangle < \infty$ can be written as $\bar{\tau}(q) < 0$ for all $q > 1$, in $\alpha_E = \infty$, or, as $W < C^{E/3}$. As a corollary, $Q < 3$.

A formal argument would consist in replacing the limits of the moments with the moments of the limit, and would suggest that $\langle \bar{\varepsilon}_1^q(x, L, 0) \rangle < \infty$ holds if and only if $\bar{\tau}_E(k) < 0$ for all $q > 1$. This formal argument is justified by a theorem in Doob 1953, p.319. Ordinarily, physicists do not feel that such justifications deserve attention but Section 4.5 will show that in this context they must be taken seriously.

A microcanonical cascade is always regular from the viewpoint of eddy averages. In addition, the condition $W < C^{E/3}$ is necessary and sufficient to insure that W be admissible as weight in a microcanonical cascade with the same E. Consider, then, both the microcanonical and the canonical cascades corresponding to a weight W of the regular class. The only effect of changing the definition of the cascade is to change the high frequency term of $\bar{\varepsilon}_r(x, L, 0)$ from 1 to some random variable having finite moments of all orders. In other words, the only difference between the full canonical random variable $\bar{\varepsilon}_r$ and its lognormal low frequency term lies in a numerical factor whose values are about the same when $\eta = 0$ and when η is small but non-zero. Such a factor is comparatively innocuous.

As a specific example, if W is binomial with

$$\Pr\{W = 1/p\} = p \text{ and } \Pr\{W = 0\} = 1 - p,$$

$\bar{\tau}_E(q) < 0$ if and only if $pC^{E/3} > 1$, i.e., $p > 1/C^{E/3}$.

The *uniformly regular class* defined by $\alpha_1 = \infty$, i.e., $D_1 > 0$. In this case, it is true for all values of E that E's, and $\bar{\varepsilon}(\mathcal{D}, L, 0)$ is a non-degenerate random variable with finite moments. In this class, $\varepsilon(\mathcal{D}, L, \eta)$ may be said to be "approximately lognormal." Yaglom has implicitly assumed that this situation prevails in practice. This may, but need not, be so. Only experiments can tell.

4.5. Degenerate classes

The degenerate class for fixed E. This class is defined by $\Pr\{\bar{\varepsilon}(\mathcal{D}), L, 0) = 0\} = 1$. A sufficient condition is $D < 0$, from which it follows that $\alpha \leq 1$.

The proof (see Section A3) consists of showing that the number of elementary subeddies of side r contributing to the bulk of $\bar{\varepsilon}_r(\mathbf{x}, L, 0)$ is roughly equal to $(L/r)^{D_E}$. From $D_E < 0$ it follows that, as $\eta \to 0$, this number tends to zero, and so does $\bar{\varepsilon}(\mathcal{D}, L, \eta)$. Two examples come to mind.

First example. $\Pr\{W = 1/p\} = p$ and $\Pr\{W = 0\} = 1 - p$ with $p < 1/C$.

Second example. Since a lognormal distribution is unbounded, *a lognormal cascade is never regular*. Because $D_E = E - \mu/2$, the cascade is degenerate when $D_E < 0$, i.e., when $\mu > 2E$. In particular, $\bar{\varepsilon}_r(\mathbf{x}, L, \eta)$ is lognormal only when $r = \eta$.

The uniformly degenerate class. When $\bar{\varepsilon}(\mathcal{D}_E, L, 0)$ is degenerate for $E = 3$, i.e., when $D < 0$, $\bar{\varepsilon}$ is also degenerate for $E = 2$ and $E = 1$.

4.6. Irregular classes

The irregular class for fixed E. This class is defined by $\Pr\{\bar{\varepsilon}(\mathcal{D}_E, L, 0) = 0\} > 0$, with $\langle \bar{\varepsilon}^q(\mathcal{D}_E, L, 0) \rangle < \infty$ for small enough $q > 1$, but $\langle \bar{\varepsilon}^q(\mathcal{D}_E, L, 0) \rangle = \infty$ for large finite q. The class is characterized by $1 < \alpha_E < \infty$, and the cut-off between finite and infinite moments is $q = \alpha_E$ (see M 1974c{N16} and Kahane 1973).

The uniformly irregular class. When $\bar{\varepsilon}(\mathcal{D}_E, L, 0)$ is non-degenerate for $E = 3$ and is irregular for $E = 1$, then it is irregular for all E.

4.7. Mixed classes

Since $\alpha_1 < \alpha_2 < \alpha_3$, it is possible that a cascade ($E = 3$) and its cross-sections ($E = 1$ and $E = 2$) belong to different classes. Neglecting the behavior for $E = 2$, three possibilities are open. We shall give one example of each.

An example from the mixed regular-degenerate class. $\Pr\{W = 1/p\} = p$ and $\Pr\{W = 0\} = 1 - p$ with $C^{-1} < p < C^{-1/3}$. Here $\bar{\tau}_3(q) < 0$ but $\bar{\tau}_1(q) > 0$ for all $q > 1$. That is, the cascade is regular for $E = 3$ but degenerate for $E = 1$.

Comments. I doubt that this mixed class is ever encountered in practice, because it implies that the spatial distribution of dissipation is extraordinarily sparse, sparser than anything I would consider as likely.

An example from the mixed regular-irregular class: lognormal W. Take $C = 27$, so that $C^{1/3} = b = 3$, with the random variable W satisfying $\Pr\{W = 3.7\} = 0.1$ and $\Pr\{W = 0.7\} = 0.9$. Since $W < C$, the resulting three-dimensional cascade is regular, and it may correspond to a canonical approximation of a microcanonical cascade. On the other hand, it is not true that $W < b$, although it is true that $\bar{\tau}'_1(1) < 0$. As a result, a one-dimensional cascade corresponding to this W is irregular.

Comments. I consider this last situation to be a very strong possibility. If and when it occurs in practice, the distribution of one-dimensional averages is not at all lognormal. *One task for the experimental study of turbulence should be to check whether or not such a mixture ever occurs. It may be that different circumstances yield either this mixture or the uniformly regular class; if so, those circumstances should be classified according to the class to which they lead.*

An example from the mixed irregular-degenerate class: lognormal W. If $2 < \mu < 6$, the full three-dimensional pattern is irregular, while one-dimensional cross-sections are degenerate.

4.8. Digression: D_E as a fractional intrinsic dimension

Select two arbitrary small thresholds ω and γ. When $D_E > 0$, it can be shown (see Section A3) that the eddies of side r can be divided into two groups. The eddies of the first group contain a proportion greater than $1 - p$ of the whole dissipation. However, their number lies between $(L/r)^{D_E - \gamma}$ and $(L/r)^{D_E + \gamma}$, which makes the group relatively very small. As a result, almost all eddies belong to the second group. But, the total dissipation they contain is at most equal to ω, which makes it negligible.

It is convenient to call D_E an intrinsic dimension; alternatively (because it need not be an integer) it can be called a fractional dimension.

The notion that a geometric figure can have a fractional dimension was conceived in 1919 by a pure mathematician, Felix Hausdorff. This concept is closely related to the Cantor set, and both have the reputation of lacking any conceivable application, and of "turning off" every natural scientist. I believe that this reputation is no longer deserved. I hope to show (elsewhere) that fractional dimension is in fact something very concrete and that different aspects of it are useful measurable physical characteristics. Examples are the degree of the wiggliness of coastlines, described in M 1967s, the degree of clustering of galaxies, and the intensity of the intermittency of turbulence.

In these applications, it is best to use a semi-formal variant of dimension called the "similarity dimension," which is of more limited validity than Hausdorff's concept, but incomparably simpler. It is rooted in some elementary features of the usual concept of dimension, as applied to segments of a straight line, to rectangles and to parallelepipeds. A line has dimension $D = 1$, and for every positive integer C, the segment where $0 \leq x < X$ exactly subdivides into C non-overlapping segments of the form $(n-1)X/C \leq x < nX/C$, where $1 \leq n \leq C$. To obtain one of these parts from the whole one performs a similarity of ratio $\rho(C) = C^{-1}$. In the same way, a plane has dimension $D = 2$, and for every integer $\sqrt{C}$, the rectangle where $0 \leq x < X$ and $0 \leq y < Y$ subdivides exactly into C non-overlapping rectangles of the form

$$\frac{(k-1)X}{\sqrt{C}} \leq x < \frac{kX}{\sqrt{C}} \quad \text{and} \quad \frac{(h-1)Y}{\sqrt{C}} \leq y < h\frac{Y}{\sqrt{C}},$$

where $1 \leq k \leq C$ and $1 \leq h \leq C$. To obtain one of these parts from the whole, one performs a similarity of ratio $\rho(C) = 1/C^{1/2} = 1/b$. More generally, a D-dimensional rectangular parallelepiped can, for every integer $C^{1/D} = 1/b$, be decomposed into C parallelepipeds. To obtain one of these parts from the whole, one performs a similarity of ratio $\rho(C) = 1/C^{1/D}$.

For each of the above figures, the dimension D satisfies the relation

$$D = \frac{-\log C}{\log \rho(C)}.$$

This suggests that the concept of dimension can be generalized to the set on which the bulk of intermittent turbulence is concentrated. Here $1/\rho(C) = L/r$, and for every $\gamma > 0$,

$$(L/r)^{D_E - \gamma} < C < (L/r)^{D_E + \gamma}, \quad \text{i.e.,} \quad D_E - \gamma < \log C / \log \rho(C) < D_E + \gamma.$$

In other words the dimension of that set is D_E. The intuitive notion that turbulence concentrates on an extremely sparse set is expressed numerically by the inequality $D_E < E$. By choosing W appropriately, the dimension D_E can take any value between 0 and E. Note also that, when $D < 3$, $D_3/3 = D/3$ is greater than $D_2/2 = (D-1)/2$, which in turn is greater than $D_1 = (D-2)$. The inequality $D < 3$ expresses that a figure does not fill the space dimensionally, and the inequality $D/3 > (D-2)$ expresses that the intersections of such a figure by straight lines are dimensionally even less filling.

I have great faith in the practical usefulness of fractional dimension and hope it will be explored further. In particular, it opens up the issue of the degree of connectedness of the volume where the dissipation concentrates. However, neither the microcanonical nor the canonical models appear to provide a satisfactory framework, because both allow the dissipation to be divided very discontinuously. Therefore connectedness should be studied in some other context, say, that of the limiting lognormal model.

4.9. Further comments on the lognormal approximation to W, and on parabolic approximations to $\Psi(q)$

Suppose that W is non-lognormal and bounded, and satisfies $\sigma^2 \log W < \infty$, and let $\overline{W}$ be its lognormal approximation and let the corresponding determining functions be $\Psi(q)$ and $\overline{\Psi}(q)$, with the obvious definitions for $\overline{\tau}_E(q)$. We have already noted that for $q \to \infty$, $\Psi(q)$ has a finite asymptotic slope $\log_C \max W$, while $\overline{\Psi}(q)$ is parabolic. Therefore, their asymptotic behaviors differ qualitatively. On the other hand, the behavior of $\Psi(q)$ and the $\overline{\tau}_E(q)$ for small q depends only on $\langle \log W \rangle$ and $\sigma^2 \log W$, therefore it remains unchanged when $\log W$ is replaced by its normal approximation. Hence the following consequences.

Since the moment of order $q = 2/3$ is likely to be covered by this approximation, the conclusions Kolmogorov and Yaglom had obtained by applying the "two thirds-law," may well be essentially unchanged.

For $q = 1$, $\overline{\Psi}(1)$ need not equal 1. Also, $\overline{\tau}'_E(1)$ need not equal $\overline{\tau}_E'(1)$. In extreme instances (see the end of Section 4.2) they may have different signs. It may happen that the "real" $\overline{\tau}'_E(1)$ is negative, meaning that the cascade is non-degenerate, while $\overline{\tau}_E'(1)$ is positive, suggesting that the cascade is degenerate. The approximations of the values of Q and of the α_E may be even poorer.

A different lognormal approximation to W, to be called $\widetilde{W}$, is achieved by approximating $\Psi(q)$ by a parabola $\widetilde{\Psi}(q)$ satisfying $\widetilde{\Psi}(0) = \widetilde{\Psi}(1) = 0$ and having the correct slope $\widetilde{\Psi}'(1)$. The mean and variance of $\log W$ are determined by the properties of $\overline{\Psi}(q)$ near $q = 0$. From the practical viewpoint, all that is of direct interest is the portion of $\overline{T}_E(q)$ that lies between $q = 1$ and $q = \alpha_E$; it follows that, when $\Psi(q)$ is smooth and the original α_E is small, the error introduced by the lognormal approximation $\widetilde{W}$ may well be acceptable. Whenever such is the case, the various properties of $\bar{\varepsilon}(\mathcal{D}, L, 0)$ linked to D_E, Q and α_E turn out to be related, after all. Given the inaccuracy inherent in experimental work, this implies that it may be necessary to return to the situation that used to prevail when the single characteristic parameter Q was believed sufficient. To the contrary, when the original α_E is large, and especially when it is infinite, the error in using $\widetilde{W}$ is very large; the process of approximation changes the class to which such a cascade belongs.

Digression. (This is another occurrence of a phenomenon also encountered in Section A.1: the moments of exp V are very sensitive to apparently slight deviation of V from normality.)

APPENDIX

A1. Approximate versus strict lognormality; the differences are deep, hence the use of approximate lognormality to calculate moments is unsafe

Let us consider the following random variables: a normal (Gaussian) random variable G, a Poisson random variable P and a bounded random variable B obtained as the sum of a large number K of random variables $B_k = \log R_{k'}$, each of which bounded by the same $\beta < \infty$. We want G, P and B to be nearly identical. Since the mean and the variance are equal in the case of P, they must be assumed equal for G and B. Finally, we want G, P and B to be near identical. Therefore, the value δ common to mean and their variance must be large. It follows that

$$\langle (e^G)^q \rangle = \exp(q\delta + \delta q^2/2) = \exp[\delta(q + q^2/2)],$$

$$\langle (e^P)^q \rangle = \exp(-\delta + \delta e^q) = \exp[\delta(e^q - 1)],$$

$$\langle (e^B)^q \rangle \leq \exp(qK\beta).$$

Thus, $\langle(e^B)^q\rangle$ increases at most exponentially with q, $\langle(e^G)^q\rangle$ increases more rapidly than any exponential, and $\langle(e^P)^q\rangle$ more rapidly still. The expectations, equal respectively to $\langle e^G \rangle = \exp(1.5\delta)$ and $\langle e^P \rangle = \exp(1.7\delta)$, are already very different. The coefficients of variation, defined as

$$\frac{\langle(e^G)^2\rangle}{\langle e^G\rangle^2} = e^\delta \quad \text{and} \quad \frac{\langle(e^P)^2\rangle}{\langle e^P\rangle^2} = e^{(e-1)2\delta} \sim e^{3\delta},$$

differ even more, and higher order moments differ strikingly. In short, it may be that B and P are nearly normal from the usual viewpoint (which is that of the so-called "weak topology"). But from the present viewpoint they provide poor approximations to the normal distribution. However, when $q < 1$ (for example, $q = 2/3$, as in the calculation of spectra) the discrepancy is smaller.

{PS 1998. *A fourth example of a manageable sequence of moments is the log-gamma distribution.* The reduced gamma r.v. of parameter δ, to be called Gamma (δ), is defined as having the density $u^{\delta-1}e^{-u}/\Gamma(\delta)$. It is well known that the sum of two independent gamma r.v. of parameters δ' and in exponent δ'' is a gamma r.v. of parameter $\delta' + \delta''$. Let the reduced log-gamma r.v. be defined as $e^{-\sigma\text{Gamma}(\delta)}/\langle e^{-\sigma\text{Gamma}(\delta)}\rangle$. The q-th moment of the log-gamma is finite if and only if $q > -1/\sigma$; when finite, its value is

$$\langle(e^{-\sigma\text{Gamma}(\delta)})^q\rangle = \int_0^\infty u^{\delta-1}e^{-u}e^{-u\sigma q}du/\Gamma(\delta) = (1+\sigma q)^{-\delta}.$$

In particular, irrespective of δ, $\sigma \leq -1$ yields $\langle e^{-\sigma\text{Gamma}(\delta)}\rangle = \infty$. That is $\langle W \rangle = 1$ cannot be insured unless $\sigma > -1$. To my knowledge, the four families listed in this paper and this PS are the only ones that are endowed with explicit analytic expressions, have some finite moments and possess the property of being closed under addition.}

A2. On Orszag's remark concerning the determination of turbulence by its moments

Homogeneous turbulence is presumed to be determined by its moments, and the bulk of the theory based on the Navier-Stokes equations is devoted to efforts to determine these moments theoretically. Is intermittent turbulence also so determined? To answer, note that a random variable can have the same moments as the lognormal distribution, without itself being lognormal. Feller 1951 vol. 2, p.227 of the 2nd edition, credits this example to Heyde 1963. {P.S. 1998. In fact, the lognormal was the very first example to be mentioned in the original (1894) paper on the moment problem; see Stieltjes 1914.}

The reason for this indeterminacy is that the moments of a lognormal e^G increase so fast that $\Sigma[\langle \exp(2qG)\rangle]^{-1/(2q)} < \infty$, which expresses that the lognormal distribution fails to satisfy a necessary condition due to Carleman. Orszag 1970 has observed that a corollary of this indeterminacy is that, if intermittent turbulence were indeed lognormal, it would not be determined by its moments. On the other hand, suppose that Yaglom's $\bar{\varepsilon}_r$ is the product of independent bounded factors. In that case, the moments of intermittent turbulence do satisfy the Carleman criterion; therefore, the indeterminacy noted by Orszag vanishes. (*Note added during revision.* Novikov 1971, p.235, contains a remark to the same effect.)

A3. The dimension exponent, as introduced through the number of eddies of side *r* within which dissipation is concentrated

The purpose of this section of the appendix is to show that among subeddies of side r, most of the dissipation is concentrated in a subset of about $(L/r)^{D_\varepsilon}$ subeddies.

Preliminary example: binomial weights. Let

$$\Pr\{W = 0\} = 1 - p \text{ and } \Pr\{W = 1/p\} = p,$$

so that $D = -\log_c p$. Then $\bar{\varepsilon}_L(\mathbf{x}, L, \eta) = 1$ factors into two terms: (a) the contents of a non-empty eddy, namely

$$(p^{-1})^{\log_c(L/\eta)} = (L/\eta)^{\log_c p} = (L/\eta)^{-D},$$

and (b) the number of non-empty eddies of side η contained in a big eddy of side L. Since $E\,\bar{\varepsilon}_L(\mathbf{x}, L, \eta) = 1$, the expectation of this last number must be $(L/\eta)^D$.

Second example: lognormal weights and cubic eddies. Let us begin with the low frequency factor $\bar{\varepsilon}_r(\mathbf{x}, L, r)$. With W lognormal as in Section 4.3, $\log \bar{\varepsilon}_r$ is Gaussian with variance $\mu \log(L/r)$ and expectation $(\mu/2)\log(L/r)$. To simplify the notation, we shall denote $\bar{\varepsilon}_r$ by V. When $L/r \gg 1$, this lognormal factor has the feature that its expectation is due almost exclusively to occasional large values. As a result, one can select a function threshold (L/r) in such a way that values of V below threshold (L/r) are negligible. Specifically, if one defines $\tilde{V}$ by

$$\tilde{V} = \begin{cases} V & \text{when } V > \text{threshold } (L/r), \\ 0 & \text{otherwise,} \end{cases}$$

then $E\tilde{V}$ is arbitrarily close to 1. Let us prove that such a result is achieved when C is a function subjected to the sole requirement $\lim_{r \to 0} C(L/r)/[\log(L/r)]^{1/2} = 0$, and when the "threshold" function is chosen to satisfy

$$\text{threshold } (L/r) = (L/r)^{\mu/2} \exp\{ - C(L/r)\sqrt{\mu \log(L/r)} \}.$$

Indeed,

$$\langle V \rangle = \frac{0.1}{\sqrt{2r\mu \log(L/r)}} \int \exp\left\{ x - \frac{[x + (\mu/2) \log(L/r)]^2}{4\mu \log(L/r)} \right\} dx,$$

with an integration range from $\log[\text{threshold }(r, L)]$ to ∞. The expression in braces transforms into

$$\frac{- [x - (\mu/2) \log(L/r)]^2}{4\mu \log(L/r)},$$

and by changing the variable of integration to

$$z = [x - (\mu/2) \log(L/r)] [2\pi \log(L/r)]^{-1/2}$$

we obtain

$$\langle \tilde{V} \rangle = (2\pi)^{-1/2} \int \exp(-z^2/2) dz,$$

with an integration range from $- C(L/r)$ to infinity. As $L/r \to \infty$, $\langle V \rangle \to 1$, which shows that the contribution of other values of V to $\bar{\varepsilon}_r$ is asymptotically negligible, and that the above choice of N was indeed appropriate to make V arbitrarily closely approximated by $\tilde{V}$. From now on, one can consider the $(L/r)^3$ cells of side r that lie within a cube of side L, and divide them into those for which $V >$ threshold (L/r), and those for which $V <$ threshold (L/r).

For the former, the expectation of their total number is

$$(L/r)^3 \Pr\{V > \text{threshold } (L/r)\}.$$

In terms of the reduced Gaussian random variable

$$[\log V + (\mu/2) \log(L/r)]\sqrt{\mu \log(L/r)} = G,$$

the above probability becomes

$$\Pr\{G > \sqrt{\mu \log(L/r)} - C(L/r)\}.$$

Using a well-known tail approximation of G, the expected number in question is approximately equal to

$$(L/r)^3 \frac{\exp[-(\mu/2)\log(L/r)]}{\sqrt{2\pi\mu \log(L/r)}} = \frac{(L/r)^{3-\mu/2}}{\sqrt{2\pi\mu \log(L/r)}} = \frac{(L/r)^D}{\sqrt{2\pi\mu \log(L/r)}}.$$

(Note that this last approximation is independent of C.)

With cubic eddies replaced by straight segments, the only change in the above formulae is that the factor $(L/r)^3$ is replaced by L/r and hence $3 - \mu/2$ by $1 - \mu/2 = D_1$.

As for the cells in which $V <$ threshold (L/r), we want to show that their total contribution is negligible. The proof involves the high frequency factors $\bar{\varepsilon}_r(\mathbf{x}, r, \eta)$ and an application of the ergodic theorem. Details need not be given here.

General weights W. The assertion is that the number of eddies that are not nearly empty is about $(L/r)^D$. The proof cannot be given, but its principle can be indicated. The quantity $\langle W \log W \rangle$ is related to Shannon's concept of entropy-information, and it enters here because our problem can be restated in terms of information-theoretical asymptotical equiprobability; see Billingsley 1967.

A4. Introduction to a model of intermittency based on the limiting lognormal processes, in which eddies are randomly generated and the partition of dissipation is continuous

My earlier paper on intermittency M 1972j{N14}, involved a departure from the assumption of Section 2.1: the grid itself was made random, and was generated by the same model as the distribution of dissipation. The purpose of this section is to provide a transition from this to the earlier work.

Our point of departure consists in a prescribed grid of eddies, and a canonical cascade with lognormal weight W: $\log \varepsilon(x, L, \eta)$ is Gaussian with variance $\mu \log(L/\eta)$ and expectation $-(\mu/2) \log(L/\eta)$. We know that the correlation of $\bar{\varepsilon}$ is approximately proportional to d^{-Q}, with $Q = \mu$. Because the eddies were prescribed, the random function $\varepsilon(x, L, \eta)$ is non-stationary and discontinuous: between an eddy and its neighbors, there may be very large discontinuities. Both non-stationarity and discontinuity are of course quite unrealistic. One may instead demand that $\log \varepsilon(x, L, \eta)$ be Gaussian and stationary, with the added restriction that it should be continuous and vary little over spans of order shorter than η. This will ensure that $\bar{\varepsilon}(x, L, \eta)$ is nearly identical to $\varepsilon(x, L, r)$. It remains to ensure that $\varepsilon(x, L, \eta)$ has a correlation proportional to d^{-Q}. The simplest way to achieve this aim is to require $\varepsilon(x, L, \eta)$ to have a truncated self-similar spectral density, namely a spectral density equal to $\eta/2\omega$ when $1/L < \omega < 1/\eta$, and equal to zero elsewhere. The resulting model may be viewed, as combining self-similarity with the maximum retrievable portion of the Kolmogorov third hypothesis.

The properties of $\bar{\varepsilon}(\mathcal{D}, L, \eta)$ relative to this model can be summarized as follows. The dimensions continue to be $D_E = E - \mu/2$ and the cascades are never regular: for $\mu < 2/E$, they are irregular with $\alpha_E = 2/\mu$, while for $\mu > 2/E$, they are degenerate. Compared with a canonical cascade with a lognormal W, the main differences involve the values of certain numerical constants.

A5. Remarks on Kolmogorov's third hypothesis of lognormality

This hypothesis states that, for every cube of center x and side $r > \eta$, $\bar{\varepsilon}_r(x, L, \eta)$ follows the lognormal distribution, the variance of $\log \bar{\varepsilon}_r$ being equal to $\mu \log(L/r)$. Within Yaglom's context of prescribed eddies, either canonical or microcanonical, it will be shown that this hypothesis cannot hold. Then, it will be shown that in a wider context this hypothesis is tenable only if one makes unlikely additional conditions.

In the context of prescribed microcanonical eddies, the difficulty is the following. To say that $\log \bar{\varepsilon}_r$ is normal is to say that a finite number of the independent random variables $\log W_k$ add to a Gaussian distribution, and it follows, by a classical theorem (Lévy-Cramèr), that the $\log W_k$ must themselves be Gaussian, i.e., unbounded. On the other hand, we know that microcanonical weights must be bounded. Thus Kolmogorov's hypothesis cannot apply strictly for any fixed value of r, not even for $r = \eta$.

In the context of prescribed canonical eddies, the source of the difficulty is different. One can show that, in a canonical context, the correlation of every pair of ε_r is positive. (To be more precise, Yaglom's rough derivation suggests that the correlation is also positive in a microcanonical context. But a careful investigation, too long to be worth reporting, shows that for some value of d it must be negative. This follows from the fact that $\langle W_s W_t \rangle < 1$; see Section 3.6.) To the contrary, we shall see momentarily that the Kolmogorov hypotheses imply that at least some of those correlations are negative. Thus, the Kolmogorov hypotheses might conceivably hold for $r = \eta$, but the hypothesis relative to several values of r are incompatible, meaning that overall the hypotheses are internally inconsistent.

More generally, the joint assumptions that the random variables $\log \bar{\varepsilon}_r(\mathbf{x}, L, \eta)$ are normal for every r, with $\sigma^2 \log \bar{\varepsilon}_r = \mu \log(L/r)$ and $\langle \log \bar{\varepsilon}_r \rangle = -(\mu/2) \log(L/r)$, are incompatible with any model that leads to a positive reduced covariance for $\bar{\varepsilon}_\eta$. Indeed it would follow from these assumptions that $r^3 \bar{\varepsilon}_r$, the mass of turbulence in a cube, satisfies

$$\langle [r^3 \bar{\varepsilon}_r(\mathbf{x}, L, \eta)]^q \rangle = r^{3q - q/2(q-1)\mu} L^{q/2(q-1)\mu}.$$

When r reaches its maximum value L, the above moment will reduce to r^{3q}, for all q. This is as it should. But the nature of convergence to this limit must be examined more closely. For example, let us subdivide our cube into (say) 2^3 portions. When $q/3 > 2\mu$, the exponent of r in the above expression is negative, implying that the value of the ratio $\langle (r^3 \bar{\varepsilon}_r)^q \rangle / \langle [(r/2)^3 \bar{\varepsilon}_r]^q \rangle$ is less than 2^3. From an elementary result of probability, this means that at least two of our subcubes must have a negative reduced correlation. This conclusion, and hence Kolmogorov's form of the lognormal hypothesis, is inconsistent with the assumed positive covariance of $\bar{\varepsilon}_\eta(\mathbf{x}, L, \eta)$. (Note added during revision. This inconsistency is observed in Novikov 1971, p. 236. The author notes the contradictory behavior of the quantities he designates as μ_ρ; however, having noted the contradiction, he does not resolve it.)

This is the moment to point out that in my limit lognormal model (M 1972j{N14} and Section A4) the above inconsistency is avoided, because every formerly misbehaving moment turns out to be infinite.

A6. Footnote added in 1972, during revision

This text appeared as footnote to Section 2.2 of the original M 1974f, but deserves being changed into an appendix and discussed in the Annotations that follow.

A referee made me aware of Novikov 1969 and 1971 which helped put [my] results in focus. Novikov 1971 p. 236 observes that the moments of $\bar{\varepsilon}$ do not tend towards those of the lognormal distribution. Yet, "in the same manner as in [a textbook by] Gnedenko, it may be shown that the limit distribution is lognormal." The puzzling discrepancy between these results appears to be due to use of conflicting approximations. Earlier, Novikov 1969, p. 105 states that "all moments (if they exist) must have a power law character." The phrase in parentheses raises the possibility that moments may not exist, but this possibility is regrettably dismissed and is not discussed again.

ACKNOWLEDGMENT

I benefited greatly from conversations with Jay M. Berger of IBM, Erik Mollo-Christensen of MIT, Robert W. Stewart, then of the University of British Columbia, and Akiva Yaglom of the Academy of Sciences of the USSR. Computer simulations were carried out in part by Gerald B. Lichtenberger, then of Yale University, and in part by Hirsh Lewitan of IBM. During the last of the revisions of this paper for publication, Robert Kraichnan helped me to sharpen the distinctions between different forms of the cascades. Several early forms of portions of this paper have been circulated privately; excerpts and abstracts have been published in the *Proceedings of the Summer Study Program in Geophysical Fluid Mechanics*, Woods Hole, Massachusetts, 1965; *Proceedings of the IUGG-IUTAM International Symposium on Boundary Layers and Turbulence including Geophysical Applications*, Kyoto, Japan, 1966 (A Supplement to *Phys. Fluids* 10, (suppl.), 1967, p. S302); and *Proceedings of the Polytechnic Institute of Brooklyn Symposium on Turbulence of Fluids and Plasmas*, New York, 1968 (John Wiley, 1969), and in M 1966k{N}, and M 1969b{N}.

&&&&&&&&&&& **ANNOTATIONS** &&&&&&&&&&&&

Editorial comments. The notation was brought close to the current usage. The letter Γ denoting a base was replaced by b; the letter h for the exponent in a moment was replaced by q; the expression $f(h)$ by $\overline{\tau}(q)$ (my old function $f(h)$ is now ordinarily denoted by $-\tau(q)$ and the notation $\overline{\tau}(q) = -\tau(q)$ avoids having to change signs throughout this paper); the D identifying a domain was replaced by $\mathscr{D}$ the letter i for the embedding dimension was replaced by E; the letter Δ for the intrinsic fractional dimension was replaced by D; the expression V^* was replaced by $\tilde{V}$, and the integer N in Section 4.8 was replaced by C. A footnote to Section 2.2 in the original was made into Appendix A.6. The other footnotes were incorporated in the text. A few lengthy ones are printed in smaller size letters.

Comment on Appendix A6. This Appendix acknowledges that E.A. Novikov and I simultaneously showed that the q-th moment of the average dissipation $\bar{\varepsilon}$ within a volume is the volume's size raised to some power $\tau(q)$. However, the derivation of this $\tau(q)$ is only the first of three stages of the theory of multifractals.

Elaborating on M 1969b {N13}, this chapter goes on to a second step and concludes that the central limit theorem is far from being sufficient from the viewpoint of the study of multifractals, that is, the distribution of $\bar{\varepsilon}$ is *not* approximately lognormal to a significant degree. M 1974c{N16} takes a further third step. It uses the Cramer large deviations theory, hence the Legendre transform, to obtain the multifractal function $f(\alpha)$ via the probability distribution of $\bar{\varepsilon}$ plotted on suitable log-log coordinates.

To the contrary, Novikov stopped at the first step. He did observe the difficulties $\tau(q)$ may present and the fact that its non-universality contradicts lognormality. But he did not resolve these difficulties. In terms of the function $f(\alpha)$ (which he did not consider), his text implied that the graph of $f(\alpha)$ is in every case a parabola.

Other scientific comments. Additional comments on this chapter are combined with comments at the end of Chapter N16.

How did M 1974f come to be written and published. This paper was hard to write, and it is hard to read because it is far too detailed for a first formal work on a new topic. That counterproductive complication was largely the unintended result of a very long refereeing process. The annotation "received 1 March 1972" refers the date of the final draft. An editor of the *Journal of Fluid Mechanics*, Keith Moffatt, invited a paper after hearing me sketch M 1967k{N12} at Kyoto in 1966. Thus, M 1974f was preceded by at least two very distinct versions ranging back to 1968. Among successive referees the *Journal* called upon, several professed utter bafflement about the problem and the solution. Other referees made nice noises, but asked not to be called again. One referee who professed competence picked endlessly at insignificant issues, and chastised me for incompleteness. (The next significant step in the theory of multifractals did not come until the mid 1980s.) All this encouraged endless rewriting that became increasingly "defensive" and counter-productive.

Fortunately, Keith Moffatt showed great foresight and kindness, but everything that sounded like "philosophy" had to be removed. As a result, the text that *J. Fluid Mech.* received on March 1, 1972 left a remainder. After revision, this became a typescript of 90 tightly packed pages, titled *The Geometry of Turbulence*. My copy indicates it was received by *J. Fluid Mech.* on August 27, 1974, but I changed my mind. One part became M1975F and a second was largely incorporated in M 1977F and M 1982F{*FGN*}. Its summary, slightly abbreviated, is reproduced as Section N4.2.

Translated from Comptes Rendus (Paris) 278A, 1974, 289-292 & 355-358   **N16**

Iterated random multiplications and invariance under randomly weighted averaging

●*Chapter foreword. First mention of the Legendre transform in the context of multifractals.* This paper's original had few readers: it was in French, was overly concise, and appeared to be a summary of M 1974f{N15}.

Actually, it was written after M 1974f{N15} and went beyond it on several accounts. Most significantly, as mentioned in Section N2.2, this paper includes for the first time an argument that became the basis of my approach to multifractals: Indeed, an argument reproduced in French in Chapter N2, Section 5.5.1, injected the function "$f(\alpha)$" and the Legendre transforms via the Cramèr theory of large deviations of sums of random variables. This method preceded the alternative approach due to Frisch & Parisi 1985 and Halsey et al. 1986. Details are found in Chapter N2 and the *Annotations on Section 21* of this chapter.   ●

✦ **Abstract.** The iteration of random multiplications yields new random functions that are interesting theoretically and practically. For example, they represent intermittent turbulence, M 1974f{N15}, and the distribution of minerals. This paper also generalizes the stable random variables: Lévy's criterion of invariance under non-random averaging is replaced by the criterion of invariance under randomly weighted averaging.   ✦

1. Construction of a multiplicative measure

Consider a sequence of "weights" W, which are independent and identically distributed random variables (i.i.d. r.v.'s). We shall write $F(w) = \Pr\{W < w\}$. The base is a given integer $b > 1$, the first b weights are denoted by $W(i_1)$, $0 \le i_1 \le b-1$, the following b^2 weights by $W(i_1, i_2)$. etc. Let t be a real number in the interval $]0, 1]$, expanded in base b in the form

$t = 0, i_1, i_2, \ldots$ Starting from $X'_0(t) \equiv 1$, the sequence of random densities $X'_n(t)$ will be defined iteratively as

$$X'_n(t) = W(i_1)W(i_1, i_2)\ldots W(i_1, i_2, \ldots, i_n).$$

Let $X_n(t) = \int_0^t X'_n(s)\,ds$. We shall primarily study the random function (r.f.) $X_\infty(t) = \lim_{n \to \infty} X_n(t)$. However, in the case where $X_\infty(t)$ itself is degenerate, we shall instead study $Y_\infty(t) = \lim_{n \to \infty} Y_n(t)$, where $Y_n(t) = X_n(t)/A_n$ and A_n is an appropriate normalizing non-random sequence.

This construction is closely related to the construction of the limit lognormal random functions, M 1972j{N14}. These functions are of the form

$$L_\infty(t) = \lim_{n \to \infty} L_n(t),$$

where $\log L'_n(t)$ is normal. In the interesting case, $L_\infty(t)$ is non-degenerate and singular. A general procedure to construct $\log L'_n(t)$ consists in decomposing it into a sum of random functions $\log L'_{n+1}(t) - \log L'_n(t)$, each with a bounded spectrum. Unfortunately, the theory of $L'_\infty(t)$ is far removed from the familiar theories of Kolmogorov and Yaglom, and the theory of $L_\infty(t)$ presents formal difficulties that the present construction is designed to avoid.

Remark. When $b = \Gamma^2$, with an integer $\Gamma > 1$, the construction generalizes to the case where t is a vector of co-ordinates $t' \in \,]0, 1]$ with $t' = 0, i'_1, i'_2, \ldots$ and $t'' \in \,]0, 1]$ with $t'' = 0, i''_1, i''_2, \ldots$ In this case, each i_n is a vector whose coordinates i'_n and i''_n are integers that range from 0 to $\Gamma - 1$.

2. Special cases: birth-and-death and symmetric binomial

The most interesting case (considering the number of applications and the precision of theorems) is when $F(0) = 0$ and $0 < EW < \infty$; in this case, we assume that $EW = 1$. Every case where $EW < 0$ can be reduced to a case where $EW > 0$ by replacing A_n by $(-1)^n A_n$.

A second case is interesting because it reduces to a classical theory. When W is binomial, with $\Pr\{W = 1\} = p > 0$ and $\Pr\{W = 0\} = 1 - p > 0$, $bX_1(1)$ is the sum of b i.i.d. r.v.'s of the form $W(i_1)$; $b^2 X_2(1)$ is obtained by replacing every term of $bX_1(1)$ by a r.v. which has the same distribution as $bX_1(1)$. Consequently, $b^n X_n(1)$ results from a birth-and-death process for which the number of descendants in each generation is given by the r.v. $bX_1(1)$. Classically, if $pb > 1$, the ratio

$$\frac{b^n X_n(1)}{[bEX_n(1)]^n} = \frac{X_n(1)}{p^n}$$

converges almost surely (a.s.) towards a non-degenerate limit.

Conclusion. If $0 < EW < \infty$, we expect to encounter problems involving a.s. convergence.

A third case that also goes back to a classical theory occurs when W is binomial with $\Pr\{W = 1\} = \Pr\{W = -1\} = 1/2$. This case brings back the central limit theorem: for Bernoulli variables the ratio $b^n X_n(1)/\sqrt{b^n}$ converges in distribution to a reduced Gaussian limit, the r.v. $X_n(1)b^{n/2}$ being independent.

Conclusion. If $EW = 0$, we expect to encounter problems involving convergence in distribution.

3. Fundamental recursion rule between distributions

One has the relation:

$$X_{n+1}(1) = b^{-1} \sum_{g=0}^{b-1} W_g X_{n,g}(1),$$

where the r.v.'s W_g and $X_{n,g}$ are independent, and, for every g,

$$\Pr\{W_g < w\} = F(w) \quad \text{and} \quad \Pr\{X_{n,g}(1) < x\} = \Pr\{X_n(1) < x\}.$$

4. Fundamental invariance (fixed point) property of the limit measure

If $X_n(1)/A_n$ converges in distribution towards $Y_\infty(1)$, one must have

$$\lim_{n \to \infty} \frac{A_{n+1}}{A_n} = A \quad (0 < A < \infty),$$

and one has the following identity *between distributions*:

$$Y_\infty(1) = \sum_{g=0}^{b-1} [Ab^{-1} W_g] Y_{\infty,g}(1),$$

where the W_g and $Y_{\infty,g}$ are independent, and, for all g,

$$\Pr\{W_g < w\} = F(w) \quad \text{and} \quad \Pr\{Y_{\infty,g}(1) < y\} = \Pr\{Y_\infty(1) < y\}.$$

It is easy to rewrite the above invariance in terms of the characteristic function of $Y_\infty(1)$.

Remark. The invariance that defines Lévy stability corresponds to the special case where the W_g are identical real numbers. The above invariance implies a functional equation in $\Pr\{Y_\infty(1) < y\}$. Its solutions depend on both $F(w)$ and b. Some are given by the construction of Section 1; the problem of the existence of other solutions remains open.

{P.S. 1998. Additional solutions were discovered in Durrett & Liggett 1983 and Guivarc'h 1987, 1990. See the *Scientific Comment on Sections 5 and 17* at the end of the Chapter.}

5. Statement of the fundamental martingale property

Let $EW = 1$; for all t, $X_n(t)$ is a martingale. For all finite n and integer $q > 1$, $EX_n^q(t) > 0$.

6. Condition for convergence of martingales

Let $EW = 1$. For every integer $h > 1$, the necessary and sufficient condition for $0 < \lim_{n \to \infty} EX_n^q(t) < \infty$ is $EW^q < b^{q-1}$.

7. Conjectured generalization of the result in Section 6

The result of Section 6 is expected to hold for all real $q > 1$.

8. Two sufficient conditions of convergence

Let $EW = 1$. In order for $X_n(t)$ to converge almost surely, two sufficient (not mutually exclusive) conditions are as follows: (a) $F(0) = 0$ (b) $EW^2 < b$ {P.S. 1996, meaning that the martingale is positive and of bounded variance}. When this second condition holds, $X_\infty(t)$ does not reduce to 0, and the equation of Section 4, with $A = 1$, has at least one non-degenerate solution, namely $X_\infty(1)$.

Proof. Theorem on convergence of martingales (see for example Doob 1953).

9. Conjectured generalization of the result in Section 8

The condition $EW = 1$ is sufficient in order for $X_n(t)$ to converge almost surely.

10. Positive weights: definitions of the exponents q_{crit} and β

When $F(0) = 0$ and $EW = 1$, let

$$q_{crit} = \max\{1, \sup [q: EW^q < b^{q-1}]\} \quad \text{and} \quad D = 1 - EW \log_b W.$$

When $F(0) = 0$, let

$$\overline{\tau}(q) = \log_b[EW^q/b^{q-1}] = \log_b EW^q - (q - 1).$$

{P.S. 1996. Today, $\overline{\tau}(q)$ is usually denoted by $-\tau(q)$ }. This function $\overline{\tau}(q)$ is convex, and $\overline{\tau}(1) = \log EW = 0$, so that the quantity q_{crit} is the larger of 1 and of the second zero of $\overline{\tau}(q)$. Formally, $D = -\overline{\tau}'(1)$.

When $F(0) = 0$ and $\overline{\tau}(0+) > 0$, let β be the value of q for which the straight line adjoining 0 to $[\beta, \overline{\tau}(\beta)[$ has no other points in common with the graph of $\overline{\tau}(q)$.

Proposition concerning q_{crit}

The condition $q_{crit} = \infty$ holds if and only if $W < b$.

Remark. We will see that, as long as $q_{crit} = \infty$, the X_∞ are regular meaning that, $E_\infty^q < \infty$ for all q. If $1 < q_{crit} < \infty$, the X_∞ are irregular. When $D < 0$, the X_∞ are degenerate. The case $D = 0$ remains to be studied.

11. Proposition concerning q_{crit} and the convergence of moments

Let $F(0) = 0$, $EW = 1$ and $q_{crit} > 2$. Then $EX_\infty^q(t) = \lim_{n \to \infty} EX_n^q(t)$ for all $q <$ the largest integer $< q_{crit}$.

Proof. Again this follows directly from the classical theorem on the convergence of martingales. (P.S. 1996. A misprint in the original was corrected}.

12. Conjectures concerning the generalization of Section 11

(A) Let $F(0) = 0$ and $EW = 1$. In order for $X_\infty(t)$ to be non-degenerate, and for the equation of Section 4 to have a non-degenerate solution with $A = 1$, it is sufficient that $q_{crit} > 1$.

(B) Furthermore, in this case, $EX_\infty^q(t) = \lim_{n \to \infty} EX_n^q(t)$ holds for *all* $q < q_{crit}$.

13. Proposition concerning divergent moments

Let $F(0) = 0$, $EW = 1$ and $q_{crit} > 1$. If $X_\infty(t)$ is nondegenerate, $EX_\infty^q(t) = \infty$ for all $q > q_{crit}$.

Proof. It suffices to take t in the form b^{-n}, and then to limit the study to $t = 1$. Then, for all $q > 1$, Section 4 gives $EX_\infty^q(1) > b^{1-q} EW^q EX_\infty^q(1)$. If $q > q_{crit}$, this requires either $EX_\infty^q(1) = 0$, which is excluded, or $EX_\infty^q(1) = \infty$, which is thereby proven.

14. Remark on related multiplicative measures

When $F(0) = 0$, $X'_\infty(t)$ generalizes the Besicovitch measure. This is the non-random singular measure {P.S. Today, it is mostly called multinomial.} one obtains when the weights $W(i_1)$ are not random, but imposed in advance, and satisfy

$$W(i_1, i_2, ..., i_n) = W(i_n).$$

It is convenient to assume that the attainable values of W are all different, the probability p_j of each value w_j being $1/b$, with $\sum_{j=0}^{b-1} w_j/b = 1$. The Besicovitch measure rules the distribution of numbers for which the "decimals" in base b have the probabilities $\pi_j = p_j w_j$.

To generalize, we shall proceed in several stages. First, while keeping the $W(i_1)$ fixed, let their sequence follow a randomly chosen permutation. Next, allow the values of the $W(i_1)$ to vary – and in particular let the number of possible values vary – while imposing on them the following sequence of "conservation relations:"

$$\sum_{i_1=0}^{b-1} W(i_1) = b, \quad \sum_{i_2=0}^{b-1} W(i_1, i_2) = b \quad \text{for all } i$$

$$\sum_{i_3=0}^{b-1} W(i_1, i_2, i_3) = b \quad \text{for all pairs } (i_1, i_2) \quad \ldots$$

The resulting measure – which has been considered by Yaglom (see M 1974f{N15}) – can be called "microcanonical." Finally, generate the W independently, thereby obtaining $X'_\infty(t)$. The possibility of degeneracy does not appear until this last step. However, when t is generalized to be multidimensional, degeneracy may already appear in one-dimensional sections.

15. A weak limit law yielding a box dimension

Let $F(0) = 0$ and $EW = 1$. For all $\varepsilon > 0$, there exists an $n_0(\varepsilon) > 1$ such that, for all integer $n > n_0(\varepsilon)$, one can write $X_n(t) = Y_n(t) + Z_n(t)$. In this representation, $Y_n(t)$ is very small, that is to say $EY_n(1) < \varepsilon$. As to $Z_n(t)$, it varies only on a small portion of the intervals of the form $kb^{-n} < t \leq (k+1)b^{-n}$, that is, a number whose expectation is much less than $(1-\varepsilon)b^{n(D+\varepsilon)}$.

Proof. It is postponed until Section 19.

16. Corollary: a condition for degeneracy

Let $F(0) = 0$, $EW = 1$ and $D < 0$. In this case, $X_\infty(t) = 0$ almost surely, and the equation of Section 4, with $A = 1$, has no non-degenerate solution that could be constructed by the method of Section 1.

Proof. If $D < 0$, and for n sufficiently large, $(1-\varepsilon)^{-1}b^{n(D+\varepsilon)} \leq \sqrt{\varepsilon}$. From this it follows that $\Pr\{Z_n(1) > 0\} < \sqrt{\varepsilon}$. Moreover, $\Pr\{Y_n(1) \geq \sqrt{\varepsilon}\} \leq \sqrt{\varepsilon}$. It follows that $\lim_{n \to \infty} \Pr\{X_n(1) > 0\} = 0$.

17. Conjecture

Let $F(0) = 0$, $EW = 1$ and $D < 0$. Then the equation of Section 4, with $A = 1$, has $X = 0$ as its only solution.

{P.S. 1998: See the P.S. at the end of section 4}.

18. Conjectured strong limit law concerning Hausdorff dimension

Let $F(0) = 1$, $EW = 1$ and $D > 0$.

(A) Define $N(a,b,t,n)$ as the number of weights W in the sequence $W(i_1), W(i_1, i_2), W(i_1, i_2, \ldots, i_n)$ that satisfy $a \leq W \leq b$. In a sense that remains to be specified, the domain of variation of $X_\infty(t)$ is characterized by $\lim_{n \to \infty} n^{-1} N(a,b,t,n) = \int_a^b fw \, dF(w)$.

(B) The Hausdorff dimension of said domain of variation is almost surely equal to D.

Remark. In the case when W has b distinct possible values w_j with the probabilities b^{-1}, the clause (A) above resembles the strong classical law for the probabilities $\pi_j = w_j/b$, and the dimension in clause (B) becomes $-\Sigma \pi_j \log_b \pi_j$, which is formally identical to the dimension of the Besicovitch measure (see Billingsley 1967).

19. Proof of the "box dimension" weak law stated in Section 15

The main idea of the proof is easily expressed in the finite case where $\Pr\{W = w_j\} = p_j$, with $\Sigma p_j = 1$ and $EW = 1$, hence $\Sigma \pi_j = 1$, with $\pi_j = p_j w_j$. Denoting by $n \psi_j$ the number of times that w_j appears in the product that defines $X'_n(t)$, one has

$$X'_n(t) = \prod w_j^{n \psi_j}.$$

The hypothesis $EX'_n(t) = (EW)^n = 1$ yields

$$\sum n! \left(\prod (n \psi_j)!\right)^{-1} \prod w_j^{n \psi_j} p_j^{n \psi_j} = 1.$$

When the π_j are interpreted as probabilities, this last equality is simply the multinomial expansion of $(\Sigma \pi_j)^n = 1$. A theorem (Billingsley 1967) that is used in some proofs of the weak law of large numbers. It states that, given $\varepsilon > 0$, there exists a $n_0(\varepsilon)$ so that for $n > n_0(\varepsilon)$, the terms of the expansion of $(\Sigma \pi_j)^n$ can be classified as follows:

- in the first class, $2|\psi_j - \pi_j| < n^{-1/2}\sqrt{\pi_j(1 - \pi_j)}$ holds for all i',
- the sum of all the terms in the second class is $< \varepsilon$.

A fortiori, the first class satisfies

$$\left|\sum (\psi_j - \pi_j) \log_b \pi_j\right| < \sum (\psi_j - \pi_j) |\log_b \pi_j| < n^{-1/2} \sum |\log_b \pi_j| \sqrt{\pi_j(1 - \pi_j)}.$$

For large enough n, this last quantity is $< \varepsilon$. It follows that the values of $X'_n(t)$ in the second class are contained between $b^{n(H-\varepsilon)}$ and $b^{n(H+\varepsilon)}$, with

$$H = \sum \pi_j \log_b w_j = EW \log_b W.$$

It is easily established that $H > 0$. The probability of each of these $X'_n(t)$ is contained between $(1-\varepsilon)b^{-n(H+\varepsilon)}$ and $(1-\varepsilon)b^{-n(H-\varepsilon)}$. Finally, noting that $]0,1]$ divides into b^n equal intervals on which $X'_n(t)$ is constant, the number of those intervals for which $X'_n(t)$ is of the second class is at most

$$(1-\varepsilon)b^{n(1-H+\varepsilon)} = (1-\varepsilon)b^{n(D+\varepsilon)}.$$

The case where W is not bounded is treated through bounded approximations. In the case where $\log W$ is Gaussian, the direct verification is easy.

20. The case $EW = 0$: choice of A_n to insure convergence in distribution

Only some formal results are available. When $EW^2 < \infty$, one can insure that $A_n = 1$ by normalizing W so that $EW^2 = b$. Define

$$\tilde{q}_{crit} = \max\{1, \sup[q : EW^q < b^{q-1}]\},$$

where q is an even integer > 2. Two cases must be distinguished.

From $\tilde{q}_{crit} = \infty$, which implies $|W| < b$, it follows that $0 < \lim_{n \to \infty} EX_n^q(t) < \infty$ when q is an even integer, and $\lim_{n \to \infty} EX_n^q(t) = 0$ when q is an odd integer. One may conjecture that $X_n \to X_\infty$, with $EX_n^q(t) \to EX_\infty^q(t)$, and that the limit X_∞ is a symmetric r.v..

When $\tilde{q}_{crit} < \infty$, the moment of order $q < \tilde{q}_{crit}$ converges either to a limit that is both > 0 and $< \infty$, or to a limit that is identically 0. The moments of even order $q > \tilde{q}_{crit}$ converge to infinity; the moments of odd order $q > \tilde{q}_{crit}$ can either tend to infinity or oscillate while moving away from 0, which raises problems. In the former case, one can conjecture that $X_n \to X_\infty$, just as above.

21. The probability distribution of X'

{P.S. 1998. This translation is limited to the middle part of Section 21 of the French original, namely to the part reproduced photographically in

Chapter N2. See the *Comments on Sections 10 and 21* in the *Annotations* appended to this chapter.} We use an inequality in Chernoff 1952 that requires $E(\log W) < \infty$, but allows $EW = \infty$. Writing $-\log_b A_n = \alpha n$ and neglecting complicating slowly ranging factors that do not affect the present argument, the Chernoff inequality takes the form

$$\Pr \{ \log [X'_n(kb^{-n})b^{-n}] \geq -\alpha n \log b \} \sim b^{-nC(\alpha)},$$

where

$$-C(\alpha) = \operatorname{Inf} \{\alpha q + \log_b E(W/b)^q\} = -1 + \operatorname{Inf} [\overline{\tau}(q) + q\alpha].$$

22. {P.S. 1998. See the *Comment on Section 10* in the *Annotations* appended to this chapter.}

23. Hyperbolically distributed weights

Let $F(0) = 0$ and $\Pr \{W > w\} = w^{-\gamma} L(w)$, where $L(w)$ is a slowly varying function for $w \to \infty$. When $\gamma > 1$, we have either $\alpha > 1$ with $\alpha \leq \gamma$, or $\beta < 1$ with $\beta \leq 1$. When $\gamma < 1$, we have $\beta < 1$ and $\beta \leq \gamma$. It was not unexpected that X_∞ is at least "as irregular" as W. But it was surprising that W_∞ could be strictly more irregular, or that X_∞ could be irregular when W is regular. An example where X_∞ and W are of precisely the same level of irregularity ($\alpha = \gamma$) occurs when $L(w)(\log w)^2$ tends very rapidly (as $w \to \infty$) towards a sufficiently small limit. {P.S. 1998. This section was translated without being thought through.}

24. Conjecture concerning the case when $F(0) = 0$ but $\overline{\tau}(0+) > 0$

The behavior of X goes beyond the above theorems and conjectures. An example is when X_n is ruled by the birth and death process of Section 2. When $p < 1/b$, one has $X_n(1)/p^n \to 0$ a.s., and there can be no sequence A_n such that $Y_n \to Y_\infty$, with a non degenerate Y_∞. It is conjectured that this conclusion holds whenever $F(0) = 0$ and $\overline{\tau}(0+) < 0$.

25. Final remark: all the features of Y_∞ investigated above are determined by the geometry of the graph of $\overline{\tau}(q)$

When $D > 0$, the moments of $X_\infty(t)$ and the dimension of its set of concentration are ruled by different features of $\phi(q)$ or of W. The same holds for the covariance of $X'_\infty(t)$, which can be shown to be ruled by $\overline{\tau}(2)$.

&&&&&&&&&&& **ANNOTATIONS** &&&&&&&&&&&

The French original is made up of two papers, presented to the Académie de Sciences on November 19, 1973 by Szolem Mandelbrojt.

Editorial changes. Section titles were overly terse in the original. To help orient the reader, many were made more explicit without special mention to that effect. Other cases where the original is not followed closely is marked as {P.S. 1998}. The notation was updated: the original denotes the present b by C, the present q by h, the present $\overline{\tau}(q)$ as $f(h)$, the present q_{crit} by α, the present α by T, and the present b^{-C} by e^{-Q}. This change of base eliminates typographical clutter and brings out the co-dimension function $C(\alpha) = 1 - f(\alpha)$, be discussed in a *Comment* directed to the Legendre transform.

How this paper came to be written. M 1972j{N13} and M 1974f{N14} include and/or imply a host of conjectures of a purely mathematical nature. For example, they involve an informal box dimension for the set on which an arbitrarily large proportion of a total multiplicative measure concentrates. (The meaning of this box dimension was not fully clarified until M 1995k). It was natural to go further and conjecture that these quantities were also the dimensions in the sense of Hausdorff-Besicovitch, but the proofs turned out to be elusive.

As described in an *Annotation* to Chapter N13, I brought the matter up with Jean-Pierre Kahane; he and his then student Jacques Peyrière soon confirmed several of my conjectures relative to the base-bound case. They also advised me to restate my work for the mathematicians. This is how the present chapter came to be written.

Literature. The most important follow-up articles are Peyrière 1974, Kahane 1974, Kahane & Peyrière 1976{N17}, Ben Nasr 1987, Durrett & Liggett 1983, and Guivarc'h 1987, 1990. When referred to jointly, the last three are denoted by DLG.

Generalization to the case when the weights W_g in the recursion of Section 3 are not independent. Dependence between weights affects neither the evaluation of $\tau(q)$ nor the derivation of $f(x)$ by using the Cramer theory of large derivation. Durrett and Liggett 1983 and Ben Nasr

1987 show that the same theorems continue to hold. This generalization is important in M 1991k, 1995k, which deal with sample estimates of $\tau(q)$.

The tail behavior $\Pr\{X_\infty > x\} = x^{-q_{crit}}$. This behavior, conjectured in M 1974f{N15}, follows directly from Sections 11 to 13. Proofs of my conjecture were provided by DLG and Guivarc'h.

Scientific comment on the quantity β defined in Section 10. Let $f(\alpha) > 0$ for some αs but $f(\alpha) < 0$ in the interval from α^*_{min} to α^*_{max}, where $\alpha^*_{min} > \alpha_{min}$ and $\alpha^*_{min} \leq \alpha_{max}$. If so, β is defined and is identical to α^*_{min}.

Alternative and equivalent definitions of α^*_{min} and α^*_{max} involve the narrowest "fan" that is contained between two half lines that start at the point of coordinates $q = 0$ and $\overline{\tau}(q) = \tau(q) = 0$ and contains the graph of $\overline{\tau}(q)$. The quantities α^*_{min} and α^*_{max} are the slopes of the half-lines that bound this narrowest fan.

M 1991k and M 1995k vindicated (belatedly) the usefulness of the quantity β by showing that the quantities α^*_{min} and α^*_{max} play an important role on their own terms. Indeed, for certain purposes, the only part of $f(\alpha)$ that matters corresponds to positive fs. Correspondingly, the part $\overline{\tau}(q)$ that matters is between the abscissas q^*_{max} and q^*_{min}, where the graph of $\overline{\tau}(q)$ touches the half lines of slopes $\alpha^*_{min} = \beta$ and α^*_{max}.

However, those roles of β were not known to me in 1974 and the role attributed to it in the French original of M 1974c was incorrect. In addition, the discussion of β was very confused, and to translate it fairly would be difficult and pointless. Therefore, Section 22 and large portions of Section 21 were omitted in this translation.

Scientific comments on Sections 4 and 17. The need to distinguish between two forms of cascade: direct (interpolative) and inverse (extrapolative). Lévy's semi-stable distributions. DLG closed the issue raised at the end of Section 4 and showed the conjecture in Section 17 to be incorrect. When $F(0) = 0$, $EW = 1$ and $D < 0$ – and also under milder restrictions – the functional equation introduced in Section 4 has additional solutions. These solutions have infinite expectations, and are *not* obtained by the measure-generating multiplicative scheme in Section 1. The functional equation of Section 4 continues to interest mathematicians.

Recall that my multiplicative scheme generates a measure proceeds to increasingly *small* eddies. Thus, it can be called an "interpolative" or "direct" cascade that *roughens* a uniform measure. To the contrary, the remaining solutions of the functional equation, as obtained by DLG, are obtained by an extrapolative inverse cascade. This cascade reinterprets the fundamental recursion rule of Section 3 as being a *smoothing* multiplicative scheme that proceeds to increasingly *large* eddies.

The inverse cascade. It is worth repeating here a few lines from M 1984e, Section 3.2.2. This and the preceding chapters show that in studying fractal measures relevant "to noise and turbulence, it is not only inevitable but essential to introduce a process of renormalization somewhat analogous to Lévy's semi-stability. And the somewhat analogous (though different) indeterminacy and complication are present in the resulting random variables, and are concretely very important. The key ingredient in this more general renormalization is to replace ordinary addition by randomly weighted addition. The weights are a semi-infinite array of independent identically distributed r.v. with row index n and column index i, namely $W(n, i)$. Now we start with $X(n_1, 1) \equiv 1$, and the first step of renormalization is to form the array $X^*(n_2, 2) = \Sigma W(n_1, 1) X(n_1, 1)$ with the sum carried over the indexes n_1 of the form n, followed by an integer between 0 and $b-1$. Of the many classes of W that have been examined, the simplest, and only class characterized by $W \geq 0$, and $\langle W \rangle = 1$ was studied [before 1984]. The proper second step in renormalization is then $X(n_2, 2) = b^{-E} X^*(n_2, 2)$, and $\langle X(n_k, k) \rangle \equiv 1$.

The first object of study is, then, the fixed-point random variable $X = \lim_{k \to \infty} X(n_k, k)$ that is invariant under renormalization.

The second object is to interpolate $X(n_k, k)$ into a random function $X(n_1, k)$, and to compare the contributions to $X(n_1, k)$ from the addends $X(n_1, k) = X(n_1, k) - X(n_n - 1, k)$ that originate in the little cubes."

On the unstable solutions discovered by DLG. The contrast between "direct" and "inverse" cascades is familiar in the study of turbulence. In statistical physics, my physicist co-authors near-always build up large structures from atoms, while mathematicians prefer an interpolative "direct" cascade. Thus, the description reproduced in the preceding paragraphs took it for granted that the inverse and the direct cascades are completely equivalent. Such is not the case in the present instance.

The existence of inverse cascade fixed points means that, given the W, one can "load" the $X(n, 1)$ in such a way that the $X(n, k)$ have the same distribution for all k. Moreover, the tail of this distribution satisfies $\Pr\{X(n, 1) > \alpha\} \sim \alpha^{-q_{crit}}$. Finally, if, and only if, the $X(n, 1)$ have essentially the same tail behavior as the fixed point, the distribution of $X(n, k)$ converges to the fixed point distribution. This tail behavior characterizes the fixed point's domain of attraction.

An important feature of the new fixed points of the smoothing transformation found by DLG is that they are unstable. When the distribution being smoothed fails to be very specifically matched to the smoothing operation, its smoothed form ceases to converge to a non trivial limit. Unstable solutions may well be of no interest in physics.

Scientific comment on Section 20 and the generalization of the multiplicative processes to multipliers that may be negative. This Section gives very formal results on an important topic that is only now coming into its own. It corresponds to the multifractal functions sketched in Chapter N1, which oscillate up and down. I hope to develop my preliminary findings further and present them in a suitable forthcoming occasion.

Scientific comment on Section 21 and the Legendre transform. Section 21 was the first statement of the thermodynamical formalism of $\tau(q)$ and $f(\alpha)$. As mentioned in Chapter N2, Frisch & Parisi 1985 and Halsey et al 1986 made this formalism familiar to many scientists. The present $\tau(q)$ is my old $-\bar{\tau}(q)$, and the present $f(\alpha)$ is my old $1-C(\alpha)$. That is (but – once again – in 1974 I did *not* state it in these terms), $C(\alpha)$ is simply the fractal co-dimension corresponding to the dimension $f(\alpha)$.

My first encounter with Chernoff 1952 and the Cramèr theory of large deviations was in a paper on coding, M 1955t, Section 4.2; the mathematics behind codes and measures is often the same.

The Cramèr theory naturally leads to the Legendre transform Inf $[\bar{\tau}(h) + h\alpha]$ but this use of the Cramèr theory in the study of the multifractals did not become fully understood until many years after this paper. The valuable core of the original Section 21 was placed among statements that concern the quantity β defined in Section 10, were not thought through, and were not translated.

Let me end by restating an important remark implicit in Chapter N2, Section 1.7: the Cramèr theory does not restrict the range of values of $C(\alpha)$. For some αs, this function may well satisfy $C(\alpha) > 1$, leading to $f(\alpha) < 0$. This remark has led to extensive discussion in M 1989c, 1989ge, 1990r, 1991k, 1995k and other papers to be collected in M1998L.

Translated from Advances in Mathematics: 22, 1976, 131-145
Post-publication appendix by J. Peyrière

"On certain martingales of Benoit Mandelbrot"
Guest contribution (Kahane & Peyrière 1976)

✦ **Abstract.** Following his critical analysis of the random model of turbulence due to A. M. Yaglom, M 1974f{N14} and M1974c{N15} introduced his own model, which he calls "canonical." It proceeds from a brick, that is subsequently divided into $b, b^2, \ldots, b^n, \ldots$ similar bricks; each brick of the n-th stage is divided into b equal bricks in the $(n+1)$-th stage. Also given is a sequence of random variables W_p, which are independent, identically distributed, positive, have mean 1 and are indexed by the bricks P under consideration. Starting from the Lebesgue measure μ_0 on the initial brick, one constructs the sequence of measures μ_n by successive stages. Thus, μ_n has a constant density on each brick P of the n-th stage, and the density of μ_n on P is the product of W_P and the density μ_{n-1} on P. The sequence of measures μ_n is a vector martingale, and it converges towards a random measure μ. M 1974c gives results and raises problems concerning the measure μ: non-degeneracy, the moments of $\|\mu\|$, the Borel sets supporting μ and their Hausdorff dimension. Some of the conjectures of Mandelbrot have been solved by Kahane 1974 or by Peyrière 1974. Here we present these results in a refined form. Theorems 1, 2 and 3 below are due to J.-P. Kahane, Theorem 4 is due to J. Peyrière. ✦

1. Introduction, definitions, main results and history

It will be convenient to take as the initial brick the interval $[0, 1]$. The "bricks" P are then the b-adic intervals for $n = 1, 2, \ldots$ and $j_k = 0, \ldots, b-1$, namely

$$I(j_1, j_2, \ldots, j_n) = \left[\sum_1^n j_k b^{-k}, \sum_1^n j_k b^{-k} + b^{-n}\right[.$$

Given an integer $b \geq 2$, and a positive random variable (r.v.) W with $E(W) = 1$, one denotes by $W(j_1, j_2, \ldots, j_n)$ a sequence of independent r.v.'s, having the same distribution as W, and one denotes by μ_n the measure defined on $[0, 1]$, whose density on the interval $I(j_1, j_2, \ldots, j_n)$ is given by $W(j_1)W(j_1, j_2) \ldots W(j_1, j_2, \ldots, j_n)$. Let

$$Y_n = \|\mu_n\| = b^{-n} \sum_{j_1, j_2, \ldots, j_n} W(j_1)W(j_1, j_2)\ldots W(j_1, j_2, \ldots, j_n). \tag{1}$$

This is a nonnegative martingale, with $E(Y_n) = 1$. Hence, it converges almost surely (a.s.) towards a r.v. Y_∞ such that $E(Y_\infty) \leq 1$. In the same fashion, for all b-adic intervals I, $\mu_n(I)$ is a martingale with expectation $|I|$ which converges a.s. to a limit $\mu(I)$. Hence μ_n tends weakly a.s. to a measure μ of total mass Y_∞.

It is convenient to write (1) in the form

$$Y_n = b^{-1} \sum_{j=0}^{b-1} W(j) Y_{n-1}(j). \tag{2}$$

The r.v. $W(j)$ and $Y_{n-1}(j)$ are mutually independent, and the $Y_{n-1}(j)$ have the same distribution as Y_{n-1}.

Consider finally the functional equation

$$Z = b^{-1} \sum_{j=0}^{b-1} W_j Z_j, \tag{3}$$

where the r.v.'s W_j and Z_j are mutually independent, the W_j having the same distribution as W, and the Z_j having the same distribution as Z. The unknown in (3) is the distribution of Z; by an abuse of language, Z will be called solution of (3). Other solutions may exist; for example, in the case $W \equiv 1$, a Cauchy variable is a solution to (3), and it cannot be of type Y_∞ because it is neither positive nor integrable.

It will be convenient to associate with W the convex function

$$\varphi(q) = \log_b E(W^q) - (q - 1). \tag{4}$$

It is always defined for $0 \leq q \leq 1$, and can be defined for values $q > 1$. The function φ is zero at the point 1, and at most at one other point, q_{crit}. The left-side derivative of φ at the point 1 is

$$\varphi'(1 - 0) = E(W \log_b W) - 1 = -D \text{ (by definition of } D\text{)}.$$

We will see the role that D plays in the non-degeneracy of μ, and in the dimension of the Borel sets supporting μ. Also, we shall see the role of q_{crit} with respect to the moments of Y_∞.

The most striking illustrations are the following. (1) $W = e^{\tau\xi - (\tau^2/2)}$, where ξ is a normal variable (this is the origin of theory); then φ is a polynomial of degree 2. (2) W has only two possible values, one of which is zero; then φ is a linear function, and $b^n Y_n$ may be interpreted as the population at time n in a birth-and-death process in which each individual gives birth to b descendants, whose probability of survival is $P(W \neq 0)$.

All these notions were introduced in M 1974f{N15} and M 1974c{N16}.

We will establish the following results.

Theorem 1. *The following statements provide equivalent conditions of non-degeneracy:*

(α) $E(Y_\infty) = 1$,
(β) $E(Y_\infty) > 0$,
(γ) equation (3) has a solution Z such that $E(Z) = 1$,
(δ) $E(W \log W) < \log b$.

Theorem 2. (Condition for the existence of finite moments). Let $q > 1$. One has $0 < E(Y_\infty^q) < \infty$ if and only if $E(W^q) < b^{q-1}$.

Theorem 3. (Case where Y_∞ has moments of every order). (1) *The following statements are equivalent:* (α_1) $0 < E(Y_\infty^q) < \infty$ for all $q > 1$; (β_1) $\|W\|_\infty =$ ess. sup adjust(u 2)$W \leq b$ and $P(W = b) < 1/b$ (strict inequality).

(2) If (β_1) holds, one has

$$\lim_{q \to \infty} \frac{\log E(Y_\infty^q)}{q \log q} = \log_b \|W\|_\infty. \tag{5}$$

Theorem 4. (Study of the measure μ). Suppose $E(Y_\infty \log Y_\infty) < \infty$. For each $x \in [0, 1[$, denote by $I_n(x)$ the b-adic interval of order n that contains x; its Lebesgue measure is $m(I_n(x)) = b^{-n}$. One has

$$\mu\text{-almost everywhere}, \lim_{n \to \infty} \frac{\log \mu(I_n(x))}{\log m(I_n(x))} = D = 1 - E(W \log_b W) \quad (6)$$

Corollary. The measure μ is a.s. supported by a Borel set of Hausdorff dimension D, while all Borel sets of Hausdorff dimension $< D$ have μ-measure zero.

Historical remarks. Condition (δ) of Theorem 1 can be written as $D > 0$. Kahane 1974 had only shown that

$$D > 0 \Rightarrow (\alpha) \Rightarrow (\beta) \Rightarrow (\gamma) \Rightarrow D \geq 0.$$

The role of D in the study of degeneracy had been guessed in M 1974c{N16}, Section 10.

Theorem 2 was conjectured in M 1974c and proved in Kahane 1974. The proof that we will give is simpler. Let us remark that the condition $E(W^q) < b^{q-1}$ can also be written as $\varphi(q) < 0$. If φ is zero for some $q_{crit} > 1$, this means $q < q_{crit}$.

Theorem 3 constitutes a critical comment on M 1974c, Proposition 10. It corresponds to $q_{crit} = \infty$. The proof will give some variants of Kahane 1974.

The Corollary of Theorem 4 confirms a conjecture of M 1974c{N16}, and improves on Peyrière 1974.

2. Proof of Theorem 1

Obviously $(\alpha) \Rightarrow (\beta) \Rightarrow (\gamma)$. Assume (γ), and let Z be a solution of (3) such that $E(Z) = 1$. There exists a sequence of independent r.v.'s $W(j_1, j_2, \ldots, j_n)$ ($n = 1, 2, \ldots ; j_k = 0, 1, \ldots, b-1$), having the same distribution as W, and a sequence of r.v.'s $Z(j_1, j_2, \ldots, j_n)$ with the same distribution as Z and independent of the $W(i_1, i_2, \ldots, i_n)$ for $k \leq n$, such that for all n

$$Z = b^{-n} \sum_{j_1, \ldots, j_n} W(j_1) W(j_1, j_2) \ldots W(j_1, j_2, \ldots, j_n) Z(j_1, j_2, \ldots, j_n). \quad (7)$$

Indeed, (7) reduces to (3) for $n = 1$ ($W(j) = W_j$ and $Z(j) = Z_j$), and equation (3), if applied to $Z(j_1, j_2, \ldots, j_n)$, gives

$$Z(j_1, j_2, \ldots, j_n) = b^{-1} \sum_{j_{n+1}} W(j_1, j_2, \ldots, j_{n+1}) Z(j_1, j_2, \ldots, j_n, j_{n+1}),$$

with the required conditions for the r.v.'s on the right hand side. The conditional expectation of Z with respect to the σ-field generated by the $W(j_1, \ldots, j_k)$ ($k \leq n$) is Y_n as defined by (1). It follows that the martingale Y_n is uniformly integrable and that a.s. $Z = Y_\infty$ (see for example Meyer 1966, Section V8). Hence (γ) $\Rightarrow$ (α), and furthermore (γ) implies $Z \geq 0$ a.s..

Assume again (γ), and consequently $Z \geq 0$. For $0 < q < 1$, the function x^q is sub-additive, hence (3) yields

$$E(b^q Z^q) \leq \sum_{j=0}^{b-1} E[(W_j Z_j)^q] = b E(W^q) E(Z^q), \tag{8}$$

with $0 < E(Z^q) \leq 1$. Therefore, the function $\varphi(q)$ as defined by (4) is nonnegative on $[0, 1]$, from which it follows that $\varphi'(1 - 0) \leq 0$, that is $D \geq 0$. To go further, (8) needs to be improved.

Lemma A. $(x + y)^q \leq x^q + q y^q$ for $x \geq y > 0$, and $0 < q < 1$.

Proof. Use the fact that $y = 1$ and the formula of finite increments.

Lemma B. Let X be a positive integrable r.v., and X' a r.v. with the same distribution as X and independent of X. There exists a number $\varepsilon_X > 0$ such that

$$E(X^q 1_{X' \geq X}) \geq \varepsilon_X E(X^q) \quad \text{for } 0 \leq q \leq 1.$$

Proof. Each of these expectations is a continuous function of h, and is strictly positive on $[0, 1]$.

Since the function x^q is sub-additive, it follows from (3)

$$b^q Z^q \leq \sum_{j=0}^{b-1} W_j^q Z_j^q \quad \text{a.s.}$$

From Lemma A,

$$b^q Z^q \leq q W_0^q Z_0^q + \sum_{j=1}^{b-1} W_j^q Z_j^q \quad \text{if} \quad W_1 Z_1 \geq W_0 Z_0,$$

hence

$$E(b^q Z^q) = \sum_{j=0}^{b-1} E(W_j^q Z_j^q) - (1-q) E(W_0^q Z_0^q 1_{W_1 Z_1 \geq W_0 Z_0}),$$

which gives, using Lemma B,

$$E(b^q Z^q) \leq b E(W^q) E(Z^q) - (1-q) \varepsilon_{WZ} E(W^q) E(Z^q). \tag{9}$$

(9) is the desired refinement of (8). Dividing by $E(Z^q)$ and taking logarithms, and writing $\varepsilon = \varepsilon_{WZ}$, one has

$$\varphi(q) + \log_b (1 - \frac{(1-q)\varepsilon}{b}) \geq 0 \quad \text{on } [0, 1],$$

from which it follows that $\varphi'(1-0) + (\varepsilon/b \log b) \leq 0$, hence $D > 0$.

We have already shown that $(\alpha) \Leftrightarrow (\beta) \Leftrightarrow (\gamma) \Rightarrow (\delta)$. We will finish the demonstration by showing that (δ) implies (β)

Lemma C. $(x + y)^q \geq x^q + y^q - 2(1-q)(xy)^{q/2}$ for $x > 0$, $y > 0$, $q_0 < q < 1$.

Proof. One verifies that the function $f(t) = e^{tq} + e^{-tq} - (e^t + e^{-t})^q$ is strictly decreasing, hence has a maximum at $t = 0$, where $f(1) = 2 - 2^q$ and $f(1) = -2\log 2 < 1$. This establishes the lemma.

Here is a corollary: one has

$$\left(\sum_1^b x_j \right)^q \geq \sum_1^b x_j^q - 2(1-q) \sum_{i<j} (x_i x_j)^{q/2} \tag{10}$$

for $x_j > 0$ ($j = 1, 2, \ldots, b$) and $q_0 < q < 1$. Indeed, (10) is obtained by induction from

$$\left(\sum_1^b x_j\right)^q \geq x_1^q + \left(\sum_2^b x_j\right)^q - 2(1-q)x_1^{q/2}\left(\sum_2^b x_j\right)^{q/2}$$

$$\geq x_1^q + \left(\sum_2^b x_j\right)^q - 2(1-q)\sum_{j>1}(x_1 x_j)^{q/2}$$

which results from Lemma C and the sub-additivity of the function $x^{q/2}$.

Let us return to formula (2) and rewrite it provisionally as

$$Y = b^{-1}\sum_{j=0}^{b-1} W_j X_j, \tag{11}$$

where Y, W_j, X_j stand for Y_n, $W(j)$, $Y_{n-1}(j)$. Suppose that $q_0 < q < 1$. Apply Lemma C in the form (10) with $x_{j+1} = W_j X_j$. One obtains

$$b^q Y^q \geq \sum_{j=0}^{b-1} W_j^q X_j^q - 2(1-q)\sum_{i<j} W_i^{q/2} W_j^{q/2} X_i^{q/2} X_j^{q/2}.$$

Taking expectations yields

$$b^q E(Y^q) \geq b E(W^q) E(X^q) - b(b-1)(1-q) E^2(W^{q/2}) E^2(X^{q/2}).$$

By returning to our initial notations,

$$E(Y_n^q) \geq b^{1-h} E(W^q) E(Y_{n-1}^q) - b^{1-q}(b-1)(1-q) E^2(W^{q/2}) E^2(Y_{n-1}^{q/2}).$$

Taking account of $E(Y_n^q) \leq E(Y_{n-1}^q)$ (inequality of super-martingales),

$$E(Y_n^q)[1 - b^{1-h} E(W^q)] \geq -b^{1-q}(b-1)(1-q) E^2(W^{q/2}) E^2(Y_{n-1}^{q/2})$$

hence

$$E(Y_n^q)(b^{\varphi(q)} - 1) \leq b^{1-q}(b-1)(1-q) E^2(Y_{n-1}^{q/2}),$$

and, by letting $q \to 1$, we get

$$D \log b \le (b-1) E^2(Y_{n-1}^{1/2}).$$

Now the r.v. $Y_n^{1/2}$ are equi-integrable, since $E(Y_n) = 1$. As they converge a.s. towards Y_∞, one has $E(Y_\infty^{1/2}) = \lim_{n \to \infty} E(Y_n^{1/2})$ (see for example Meyer 1966, II.21), so that $E(Y_\infty^{1/2}) \ne 0$. This implies (β), which concludes the demonstration of Theorem 1.

3. Proof of Theorem 2

First, suppose that (3) has a positive solution Z such that $0 < E(Z^q) < \infty$, with given $q > 1$. Because the function x^q is super-additive, one has

$$b^q Z^q \ge \sum_{j=0}^{b-1} (W_j Z_j)^q,$$

and the strict inequality holds with with positive probability, so that

$$b^q E(Z^q) > b E(W^q) E(Z^q),$$

that is $E(W^q) < b^{q-1}$.

Conversely, suppose that $E(W^q) < b^{q-1}$, so that $\varphi(q) < 0$, and let k be an integer with $k < q \le k+1$. As the function $x^{q/(k+1)}$ is sub-additive, one has, for $x_j \ge 0$ ($j = 1, 2, \ldots, b$),

$$(x_1 + x_2 + \ldots + x_b)^q \le (x_1^{q/k+1} + \ldots + x_b^{q/k+1})^{k+1}$$
$$= x_1^q + \ldots + x_b^q + \sum \gamma_{\alpha_1, \ldots, \alpha_b} (x_1^{\alpha_1} \ldots x_b^{\alpha_b})^{q/(k+1)}.$$

In the last sum, the exponents of x_j do not exceed k, the coefficients are positive, and $\sum \gamma_{\alpha_1, \alpha_2, \ldots, \alpha_b} = b^{k+1} - b$.

Reconsider formula (2) in the form (11), and remark that $E(U^{q/(k+1)}) \le E(U)$ if $U \ge 0$ and that $\prod_j E(U^{\alpha_j}) \le E(U) E(U^k)$ if the α_j are non-negative integers such that $\sum \alpha_j = k+1$, and at least two of the α_j be different from 0. We obtain

$$b^q E(Y^q) \le b E(W^q) E(X^q) + (b^{k+1} - b)[E(W^k) E(X^k)]^{q/k}.$$

Hence

$$E(Y_n^q) \leq b^{1-q}E(W^q)E(Y_{n-1}^q) + b[E(W^k)E(Y_{n-1}^k)]^{q/k}.$$

Taking into account the sub-martingale inequality $E(Y_n^q) \geq E(Y_{n-1}^q)$

$$E(Y_n^q)(1 - b^{1-q}E(W^q)) \leq b[E(W^k)E(Y_n^k)]^{q/k}.$$

Letting n go to infinity, one can see that

$$E(Y_\infty^k) < \infty \Rightarrow E(Y_\infty^q) < \infty.$$

This establishes the desired result for $1 < q \leq 2$. Now assume that $q > 2$. As the hypothesis $\varphi(q) < 0$ implies $\varphi(l) < 0$ for all integer $l \leq q$, one also has

$$E(Y_\infty^{l-1}) < \infty \Rightarrow E(Y_\infty^l) < \infty$$

for $l = 2, \dots, k$. It results from the above implications that $E(Y_\infty^q) < \infty$. This concludes the proof of Theorem 2.

4. Proof of Theorem 3

Part 1. According to Theorem 2, (α_1) implies $E(W^q) < b^{q-1}$ for all $q > 1$. This implies (β_1). Conversely, if (β_1) holds, condition (δ) of Theorem 1 is satisfied, hence $E(Y_\infty^q) > 0$. Furthermore $E(W^q) \leq b^q$ i.e. $\varphi(q) \leq 1$ for all $q > 0$. Since $\varphi(1) = 0$ and φ is convex, this means that $\varphi(q) < 0$ for all $q > 1$, that is $E(W^q) < b^{q-1}$, or $\varphi \equiv 0$, i.e. $W \equiv 1$. Hence $E(Y_\infty^q) < \infty$ for all $q > 0$. Therefore (α_1) $\Leftrightarrow$ (β_1).

Part 2. (β_1). Hence (by Theorem 1) $E(Y_\infty) = 1$. On the other hand, there exists an $\varepsilon > 0$ such that $\varphi(q) < \log_b(1 - \varepsilon)$ for $q \geq 2$. That is,

$$E(W^q) \leq (1 - \varepsilon)b^{q-1} \quad (q \geq 2). \tag{12}$$

Consider formula (3), with $Z = Y_\infty$, and let h be an integer ≥ 2. One has

$$b^q Z^q = \left(\sum_{j=0}^{b-1} W_j Z_j\right)^q,$$

from where

$$b^q E(Z^q) = bE(W^q)E(Z^q) + \sum_{\substack{q_1 + \ldots + q_b = q \\ q_i \leq q-1}} \frac{q!}{q_1! \ldots q_b!} \prod_{j=1}^{b} E(W^{q_j}) \prod_{j=1}^{b} E(Z^{q_j}). \qquad (13)$$

(13) together with (12) gives

$$\varepsilon b^q E(Z^q) \leq \sum_{\substack{q_1 + \ldots + q_b = q \\ q_i \leq q-1}} \frac{q!}{q_1! \ldots q_b!} \prod E(W^{q_j}) \prod E(Z^{q_j}), \qquad (14)$$

hence

$$E(Z^q) \leq \frac{1}{\varepsilon} \sum_{\substack{q_1 + \ldots + q_b = q \\ q_i \leq q-1}} \frac{q!}{q_1! \ldots q_b!} \prod E(Z^{q_j}).$$

Lemma D. *For all $\alpha > 0$, one has*

$$\sum_{\substack{q_1 + \ldots + q_b = q \\ q_i \leq q-1}} (q_1! \ldots q_b!)^\alpha = o((q!)^\alpha) \quad (q \to \infty).$$

The proof follows immediately for $b = 2$, and continues through recurrence over b.

$c > 0$ being given, write $A_q = \sup_{l < q} (E(Z^l)/(l!)^{1+c})^{1/l}$. Then

$$A_{q+1}^q \leq \sup\left(\frac{1}{\varepsilon} \frac{\sum (q_1! \ldots q_b!)^c}{(q!)^\alpha} A_q^q, A_q^q \right).$$

According to Lemma D, the sequence A_q is bounded, therefore

$$E(Z^q) \leq A(c)^q (q!)^{1+c}, \text{ with } A(c) < \infty.$$

As a consequence,

$$\limsup_{q \to \infty} \frac{\log E(Y_\infty^q)}{q \log q} \leq 1. \qquad (15)$$

If we suppose that $\|W\|_\infty = \gamma < b$, (14) gives

$$E(Z^q) \leq \frac{1}{\varepsilon} \left(\frac{\gamma}{b}\right)^q \sum \frac{q!}{q_1!\ldots q_b!} \prod E(Z^{q_i}).$$

Set $B_q = \sup_{l \leq q}(E(Z^l)/l!)^{1/l}$.

As the number of terms in the sum Σ does not exceed q^b, one obtains

$$B_{q+1}^q \leq \sup\left(\frac{1}{\varepsilon}\left(\frac{\gamma}{b}\right)^q q^b B_q^q, B_q^q\right)$$

which means that the sequence B_q is bounded. It results that $E(e^{tZ}) < \infty$ for all small enough $t > 0$.

Now set $e^{\chi(t)} = E(e^{tZ})$. Formula (3) then becomes

$$e^{\chi(bt)} = E^b(e^{\chi(Wt)}). \tag{16}$$

The hypothesis $\|W\|_\infty = \gamma < b$ implies $\chi(bt) \leq b\chi(\gamma t)$, hence $\chi((b/\gamma)^n) = O(b^n)$ for $n \to \infty$. Setting $(b/\gamma)^K = b$, one has

$$\chi(t) = O(t^K) \quad (t \to \infty). \tag{17}$$

It is an easy exercise to show that (17) is equivalent to the existence of a positive real B such that

$$E(Z^q) \leq B^q(q!)^{1-(1/K)}.$$

However, $1 - (1/K) = \log_b \gamma$. Hence one has

$$\limsup_{q \to \infty} \frac{\log E(Y^q)}{q \log q} \leq \log_b \gamma. \tag{18}$$

Let us now choose $1 < \gamma_1 < \|W\|_\infty$ (the case $\|W\|_\infty = 1$ is straightforward). There exists an $\varepsilon > 0$ such that $E(W^q) \geq \varepsilon \gamma_1^q$. Reexamine formula (13). Since

$$\sum_{\substack{q_1+\ldots+q_b=q \\ q_i \leq q-1}} \frac{q!}{q_1!\ldots q_b!} = b^q - b,$$

one has

$$E(Z^q) \geq \frac{b^q - b}{b^q} \inf \prod_{j=1}^{b} E(W^{q_j}) E(Z^{q_j})$$

$$\geq |\frac{1}{2}| \varepsilon^b \gamma_1^q \inf \prod_{j=1}^{b} E(Z^{q_j})$$

the infimum being taken over all b-tuples $(q_1, q_2, \ldots, q_b)$ such that $q_1 + q_2 + \cdots + q_b = q$ and $\sup_j q_j \leq q - 1$. Let q be a multiple of b.

The infimum is then $E^b(Z^q/b)$. Hence

$$\frac{\log E(Z^{bq})}{bq} \geq \frac{\log E(Z^q)}{q} + \log \gamma_1 + O\left(\frac{1}{q}\right),$$

with the consequence

$$\log E(Z^q) \geq \eta q \log q + O(q), \quad \eta = \log_b \gamma_1 \qquad (19)$$

from where

$$\liminf_{q \to \infty} \frac{\log E(Y_\infty^q)}{q \log q} \geq \log_b \gamma_1 \qquad (20)$$

(15), (18) and (20) lead to

$$\lim_{q \to \infty} \frac{\log E(Y_\infty^q)}{q \log q} = \log_b \|W\|_\infty.$$

This terminates the proof of Theorem 3.

Remark. In the case $0 < P(W = b) < 1/b$, one has $\gamma_1 = b$ in (19), from which follows that $E(e^{tZ}) = \infty$ for large enough $t > 0$.

5. Proof of Theorem 4

Let Ω be the space on which the random variables $W(j_1, j_2, \ldots, j_n)$ are defined. On the product space $\Omega \times [0, 1]$, consider the probability Q defined by

$$Q(A) = E\left(\int 1_A d\mu\right).$$

Let $X_n = \sum_{j_1,\ldots,j_n} W(j_1,\ldots,j_n) 1_{I(j_1,\ldots,j_n)}$. One has then

$$\mu_n = b^{-n} \prod_{1 \le j \le n} X_n$$

(with an evident abuse of notation).

Write $\mu = \mu_n \nu_n$; here ν_n is a measure whose restriction to each interval of the n-th stage is defined in an analogous fashion as for μ.

Observe that the variables $\nu_n(I(j_1,\ldots,j_n))$ have the same distribution as Y_∞ and that, for fixed n, they are mutually independent. Moreover, the variables $\nu_n(I(j_1,\ldots,j_n))$ and $W(k_1,\ldots,k_p)$ are independent as long as the interval $I(k_1,\ldots,k_p)$ is not strictly contained in the interval $I(j_1,\ldots,j_n)$.

It is convenient to consider the random function

$$T_n = \sum_{j_1,\ldots,j_n} b^n \nu_n(I(j_1,\ldots,j_n)) 1_{I(j_1,\ldots,j_n)}.$$

If u is a function defined on $[0, 1]$, constant on the intervals of the n-th stage, one has

$$\int u d\mu = \int_0^1 u(x) \mu_n(x) T_n(x) dx. \tag{21}$$

The theorem results from the two following Lemmas.

Lemma E. *If $E(W \log_b W) < 1$, then almost surely μ-almost everywhere $(1/n) \log \mu_n$ tends towards $E(W \log W)$ when $n \to +\infty$.*

Proof. We will show that

$$\sum_{n \ge 1} Q(\{X_n > e^{n-1}\}) < \infty \tag{22}$$

holds and that the series

$$\sum_{n\geq 1} \frac{1}{n} \{ \log \inf (X_n, e^{n-1}) - E[W \log(\inf (W, e^{n-1}))]\} \qquad (23)$$

converges Q-almost surely. Therefore, one will have Q-almost surely $X_n \leq e^{j-1}$ beyond a certain rank and

$$\lim_{n\to\infty} \frac{1}{n} \sum_{j=1}^n \log \inf(X_j, e^{n-1}) = E(W \log W),$$

from which the lemma follows.

Let us begin by evaluating $Q([X_n > e^{n-1}])$. One has

$$Q([X_n > e^{n-1}]) = E\left[\int 1_{[X_n > e^{n-1}]} d\mu\right].$$

Taking account of (21) and the properties of independence of variables, we obtain

$$\begin{aligned}
Q(\{X_n > e^{n-1}\}) &= E\int_0^1 1_{\{X_n(x) > e^{n-1}\}} \mu_n(x) T_n(x) dx \\
&= \int_0^1 E(X_n(x) 1_{\{X_n(x) > e^{n-1}\}}) E(\mu_{n-1}(x)) E(T_n(x)) dx \\
&= E(W 1_{\{W > e^{n-1}\}}),
\end{aligned}$$

from where we have

$$\sum_{n\geq 1} Q(\{X_n > e^{n-1}\}) \leq E\left\{W \sum_{n\geq 1} 1_{\{W > e^{n-1}\}}\right\} \leq E(W(1 + \log^+ W)),$$

which proves (22).

For ease of writing, set $X'_n = \log \inf(X_n, e^{n-1})$. We will calculate $E_Q(X'_n | X_1, \ldots, X_{n-1})$. Let u be a bounded Borel function from $\mathbb{R}^{n-1}$ into $\mathbb{R}$. One has

$$\int u(X_1, \ldots, X_{n-1}) X'_n dQ$$

$$= E \int_0^1 u(X_1(x), \ldots, X_{n-1}(x)) X'_n(x) \mu_n(x) T_n(x) dx$$

$$= \int_0^1 E[u(X_1(x), \ldots, X_{n-1}(x)) \mu_{n-1}(x)] E[X'_n(x) X_n(x)] dx$$

$$= E[W \log \inf(W, e^{n-1})] \int_0^1 E[u(X_1(x), \ldots, X_{n-1}(x)) \mu_{n-1}(x) T_{n-1}(x)] dx.$$

This proves that $E_Q(X'_n | X_1, \ldots, X_{n-1}) = E[W \log \inf(W, e^{n-1})]$. Therefore

$$\int (X'_n)^2 dQ = E \int_0^1 (X'_n(x))^2 \mu_n(x) T_n(x) dx = E[W(\log \inf(W, e^{n-1}))^2],$$

hence

$$\sum_{n \geq 2} \frac{1}{n^2} \int (X'_n)^2 dQ$$

$$= E\left[W \sum_{n \geq 2} \frac{1}{n^2} (\log \inf(W, e^{n-1}))^2\right]$$

$$\leq E\left[W(\log W)^2 \sum_{n \geq \sup(2, 1 + \log W)} n^{-2} + W \sum_{2 \leq n < 1 + \log W} \left(\frac{n-1}{n}\right)^2\right]$$

$$\leq E\left[W\left(\log^+ W + \frac{(\log W)^2}{\sup(1, \log W)}\right)\right].$$

The theorem on the convergence of L^2-martingales gives (23).

Lemma F. *Suppose that $E(W \log_b W) < 1$ and that $E(Y_\infty \log Y_\infty) < \infty$. Then almost surely μ-almost everywhere $(1/n) \log v_n(I_n(x))$ tends towards $-\log b$.*

Proof. One has

$$\int T_n^{-1/2} dq = E \int_0^1 \mu_n(x) (T_n(x))^{1/2} dx = E(Y_\infty^{1/2}),$$

from where

$$\int \left(\sum_{n \geq 1} \frac{1}{n^2} T_n^{-1/2}\right) dQ < \infty.$$

As a consequence, Q-almost surely, $T_n^{-1/2} \leq n^2$ holds above a certain n, hence $\liminf_{n \to \infty} (1/n) \log T_n \geq 0$. Up to now, we have not used the second hypothesis. Now let us show that, Q almost surely, one has lim sup adjust(u 2)$_{n \to \infty} (1/n) \log T_n \leq 0$. Take a number $\alpha > 1$. Using (21),

$$E\int 1_{\{T_n > \alpha^n\}} d\mu = E(Y_\infty 1_{\{Y_\infty > \alpha^n\}}).$$

Then

$$\sum_{n \geq 1} Q(\{T_n > \alpha^n\}) = E\left(Y_\infty \sum_{n \geq 1} 1_{\{Y_\infty > \alpha^n\}}\right) \leq E(Y_\infty \log_\alpha^+ Y_\infty)$$

follows, proving that, for all $\alpha > 1$, it is Q-almost sure that lim sup adjust(u 2)$_{n \to \infty} (1/n) \log T_n \leq \log \alpha$.

The desired result follows. As $v_n(I_n(x)) = b^{-n} T_n(x)$, the lemma is proven. To prove the corollary, use Billingsley 1967, p. 136-145.

Remark. Under the only hypothesis $E(W \log_b W) < 1$, one obtains that almost surely every Borel set of dimension $< D$ is of vanishing μ-measure.

&&&&& POST-PUBLICATION APPENDIX &&&&&

AN ELEMENTARY EXPLICIT SOLUTION OF MANDELBROT'S FUNCTIONAL EQUATION $bY = \sum_{j=1}^{b} W_j Y_j$, BY J. PEYRIÈRE (1998)

Mandelbrot's equation starts with an integer base $b > 1$ and a random variable $W \geq 0$ satisfying $EW = 1$ and $E(W \log W) < \log b$. The unknown is a random variable Y satisfying $EY = 1$. This appendix proposes to show that an elementary explicit solution exists in the special case when $0 < W < b$ and $\Pr\{W < w\} = (w/b)^\alpha$ for $w < b$, with $\alpha = 1/(b-1)$. In that case, the generating function of Y, written as $g(t) = E(e^{-tY})$, verifies the functional equation

$$g(t) = \left(\alpha \int_0^1 g(tx) x^{\alpha - 1} dx \right)^b,$$

with $g(0) = -g'(0) = 1$. Defining $h(t) = g(t)^{(1/b)}$, one has $h(t) = \alpha \int_0^1 h(tx)^b x^{\alpha - 1} dx$, $h(0) = 1$ and $h'(0) = -1/b$. Differentiating under the integral sign and integrating by parts yields the differential equation

$$(b-1)h'(t) = \frac{h(t)^b - h(t)}{t},$$

from which it follows that $\dfrac{1 - h(t)^{1-b}}{t}$ is a constant. The final result is

$$g(t) = \left(1 + \frac{b-1}{b} t \right)^{\frac{-b}{b-1}}.$$

That is, the random variable $Yb/(b-1)$ follows the gamma distribution of parameter $b/(b-1) = \alpha + 1$.

&&&&&&&&&&& **ANNOTATIONS** &&&&&&&&&&&

Source and acknowledgement. J. Peyrière proofread the translation and R. Vojak corrected a few slips in the original.

Editorial changes. The text was divided into sections and the notation changed to fit the present usage. The original denoted b by c, q by h, q_{crit} by α_0, and c by α.

More recent developments. The decision to translate and reproduce this text was motivated by the desire to correct an injustice: indeed, the results of this paper are often credited to later authors. Those results, with additional ones, were restated in Kahane 1987b, which summarizes Kahane 1987a, and other papers written in French. In particular, the corollary of Theorem 4 is shown to hold under wider and more natural conditions. There are further extensions in Kahane 1989, 1991a.

Intermittent turbulence and fractal dimension: kurtosis and the spectral exponent 5/3 + B

• *Chapter foreword.* Beyond the earliest specific applications of multifractals, which concerned turbulence, this chapter contributes to the general theory. It illustrates the remark made in Section 25 of M 1974c {N16}, that the multifractal properties of a measure are determined by the geometry of the graph of the function $\tau(q)$ also denoted as $-\overline{\tau}(q)$. As observed in Chapter N2, $\tau(q)$ itself is an old notion. In the 1960s, the Russian school interpreted $\tau(2)$ as the correlation exponent. Later, M1974f {N15} gave a fundamental meaning to $\tau'(1)$ as a fractal dimension, and to the root other than 1 of the equation $\tau(q) = 0$ as a "critical q". This paper gave a fundamental meaning to $\tau(2/3)$ and $\tau(4)$. Concerning terminology, Chapter N2 told how the term "generalized dimension" became attached (to my dismay) to the ratio $D(q) = \tau(q)/(q-1)$. Since $\tau(2) = D(2)$, the exponent $\tau(2)$ became known as the "correlation dimension". In the same terminology, the L'Hospital rule interpreted $\tau'(1)$ as $D(1)$ and called it "information dimension". •

✦ **Abstract.** Various distinct aspects of the geometry of turbulence can be studied with the help of a wide family of shapes for which I have recently coined the neologism "fractals." These shapes are loosely characterized as being violently convoluted and broken up, a feature denoted in Latin by the adjective "fractus." Fractal geometry approaches the loose notion of "form" in a manner different and almost wholly separate from the approach used by topology. Until recently, it was believed that fractals could not be of use in concrete applications, but I have shown them to be useful in a variety of fields. In particular, they play a central role in the study of three aspects of turbulence: (a) homogeneous turbulence, through the shape of the iso-surfaces of scalars (M 1975f{H}), (b) turbulent dispersion (M 1976c{N20}), and especially c) the intermittency of turbulent dissipation (M 1972j {N14}, M1974f{N15} and M1974c{N16}).

The present paper will sketch a number of links between the new concern with fractal geometry and the traditional concerns with various spectra of turbulence and the kurtosis of dissipation. Some of the results to be presented will improve and/or correct results found in the literature, for example, will further refine M 1974f,{N15}, and M1974c{N16}.

One result described in Section 4 deserves special emphasis: It confirms that one effect of the intermittency of dissipation is to replace the classical spectral exponent $5/3$ by $5/3 + B$. However, it turns out that in the general case the value of B is different from the value accepted in the literature, for example in Monin & Yaglom 1975. The accepted value, derived by Kolmogorov, Obukhov, and Yaglom, is linked to a separate *Ansatz*, called lognormal hypothesis, and is shown in M 1974f{N15} to be highly questionable.

Other results in this paper are harder to state precisely in a few words. They demonstrate the convenience and heuristic usefulness of the fractally homogeneous approximation to intermittency, originating in Berger & M 1963{N6} and Novikov & Stewart 1964. They also demonstrate the awkwardness of the lognormal hypothesis. Many writers of the Russian school have noted that the latter is only an approximation, but it becomes increasingly clear that even they underestimated its propensity to generate paradoxes and to hide complexities.

M 1975O describes numerous other concrete applications of fractals. It can also serve as a general background reference, but its chapter on turbulence is too skimpy to be of use here. This deficiency should soon be corrected in the English version, M 1977F, meant to serve as a preface to technical works such as the present one. Nevertheless, in its main points, the present text is self-contained. ✦

1. CURDLING AND FRACTAL HOMOGENEITY. ROLE OF THE FRACTAL DIMENSION $D = 2$

"Curdling" is a convenient term I use to denote any of several cascades through which dissipation concentrates in a small portion of space. Absolute curdling is described by the Novikov & Stewart 1964 cascade. Its outcome was described independently, without any generating mechanism, in Berger & M1963{N6} and M1965c{N7}. Weighted curdling is described by the cascades of Yaglom 1966, M1972j{N14}, M1974f{N15} and

M1974c{N16}. It turns out that absolute curdling is a more realistic model than its extreme simplicity would suggest and in addition it provides an intrinsic point of reference to all other models. Therefore, it deserves continuing attention. Weighted curdling, which we examine next, turns out to mostly add complications, and its apparent greater generality is in part illusory.

1.1. Absolute curdling

Before each stage, dissipation is assumed uniform over a certain number of spatial cells, and zero elsewhere. In the process of curdling, each of the initial cells breaks into $C = b^3$ sub-cells and dissipation concentrates within $N \geq 2$ of these, called "curds." The quantity $\rho = 1/b$ is the ratio of similarity of sub-cells with respect to the "parent" cells.

After a finite number of stages of absolute curdling, dissipation concentrates with uniform density in a closed set, whose outer and inner scales are L and ε. This set constitutes an approximation to a fractal. Figure 1 represents such an approximate fractal in the plane. (We shall soon see that it is very close to being a plane cut through an approximate spatial fractal.)

The most important characteristic number associated with a fractal is its fractal (Hausdorff-Besicovitch) dimension, which in the case of absolute curdling is most directly defined as

$$D = \frac{\log N}{\log(\frac{1}{\rho})}.$$

D is always positive (because of the condition $N \geq 2$) and is ordinarily a fraction. In Figure 1, $b = 5 = 1/\rho$ and $N = 15$, so that $D = 1.6826$. (The values of N and ρ were chosen to make this D as close as conveniently feasible to 5/3.) One reason for calling D a dimension is explained in the legend of Figure 2.

When dissipation is uniform over a fractal of dimension $D < 3$, turbulence will be called *fractally homogeneous*. The modifier is of course meant to contrast it with G. I. Taylor's classical concept of homogeneous turbulence, which can be viewed as the special limit case of fractally homogeneous turbulence for $D \rightarrow 3$. The salient fact is that fractally homogeneous generalization allows $D - 3$ to be negative.

The value of D characterizes one among many mathematical structures present in a set. It follows that the same D can be encountered in sets that differ greatly from the viewpoint of other structures; for example, sets that are topologically distinct. Nevertheless, many aspects of fractally homogeneous turbulence turn out to depend solely upon D. In an approximate fractal of dimension D and scales L and η, that dissipation concentrates in $(L/\eta)^D$ out of $(L/\eta)^3$ cells of side η. The total volume of these cells is $(L/\eta)^D \eta^3$. The relative occupancy ratio of the region of dissipation (measured by the relative number of curds of side ε within a cell of side L) is $(\eta/L)^{D-3}$. Therefore the uniform density of dissipation in a curd must be equal to $(L/\eta)^{3-D}$ times the overall density of dissipation.

The codimension will be defined as the difference $3 - D$. More generally, given a set in E-dimensional Euclidean space with $E \neq 3$, the

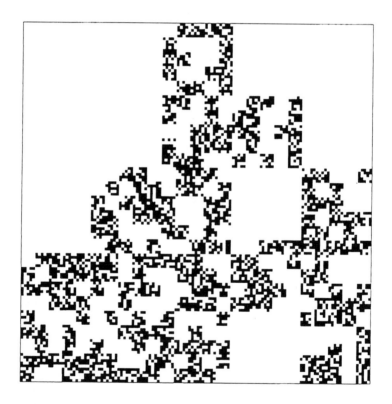

FIGURE N18-1. A fractal obtained by random absolute curdling on a square grid of base $b = 5$. We show the effect of four stages, each of which begins by dividing the cells of the previous stage into $5^2 = 25$ subcells, then "erases" 10 of them to leave the remaining 15 as "curds."

N18 ◊ ◊ KURTOSIS AND THE SPECTRAL EXPONENT 393

codimension is the difference $E - D$. This usage is consistent with that prevailing in the theory of vector spaces.

The quantities evaluated in the preceding paragraphs only concern only the way blobs of intermittent turbulence are *spread around*. Therefore D is a called a *metric* characteristic. It is conceptually distinct from the *topological* characteristics; topology is only concerned with the way in which blobs are *connected*. As exemplified by $D > D_T$, most relations between fractal and topological structures are expressed by inequalities. Topological structures prove more difficult to investigate than fractal structures. It is therefore fortunate that several structures which a casual examination would classify as topological actually turn out to be exclusively or predominantly metric. Specifically, they turn out to be fractal. One example of this is the degree of intermittency as measured by the kurtosis. Another is the intermittency correction to the 2/3 and 5/3 laws, even though intermittency may conceivably have a distinct topological facet.

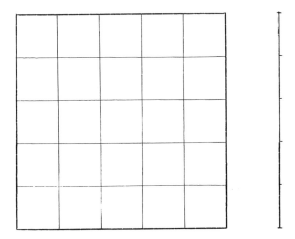

FIGURE N18-2. Definition of the similarity dimension. A segment of line can be paved with – and is therefore equivalent to – $N = 5$ replicas of itself reduced to the ratio $r = 1/5$. A square is equivalent to $N = 25$ replicas of itself reduced in the ratio $r = 1/5$. The property of self-similarity is of wider scope: for example the pattern of Figure 1 is equivalent to $N = 15$ replicas of itself reduced in the ratio $r = 1/5$. In each of the classical cases, the dimension, is $D = \log N / \log(1/r)$. The point of departure of fractal geometry is that this last expression a) remains well defined and b) happens to be useful for all self-similar sets such as the pattern of Figure 1, and that it is not excluded for D to be a fraction.

1.2 Weighted curdling

This process is more general than absolute curdling, and proceeds as follows. Each stage starts with dissipation being uniform within each cell. When it ends, the density of dissipation in each subcell is equal to the density within the whole cell multiplied by a random factor W, with $\langle W \rangle = 1$. The multiplying weight W is related to the Yaglom multiplier, but is not identical to it. The cascade underlying weighted curdling is a generalization and a conceptual tightening-up of various arguments concerning the lognormal distribution. (In the absence of viscosity cutout, weighted curdling leads asymptotically to an everywhere dense fractal that is – topologically – open rather than closed. However, this distinctive feature vanishes when high frequencies are cut out by viscosity.)

1.3 The inner scale

The value of the inner scale η, is determined by the dissipation and the viscosity ν. Its value in the Taylor homogeneous case is well known. In the fractally homogeneous case, it continues to be well-defined, and it depends on D. In the general case, the notion of η involves great complications. They will be avoided in Sections 3 and 4, and confronted in Section 5.

1.4 Behavior of linear cross-sections

Most conveniently, the fractal dimensions of the linear and planar cross-sections of a fractal are given by the same formulas as the Euclidean dimension of the corresponding cross-sections of an elementary geometric shape. Using this rule, we shall show that evidence strongly suggests that the D of turbulent diffusion must be greater than 2.

Rule: When the fractal or Euclidean dimension D of a shape is above 2, its cross-section by an arbitrarily chosen straight line has a positive probability of being non-empty with the dimension $D - 2$. Otherwise, the cross-section is empty. Analogous results apply for planar sections, except that instead of subtracting 2, one must subtract 1 (see Mattila 1975).

Finite ε-approximations to fractals. Start with $(L/\eta)^D$ curds of side η. When $D > 2$, the typical linear cross-section will either be near-empty, or (with a positive probability, which is nearly independent of η) it will include about $(L/\eta)^{D-2}$ segments of side about η. When $D < 2$, on the contrary, the probability of hitting more than a small number of curds, (say two curds or more) will greatly depend on η and will tend to zero with η/L. At the limit, suppose that $D = \log 2/\log(L/\varepsilon)$ (among possible values of D, this is the closest to $D = 0$); then everything concentrates in two curds; the probability of hitting either by an arbitrarily selected line or a plane is minute.

Illustration. Figure 1, which represents a fractal of dimension log15/log5, has the same dimension as the typical planar section of an approximate spatial fractal with $D = 1 + \log 15/\log 5$.

There is a deep but elementary experimental reason to believe that turbulent dissipation satisfies D>2. By necessity, turbulence is ordinarily studied through linear cross-sections in space-time. Under Taylor's frozen turbulence assumption, they are the same as linear cross-sections through space. Turbulence is a highly prevalent phenomenon, in the sense that the typical cross-section hits it with no effort and repeatedly. This would not happen if $D < 2$, because in that case the probability of hitting anything is practically zero. Hence, the intersection rule has given us an elementary reason (an especially profound one) for believing that the fractal dimension D of turbulent dissipation satisfies $D > 2$.

1.5 Digression: Possible relevance of fractal geometry to the study of the Navier-Stokes and Euler equations

My approach to the geometry of turbulence is to a large extent "phenomenological," as was Kolmogorov's approach, and it is geometric rather than dynamic. From these viewpoints, the study of the Navier-Stokes and Euler equations has not yet yielded anything that could be of help.

On the other hand, any success the fractal approach may be able to achieve should assist in the notoriously difficult search for turbulent solutions of the Navier- Stokes and Euler equations. I think, indeed, that the greatest roadblock to this search has been due to the lack of an intrinsic characterization of what was being sought. It could even be argued that no one could be sure he would recognize such a solution if one presented itself. In the study of other equations of physics, knowing which singularities should be expected has often made it less difficult to find the whole solution. This approach has not yet worked for turbulence. Von Neumann 1949-1963 has noted that "its mathematical peculiarities are best described as new types of mathematical singularities," but he made no progress in identifying them.

In this vein, I propose to infer from empirical evidence that, for non-linear partial differential equations like Euler's system (when viscosity is absent) or the Navier-Stokes system (when viscosity $\rightarrow$ 0, or possibly even at a small positive viscosity), the singularities of sufficiently "mature" solutions are likely to become fractal.

The singularities of Euler solutions should be viewed as associated with curdling, as discussed above and in the body of this paper. As to the Navier-Stokes equations, the notion that the solution can possess singularities remains unproven and controversial. But if singularities in the Oseen-Leray sense do in fact exist, they must be very much "sparser" than the Eulerian ones; possibly a proper subset.

Assuming that singularities do indeed exist, Scheffer 1975 has been successful in restating some of my rough hunches into precise conjectures, and has proved several of them. His results are related to the results in Leray 1934 and open new vistas on this ancient problem. See also Scheffer 1977.

Closely related forms of intermittency occur in phenomena ruled by diverse other equations. Therefore, the "fractality" of the solution may be due to some characteristic of the Navier-Stokes equation counterpart in broad classes of other equations; it may well be more useful to study fractality within a broader mathematical context.

Digression. A second possible connection between fractality and Navier-Stokes equations involves the shapes of coastlines. A priori, it may well be that fractality is *wholly* related to the Oseen-Leray argument that a solution with good initial data may, after a certain time, have large velocity gradients. Alternatively, we have the Batchelor & Townsend 1949 argument "that the distribution of vorticity is made 'spotty' in the early stages of the decay by some intrinsic instability and is kept 'spotty' throughout the decay by the action of the quadratic terms of the Navier-Stokes equations." However, "spottiness" may also be affected by a third factor. Indeed, the study of partial differential equations, while stressing the respective roles of the equation itself and of boundary conditions, usually fails to consider the possible effects of the shape of that boundary. More precisely, the boundary is nearly always assumed smooth, for example it is taken to be a cube. For atmospheric and ocean turbulence, this approximation may well be unrealistic. The fact is that the shapes of coastlines contain features whose "typical lengths" cover a wide span, and I argue in M 1967s, 1977f that coastlines' fractal dimensions are greater than 1. An analogous statement can be made concerning the rough surface of the Earth, and both factors may well combine with intrinsic instabilities as a third contribution to the roughness of observed flow.

2. THE FUNCTION $\Psi(q)$: THE FRACTAL DIMENSIONS ARE DETERMINED BY $\Psi'(1)$

All the aspects of intermittency to be studied in this paper are ruled by power laws. If certain further assumptions are added, the exponents of these scaling laws are not independent, but instead are linked to one another; for example (through the fractal dimension D of the carrier or the parameter "μ" of the Kolmogorov theory). In the general case, however, these exponents are distinct. As Novikov 1969 had observed in the case of spectra and moments, each power law is merely a symptom of self-similarity. The multiplicity of different exponents shows the self-similarity syndrome to be complex and multifarious.

Nevertheless, the exponents that enter in my previous papers and in the present one can all be derived from various distinct properties of the determining function defined as

$$\Psi(q) = \log_C \langle W^q \rangle.$$

Recall that $C = b^3$ is the number of subcells per cell. In absolute curdling, W is a binomial random variable: it is either 0, will probability $1 - p$, or $1/p$, with probability p. It follows that $\langle W \rangle = 1$. In weighted curdling, W is a more general random variable, still satisfying $\langle W \rangle = 1$. And the limit lognormal model of M 1972j{N14}, fits into the same scheme by appropriate interpolation of W. By a general theorem of probability (Feller 1971, p.155) $\Psi(q)$ is a convex function; it obviously satisfies $\Psi(1) = 0$. Furthermore, whenever $\Pr\{W > 0\} = 1$, it follows that $\Psi(0) = 0$.

Two examples stand out. First, the graph of $\Psi(q)$ is a straight line if, and only if, curdling is absolute. This graph passes through $\Psi(1) = 0$ but not through $\Psi(0) = 0$. As a second example, $\Psi(q)$ is a parabola when W is lognormal. These two cases are drawn on Figure 3, which also illustrates other features of $\Psi(q)$.

My past and present papers show that the following characteristics of the function $\Psi(q)$ are of interest: $\Psi(2/3)$, $\Psi'(1)$, $\Psi'''(1)$, $\Psi(2)$, $\Psi(q)$ for q integer > 2, and α_1, α_2, $\alpha_3 = \alpha$, where α_m is defined as the root other than $q = 1$ of the equation $\varphi_m(q) = 3\Psi(q) - m(q - 1) = 0$. The functions φ_m being convex, each α_m is unique, but of course one or more among them can be infinite. Since the condition $\varphi_1(q) < 0$ is at least as demanding as $\varphi_2(q) < 0$, and, a fortiori, as $\varphi_3(q) < 0$, we see that if $\alpha_1 > 1$, then $\alpha_1 \leq \alpha_2 \leq \alpha_3 = \alpha$.

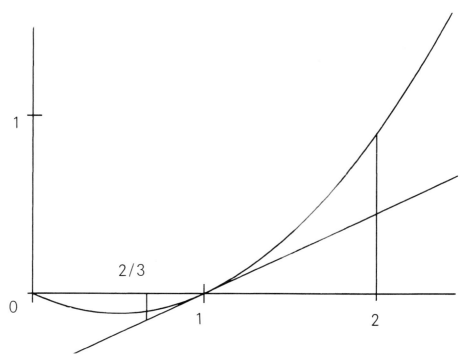

FIGURE N18-3. The determining function $\Psi(q)$. The two lines represent two determining functions $3\Psi(q)$ which yield the same value of $3 - D = 0.45$. The present paper concentrates upon the roles played in the theory of intermittency by the quantities $\Psi(2/3)$, $\Psi(2)$ and $\Psi''(1)$. Earlier work, M 1974f,c{N16}, had concentrated on the role played by $\Psi'(1)$ and the α's (the latter are not shown here). The lower line in the present Figure is straight of equation $3\Psi(q) = 0.45(q - 1)$. It corresponds to fractally homogeneous turbulence and is the lowest compatible with the given D. The upper line, which is the parabola $3\Psi(q) = 0.45q(q - 1)$, corresponds to lognormal intermittency with $\mu = 0.9$. For other forms of curdling, the determining function can lie between the above lines or even higher than the parabola. Two examples are of interest. The fractally homogeneous case can be changed so that the value 0 is replaced by some scatter of values slightly above it, while the value of $1/p$ is slightly changed to keep D invariant. Alternatively, the lognormal can be truncated sharply. In either case, the resulting line $\Psi(q)$ will be approximately straight for abscissas to the right of $q = 1$ and approximately parabolic to the left.

The concept of "weak" or "vague" approximation, which is used in the bulk of probability theory, is of little value in the present context. A random variable W' may be a close approximation to W and still lead to a markedly different determining function and hence to markedly different form of intermittency.

The order in which these various characteristics have been listed in the preceding paragraph is that of increasing sensitivity to the distribution of W. The first and least sensitive – and, in my opinion, the most basic – is

$$\Psi'(1) = \langle W \log_C W \rangle.$$

This last quantity was first considered in M 1974f,c{N15, N16}, then in Kahane 1974 and Kahane & Peyrière 1976{N17}. When $3\Psi'(1) < 3$, the carrier of intermittency is nondegenerate and its fractal dimension is $D = 3 - 3\Psi'(1)$. Thus, $3\Psi'(1) = \langle W\log_b W \rangle$ is the codimension $3 - D$. When $3\Psi'(1) < 2$ and when $\Psi'(1) < 1$, planar and linear intersections are nondegenerate with fractal dimensions $D_2 = 2 - 3\Psi'(1)$ and $D_1 = 1 - 3\Psi'(1)$. As expected, $D_2 = D - 1$ and $D_1 = D - 2$. In the case of a lognormal W of basic parameter μ, one has $\Psi(q) = (q-1)q\mu/6$. Hence $D = 3 - \mu/2$ and μ is merely twice the codimension.

Next, as the present paper will show, the traditional concerns with second order (spectral) properties are ruled by different characteristics of $\Psi(q)$, which are more sensitive to details of W. One must distinguish an inertial and a dissipative range (these are probably misnomers). In the former, the value of $\Psi(2/3)$ rules the corrective term B to be added to the exponent in the Kolmogorov $k^{-5/3}$ law (this will be shown in Section 4). Similarly, the value of $\Psi(2)$ rules the variance, the kurtosis and the exponent of the spectrum of dissipation (this will be shown in Section 3). The next simplest characteristic of $\Psi(q)$ is $\Psi''(1)$. Our last result will be that $\Psi'''(1)$ determines the width of the dissipative range. When $\Psi'''(1) = 0$, an equality characteristic of the fractally homogeneous case, the dissipative range is vanishingly narrow. Otherwise, it is most significant, especially when L much exceeds the Kolmogorov inner scale.

Digression. Each property that involves a moment of higher order $\Psi(q)$ is ruled by the corresponding $\Psi(q)$. As q increases, $\Psi(q)$ becomes increasingly sensitive to details of the distribution of W. This explains why the moments computed from the lognormal assumption appear inconsistent (see Novikov 1969). This difficulty was eliminated when M 1972j, 1974f,c{N14, N15 and N16}, showed that – in the absence of viscosity cutoff – the population moments above a certain order, namely α_1, α_2, or α_3, are in fact infinite. In my opinion, this feature explains why the experimentalists have found empirical moments of higher order to be so elusive.

3. COVARIANCE, FLATNESS AND KURTOSIS OF DISSIPATION: EXPONENTS DETERMINED BY $\Psi(2)$

3.1 The covariance exponent of the dissipation

Take two domains Ω' and Ω'' whose diameters are small compared to the shortest distance r between them, and large compared to ε. Define the covariance of the dissipation density $\varepsilon(\mathbf{x})$ as the expectation of the product of the average of ε within these domains. Without entering into details, let it be stated that in the fractally homogeneous case this covariance is approximately $(r/L)^{D-3}$, and in the general case it is $(r/L)^{-3f(2)}$. The proof closely follows that of Yaglom (see Monin & Yaglom 1975, p. 614), but stops before the point where those authors approximate the product of many W's by a lognormal variable.

Since $\Psi(q)$ is convex, we have $\Psi(2) \geq f(1) + (2-1)\Psi'(1) = \Psi'(1)$. Thus $3\Psi(2) \geq 3\Psi'(1) = 3 - D$. Equality prevails if and only if $\Psi(q)$ is rectilinear, i.e., the curdling is absolute and turbulence is fractally homogeneous. This also happens to be the sole case where r can be as low as λ. In every other case, an evaluation of the codimension $3 - D$ through the observed exponent $\Psi(2)$ would lead to overestimation. For example, the strictly lognormal W yields $3\Psi(2) = \mu$, which is the *double* of the estimate obtained by using the dimension, namely $3\Psi'(1) = \mu/2$.

The same exponents play an equally central role in the study of the kurtosis of dissipation, if it is observed after averaging over small domains of side r. In other words, at least in intermittency generated by curdling, the covariance and the kurtosis of dissipation are conceptually identical.

3.2 Kurtosis in the fractally homogeneous case

Our point of departure is that the dissipation vanishes, except in a region of relative size $(L/r)^{D-3}$, in which it equals $(L/r)^{3-D}$. Hence, it is readily shown that the kurtosis is simply $(L/r)^{3-D}$. It increases as r becomes smaller, and when r takes its minimum value λ (as announced, we shall show that ε is well-defined in the fractally homogeneous case), the kurtosis reaches its maximum value $(L/\lambda)^{3-D}$. The measure of degree of intermittency depends both on the intrinsic characteristic of the fluid, as expressed by D, and on outer and inner scale constraints, as expressed by L/λ, which is related to Reynolds number. Therefore, it is better to measure the degree of intermittency by D itself. The empirical value of the exponent is 0.4 (Kuo and Corrsin 1972); this suggests $D = 2.6$.

In order to explore the significance of these findings, let me begin by sketching the results of previous studies of the kurtosis by Corrsin 1962 and by Tennekes 1967. These and other authors took it for granted that the exponent of kurtosis depends mainly upon whether the support of intermittency is a "blob," a "slab," or a "sheet." While it was recognized that other factors are also involved in each model, they were felt to be secondary. This turns out to be unwarranted. The crucial fact is that each of these models leads to fractally homogeneous intermittency, whose dimension D is affected by *all* the assumptions made, and determines the exponent of the kurtosis.

In the Corrsin model, the exponent's value is $3 - D = 1$ (his formula 10), hence $D = 2$. This fractal dimension is experimentally wrong, it even fails to satisfy the basic requirement that $D > 2$. It is interesting to note that $D = 2$ is the smallest fractal dimension compatible with Corrsin's featured assumption, that turbulent dissipation concentrates with uniform density within sheets of thickness ε enclosing eddies of size L. In other words, Corrsin's additional assumptions cancel out: there was no surreptitious increase of D, and he worked with a classical shape, not a fractal.

On the other hand, the Tennekes model implies $D = 7/3$. This value *does* satisfy $D > 2$, is reasonably close to observations, and safely exceeds the minimum fractal dimension, namely $D = 1$, which topology imposes on a shape including ropes. However, Tennekes was mistaken in believing his argument was testing the assumption that dissipation occurs in vortex tubes of diameter λ. The more vital assumption was that the average distance between tubes is the Taylor microscale. The fact that the Tennekes value $D = 7/3$ is even higher than the Corrsin value $D = 2$ is strong evidence that a tube, if sufficiently convoluted, ends up by ceasing to be a tube from a metric-fractal viewpoint, and becomes a fractal.

Now, back to data: the experimental $D = 2.6$ excludes blobs, *does not exclude* ropes or sheets, but *does not require* either.

3.3 Kurtosis of nonfractally homogeneous intermittency generated by weighted curdling

The kurtosis $\langle \lambda^2 \rangle / \langle \lambda \rangle^2 = \langle \lambda^2 \rangle$ is equal to

$$\langle W^2 \rangle^{\log_b (L/r)} = (L/r)^{\log_b (L/r) \log_b \langle W^2 \rangle} = (L/r)^{3\Psi(2)}.$$

We know that $3\Psi(2) \geq 3 - D$. Hence, among all forms of curdling having a given D, the fractally homogeneous case is the one where the kurtosis is smallest. Therefore, $3\Psi(2) = 0.6$ yields only $D \geq 2.6$.

3.4. Digression concerning the fractally homogeneous case: the behavior of the Fourier transform

Fourier transforms do not deserve the near exclusive attention which the study of turbulence gave them at one time, because of the importance that was attached to spectra. Nonetheless they are important, hence it is useful to mention that in the fractally homogeneous case, the Fourier properties of a fluid also involve fractal dimension. The nature of the relationship between D and Fourier analysis has long been central to the fine mathematical aspects of trigonometric series (see Kahane & Salem 1963), but the resulting theory is little known and little used beyond its original context.

This theory is concerned with functions that are constant except over a fractal of dimension D. Such functions are called "singular"; they have no ordinary derivatives but have generalized derivatives which are measures carried by the fractal in question. In a very rough first approximation, the squared moduli of the Fourier coefficients of the measure in question "tend" to be $\sim k^{-D}$.

A finer approximation, however, confirms an assertion made earlier, that the consequences of self-similarity split into conceptually distinct aspects: dimensional, spectral and others. Each aspect is governed by an exponent of self-similarity, and the different exponents are related by inequalities.

If, as I hope, the importance of fractal shapes in turbulence is recognized, the spectral analysis of the motion of fluids may at long last benefit from a number of pure mathematical results in harmonic analysis.

4. MODIFICATIONS TO THE 2/3 AND 5/3 LAWS; A SPECTRAL EXPONENT CHANGE DETERMINED BY $\Psi(2/3)$

Kolmogorov 1962 and Obukhov 1962 have noted that intermittency modifies the classical exponents 5/3 and 2/3 by adding a positive factor to be denoted by B. A more careful examination of the problem, to which we now proceed, confirms this conclusion but yields values of B that do not generally agree with those asserted by the Russian school.

Consider two points P' and P'' separated by a distance r. Denote the vector $P'P''$ by $\mathbf{x}$, the distance from P' to P'' by r, the vectorial velocities at P' and P'' by $\mathbf{u}(P')$ and $\mathbf{u}(P'')$, and the difference $\mathbf{u}(P'') - \mathbf{u}(P')$ by $\Delta \mathbf{u}$. In Taylor homogeneous turbulence of uniform dissipation ε, one has $\langle (\Delta \mathbf{u})^2 \rangle = (\varepsilon r)^{2/3}$. In the intermittent case, the nonrandom ε is replaced by a random field $\varepsilon(\mathbf{x})$ with $\langle \varepsilon(\mathbf{x}) \rangle = \varepsilon$. Therefore, the expression $(\varepsilon r)^{2/3}$ must be replaced by some quantity that is characteristic of the random field and also of P' and of P''. There are several different ways of selecting this quantity, but no harm in always denoting it by ε_r.

Like Yaglom, we shall at first closely follow the approach of Obukhov 1962 and Kolmogorov 1962, which defines ε_r as the average of $\varepsilon(\mathbf{x})$ over the "Obukhov sphere" $- \Omega(P', P'')$, whose poles are P' and P''. In practice, in the case of curdling within cubic cells, it is more convenient to select for Ω the smallest cell containing both P' and P''. We find, as have Kolmogorov and Obukhov, that intermittency requires the replacement of the classical spectral density $E(k) = E_0 \varepsilon^{2/3} k^{-5/3}$ by an expression of the form $E(k) = E_0 \varepsilon^{2/3} k^{-5/3} (k/L)^{-B}$. On the other hand, we disagree with them on the basic points described in the following two paragraphs.

Summary of results concerning the value of the exponent B. When the correction B is expressed in terms of the dimension D, the Kolmogorov-Obukhov corrections come out as $B = (3 - D)/4.5$. But this value is due to very specific and arguable features of the lognormal assumption, which is part of their model. Fractally homogeneous intermittency, yields the different and *larger* value $B = (3 - D)/3$. The general result it $B = -3\Psi(2/3)$, which can lie anywhere between the bounds 0 and $(3 - D)/3$. Thus $(3 - D)/4.5$ is a compromise value that is perfectly admissible but by no means necessary.

The highest value of k for which a spectral density with the exponent $5/3 + B$ is conceivable. The fractally homogeneous case is unique in that it allows a widely-liked approximation with the following properties: the dissipative range reduces to one value of r, $E(k) \sim k^{-5/3-B}$ all the way to $k = 1/\lambda$, and $E(K) = 0$ for $k > 1/\lambda$. Weighted curdling yields an altogether different result: the traditional approximation of $E(k)$ by a truncated power law expression leads to a paradox. This Section will evaluate B and present the paradox; Section 5 will solve this paradox.

The remainder of this Section will derive the above result, then subject the Obukhov-Kolmogorov approach to critical analysis. Their chosen definition of ε_r, as the average of $\varepsilon(\mathbf{x})$ over the Obukhov sphere, is to a large extent suggested less by physics than by convenience. Other definitions of

Ω are therefore worth considering. The first substitute Ω is the interval from P' to P''. When $D > 2$ (as we believe is the case for turbulence), the expression for B is unchanged and the coefficient E_0 is modified, but remains positive and finite. When $D < 2$, on the contrary, E_0 vanishes and B becomes meaningless. A second substitute Ω will be determined by the distribution of intermittency, and will inject topology. The argument will give reasons for believing that the carrier of turbulence, must not only satisfy the metric inequality $D > 2$ divided in Section 1, but must, in some topologic sense, be "at least surface-like." However, this choice of Ω is very tentative, and so are the conclusions to be drawn from it.

4.1. The fractally homogeneous case when ε_r is the average of $\varepsilon(x)$ over the Obukhov sphere Ω, or an approximating cube

By the theorem of conditional probabilities, one can factor $\langle \varepsilon_r^{2/3} \rangle$ as the product of two factors: the probability of hitting dissipation in Ω, and the conditional expectation of $\varepsilon_r^{2/3}$ where "conditional" means that averaging is restricted to the cases where $\varepsilon_r > 0$.

When Ω is the Obukhov sphere, or is the smallest cubic eddy of side Lb^{-k} that contains both P' and P'', it can be shown that, as $k \to \infty$, the hitting probability becomes at least approximately equal to $(r/L)^{3-D}$.

The product of the hitting probability and the conditional expectation of $\varepsilon_r r$ is simply the nonconditional expectation ε. Hence, the conditional expectation must be equal to $\varepsilon_r r(r/L)^{D-3}$. A stronger statement, in fact, holds true. Assuming fractal homogeneity, a positive value of ε_r is the product of $\varepsilon_r^{D-3} L^{3-D}$ and a random variable having positive and finite moments of every order. Consequently,

$$\langle (\varepsilon_r r)^{2/3} \rangle = V_{1/3} (r/L)^{3-D} \varepsilon^{2/3} r^{2/3} (r/L)^{-2(3-D)/3}$$

$$= V_{1/3} \varepsilon^{2/3} L^{-(3-D)/3} r^{2/3 + (3-D)/3}$$

$$= V_{1/3} \varepsilon^{2/3} L^{-(3-D)/3} r^{1 - (D-2)/3}.$$

The corresponding spectral density is

$$E(k) = E_0 \varepsilon^{2/3} L^{-(3-D)/3} k^{-5/3 + (D-3)/3}$$

$$= E_0 \varepsilon^{2/3} L^{-(3-D)/3} k^{-2 + (D-2)/3}.$$

It is important to know that the numerical coefficients E_0 and $V_{1/3}$ are positive, but their actual values will not be needed.

These expressions show that intermittency has two distinct effects: it injects L, and changes the exponent of k from 5/3 to $5/3 + B$, with $B = (3 - D)/3$.

Since $B \geq 0$, the exponent $5/3 + B$ always exceeds the 1941 Kolmogorov value 5/3. As expected, $B = 0$ corresponds to the limit case $D = 3$, when dissipation is distributed uniformly over space.

The point where $5/3 + B$ goes through the value 2 occurs when $D = 2$, a relationship which we shall return to in Section 4.4.

Even if it is confirmed that (as inferred in Section 1.4) ordinary turbulence satisfies $D > 2$, values $D < 2$ are worth including for the sake of completeness. Since $B \leq 1$, the exponent $5/3 + B$ always lies below the value 8/3. This value corresponds to the limit $D = 0$, where dissipation concentrates in a small number of blobs. (In absolute curdling, we saw that D is at least $\log 2/\log(L/\lambda)$, the value corresponding to dissipation concentrated into two curds. However, variants of curdling yield a more relaxed relationship between the limit value $D = 0$ and dissipation that concentrates in few blobs.) We shall see in Section 4.3 that among curdling processes of given D, the value of B is greatest in the fractally homogeneous case. Hence, the bound $B \leq 1$ is of wide generality. Sulem & Frisch 1975 were able to rederive it by an entirely different argument from the characteristic that for $D = 0$ all dissipation concentrates in a small number of blobs.

4.2. The fractally homogeneous case with other prescribed domains Ω

Sections 4.2 and 4.3 will generalize Obukhov's specification of Ω. To do so, we must first analyze it in greater detail. A cumbersome but useful preliminary is to decompose it into several parts of increasing degrees of arbitrariness. (a) An assumption we shall not question is that one has to replace ε by the average ε_r of the local dissipation rate $\varepsilon(\mathbf{x})$, taken over an appropriate domain Ω. (b) The domain Ω should be independent of $\varepsilon(\mathbf{x})$. (c) Ω should be three-dimensional, even if it is not the Obukhov sphere whose poles are P' and P''. In Section 4.5, we shall question both (b) and (c). In the present Section we keep b) and question (c).

Specifically, we suppose that Ω is *either* fixed but nearly one-dimensional, namely we choose for it a cylinder of radius 2λ and axis $P'P''$, *or* is strictly one-dimensional, namely (for reasons of symmetry) the segment $P'P''$.

When Ω is the line $P'P''$, the results are more complex than when Ω is Obukhov's sphere, because in the limit $\lambda \to 0$, the probability of $P'P''$ hitting turbulence depends on the value of D. When $D < 2$, we know this probability is zero. When $D > 2$, we know it to be positive because of the nondegeneracy of linear cross-sections, and it turns out that the hitting probability is approximately equal to the expression familiar from Section 4.1, namely $(r/L)^{3-D}$. As a result, the dependence of $\langle (\Delta \mathbf{u})^2 \rangle$ on L/r and of $E(k)$ upon Lk is also as in Section 4.1. But there is a single vital change: the coefficients E_0 and $V_{1/3}$ remain positive if $D > 2$, but vanish if $D \leq 2$. In particular, the exponent of r is restricted to the narrower range of values between $2/3$ and 1, and the exponent of k^{-1} always lies above the Kolmogorov value $5/3$, but below 2. This bound 2, which is familiar from the Burger's theory, is when all turbulent diffusion *within the segment $P'P''$* reduces to a few blobs.

4.3. Dissipation generated by weighted curdling

As was the case for the correlation in Section 3, the formal argument can be borrowed from Monin & Yaglom 1975, with one fundamental exception. As in Section 3, one *must not*, and we *shall not*, replace log W by its Gaussian approximation. The exact result, supposing that $r \ll \lambda$, is as follows:

$$\langle \varepsilon_r r^{2/3} \rangle = V r^{2/3} [\langle W^{2/3} \rangle]^{\log_b(L/r)} = V r^{2/3} (L/r)^{-B},$$

with $B = -3\Psi(2/3)$, and

$$E(k) = E_0 L^{-B} k^{-5/3 - B}.$$

To evaluate $\Psi(2/3)$, we shall return to the determining function $\Psi(q)$. By convexity, the $0 \leq q \leq 1$ portion of the graph of $\Psi(q)$ lies between the q axis and the tangent to $\Psi(q)$ at $q = 1$; this tangent's slope is equal to $\Psi'(1) = -(3-D)/3$. As a result, given any value $D < 3$, B can range from the maximum value $B = 3\Psi'(1)/3 = (3-D)/3$ (obtained in the fractally homogeneous case) down to 0. It is possible to show that this last value cannot be attained, but can be approached arbitrarily closely. So it is conceivable, however unlikely, that intermittency should bring no change to the $k^{-5/3}$ spectral density.

The inequality $B \leq (3-D)/3$ generalizes the equality $B = (3-D)/3$, which is only valid in the fractally homogeneous case. The value of B corresponding to a lognormal W is $\mu/9$, as is shown in Kolmogorov 1962. Written in terms of $D = 3 - \mu/2$, it yields $B = (3-D)/4.5$. This value confirms that the changes in the 2/3 exponent can be smaller than $(3-D)/3$.

4.4. Paradoxically, the "Burgers" threshold spectrum k^{-2} and $D = 2$ are mutually related in absolute but not in weighted curdling

The preceding argument easily generalizes to Burgers turbulence and more generally to all models leading to $\langle u^2 \rangle = |P'P''|^{2H}$, where H is a constant: the Hölder exponent. One consequence is that the value $B = -3f(2/3)$ is replaced by $B = -3f(2H)$. It follows that the Burgers case $H = 1/2$, and this case alone, has the remarkable property that $B = 0$. The value of the spectral exponent is not only independent of D, but also of the random variable W. In other words, even after Burgers turbulence is made intermittent as a result of curdling, its spectral density continues to take the familiar form k^{-2}.

More generally, the "Burgers threshold" will be defined as the point where the intermittency has the intensity needed for the spectrum to become k^{-2}. It is a well-known fact (exploited in M 1975c{N16}) that the k^{-2} spectrum prevails when the turbulent velocity change is due to a finite number of two-dimensional shocks of finite strength. Hence it was expected that the spectrum is k^{-2} should be found in the case of fractal homogeneity with $D = 2$. This dimension marks the borderline between the cases when the probability of the segment $P'P''$ hitting dissipation is > 0 and the cases when it is $= 0$.

On the other hand, it seems that the logical correspondence between $D = 2$ and $B = 1/3$ fails in the case of weighted curdling. Example: for $D = 2$, the lognormal approximation combined with the choice of Obukhov sphere for Ω yields $E(k) = E_0 L^{-2/9} k^{-17/9}$, with $E_0 > 0$. Even though (assuming it is confirmed that turbulence satisfies $D > 2$) the behavior of the spectrum about $D = 2$ has no practical effect, the inequality $17/9 < 2$ is a paradox that must be resolved. We shall postpone this task to Section 5.

4.5. Non-prescribed domains Ω in fractally homogeneous turbulence and the issue of topological connectedness

Let us resume the discussion of the choice of Ω, that was started in Section 4.3. The use of any *fixed* Ω implies the belief that the mutual interaction between $\mathbf{u}(P')$ and $\mathbf{u}(P'')$ is on the average independent of the fluid flow between the points P' and P''. However, let us briefly envision interactions that propagate along, say, lines of least resistance. In the all-or-nothing fractally homogeneous case, it may well be possible to join P' and P'' by a line Λ such that the integral taken along this line satisfies $\int \varepsilon(\mathbf{x})d\mathbf{x} = 0$. If so, it could well be argued that $\langle \mathbf{u} \rangle$ should vanish.

In this spirit, let us replace ε_r, by the product of $(1/r)$ by the greatest lower bound of all the integrals of the form $\int \varepsilon(\mathbf{x})d\mathbf{x}$ taken along all the curves Λ joining P' to P''. Denote this product by $(1/r)$ glb $\int$. The principle of the new specification of Ω is radically different from $\Lambda = P'P''$, because, if accepted, it would open the door to topology. In particular, two of the shapes studied in M1977F{P.S.1998, and also in M1982F} (namely, the Sierpiński sponge and pastry shell) have the same $D > 2$ but very different topology. For the sponge glb $\int = 0$ for any P' and P'', while for the pastry glb $\int = 0$ if P' and P'' lie in the same cutout, and glb $\int > 0$ otherwise. Since turbulence does in fact exist, so that $\langle (\Delta \mathbf{u})^2 \rangle > 0$, accepting the Λ that minimizes $\int$ would lead to the following tentative inference.

Among all point doublets P' and P'', selected at random under the constraint that $|P'P''| = r$, doublets such that every line from P' to P'' hits the carrier of turbulence must have a positive probability. In other words, the probability of P' and P'' being separated by "sheets" of turbulence must be non-zero. The mathematical nature of this tentative inference is entirely distinct from the fractal inequality $D > 2$; the latter is purely metric, while the present one combines topology with probability. A mixture of theoretical argument with computer simulations shows that there exists a critical dimension D_0 such that when $D < D_0$ the probability of the set generated by absolute curdling being sheet-like is zero, while when $D > D_0$ this probability is > 0. This D_0 is much closer to 3 than to 2.

In this line of thought, it is tempting to constrain Λ to stay in the Obukhov sphere and denote the restricted glb by glb' $\int$. In this case, the inference that $\langle \text{glb}' \int \rangle > 0$ would involve a combination of topological, probabilistic and metric features; this lead can not be developed further.

The preceding reference to topology is extremely tentative, by far less firmly established than the fractal inequality $D > 2$. It fails to question the validity of two logical steps taken by Kolmogorov and Obukhov: the link they postulated in 1941 between $\langle (\Delta \mathbf{u})^2 \rangle$ and a uniform ε, and the link they postulated in 1962 between $\langle (\Delta \mathbf{u})^2 \rangle$ and the expectation of $\varepsilon^{2/3}$. Moreover, the all-or-nothing fractal homogeneity may well be too flimsy a model to support extensive theorizing.

5. THE INNER SCALE AND THE DISSIPATIVE RANGE

Thus far the existence of actual dissipation was acknowledged only indirectly, because an inner scale λ, like L, was arbitrarily imposed from the outside. The idea is that for Taylor homogeneous turbulence with viscosity ν and uniform rate of dissipation ε, the dissipative range is vanishingly narrow around the inverse of $\eta_3 = \nu^{3/4} \varepsilon^{-1/4}$. The quantity η_3 is defined to insure that, if the spectrum $E(k) = E_0 \varepsilon^{2/3} k^{-5/3}$ is truncated at an appropriate numerical multiple of $k = 1/\eta_3$, one has the identity $\varepsilon = \nu \int k^2 E(k) dk$, where the integral is carried from $1/L$ to $1/\eta_3$.

This Section shall first examine formally how the idea of an internal scale is affected by intermittency. Then we shall proceed to an actual analysis of the inner scale of curdling. In the fractally homogeneous case, the inner scale will continue to be defined as the inverse of the spectrum's truncation points. This result was already obtained by Novikov & Stewart, but it deserves a more careful analysis. In all other cases the result is *quite different*. The analysis will show the necessity of a dissipative range that does *not* reduce to the neighborhood of any single value $1/\eta$, but has a definite width determined by the value of $\Psi''(1)$.

5.1. Truncation point for the power-law spectrum

As was shown in Section 4, the spectral density of velocity in intermittent turbulence is of the form $E(k) = E_0 \varepsilon^{2/3} k^{-5/3} (Lk)^{-B}$. Suppose we want the identity $\varepsilon = \nu \int k^2 E(k) dk$, for ε to continue to hold. If so, neglecting numerical factors, the truncation value η'_D becomes defined as

$$\varepsilon = \nu \varepsilon^{2/3} L^{-B} \eta'_D{}^{-4/3-B}$$
$$\eta'_D = [(\nu^3/\varepsilon) L^{-3B}]^{1/(4-3B)}$$
$$\eta'_D / L = (\eta_3 / L)^{1/(1-3B/4)}.$$

Since $0 \leq B \leq 1$, we have $\eta'_D \ll \eta_3$. When D is fixed, η'_D is a monotone function of B. Thus, when B reaches its maximum value $B = (3-D)/3$, η'_D reaches its minimum value $L(\eta_3/L)^{1/(D+1)}$ and $1/\eta'_D$ reaches its maximum. Note that the value of η'_D depends not only on ν and ε, but also on L.

More generally, if one stays within a sub-domain of length scale $r \ll L$, in which the average dissipation is ε_r, one will have the new inner scale $\eta'_D(r)$ such that

$$v^{3/4}\langle\varepsilon_r\rangle^{-1/4}/r = [\eta'_D(r)/r]^{1-3B/4}.$$

5.2. Critique of an inner scale of intermittent turbulence that was tentatively suggested by Kolmogorov

The need to reexamine the concept of inner scale had been felt previously. Kolmogorov 1962, p. 83 (seventh formula) suggests for this role the expression $v^{3/4}\langle\varepsilon_r\rangle^{-1/4}$ which occurs in the left-hand side of the last formula of the preceding Section. This choice is not explained, no further use is made of it and it is difficult to understand. The first odd feature of his definition is that when $r = L$, his modified inner scale reduces to η_3. Hence, contrary to η_D, it is independent of the degree of intermittency. A second odd feature relates to $r \to 0$. To describe it, let us follow Kolmogorov in assuming $\log \eta_r$ to be lognormal, with the variance $\mu \log(L/r)$ and a mean adjusted to insure that $\langle\varepsilon_r\rangle = \varepsilon$. It follows that

$$\langle\varepsilon_r^{-1/4}\rangle = \varepsilon^{-1/4} \exp[(1/2)(-1/4)(-5/4)\mu \log(L/r)] = \varepsilon^{-1/4}(L/r)^{5\mu/32}.$$

Hence, as $r \to 0$, the modified Kolmogorov scale *increases* on the average and may exceed r. We shall not attempt to unscramble this concept.

5.3. Inner scale of curdling in the fractally homogeneous case

The truncation of $E(k)$ shows that the energy cascade must stop when it reaches eddies of the order of magnitude of η'_D. But what about the curdling cascade? It too must have an end, to be followed by dissipation. We shall now identify the scale η_D at which it stops.

In the fractally homogeneous case, let us show that η_D is identical to the $\eta_{D'}$ defined through $E(k)$. Consider a cube of side L filled with a Taylor homogeneous turbulent fluid of viscosity v, dissipation ε and inner scale η_3 Since we assume that the increasingly small curds created by a Novikov-Stewart cascade are themselves Taylor homogeneous, these curds are endowed with a classical Kolmogorov scale varying with the cascade stage. Moreover, we assume that the instability and breakdown leading to curdling are encountered if and only if the curd size exceeds the Kolmogorov scale. (This assumption is equivalent to a little-noticed condition of Novikov & Stewart, as reported in Monin & Yaglom 1975, p. 611.)

The first curdling stage leads to curds of side L/b in which dissipation is equal to either 0 or εb^{3-D}. In the empty cells, the Kolmogorov scale is infinite, and further curdling is, of course, impossible. In the first stage curds, the inner scale is $\eta^{(1)} = \eta_3 b^{-(3-D)/4}$. In the m-th stage curds, the average dissipation is $\varepsilon b^{m(3-D)}$, the curd size is Lb^{-m}, and the inner scale is therefore $\eta^{(m)} = \eta_3 b^{-m(3-D)/4}$. We see that the inner scale and the curd size both decrease with $1/m$. Since we postulate that there is no further curdling after these two scales meet, we are left with the criterion $\eta_3 b^{-m(3-D)/4} \sim b^{-m}L$, i.e., $\eta_3/L = [b^{1-(3-D)/4}]^{-m}$. The solution yields $b^{-m}L = \eta_D$, with η_D identical to the η'_D obtained earlier in this Section through the truncation of $E(k)$. Hence the fractal dimension not only rules the manner in which Novikov-Stewart curdling proceeds, but the point where it stops. In addition, it is reasonable to assume that the cutoff of $E(k)$ near $1/\eta_D$ is very sharp.

Digression concerning curdling in spaces of Euclidean dimension $E>3$. The derivation of η_D has relied on the fact that, in a space of Euclidean dimension $E = 3$, the decrease in η_m is less rapid than the decrease in curd size. However, this last feature is highly dependent upon $E - D$, and therefore upon the value of E. As in many other fields of physics, a qualitative change may be observed when $E \neq 3$. Indeed, our stability criterion readily yields the result that a nonvanishing inner scale *need not exist*. It exists if and only if $E - D < 4$. Its value is given by the relation

$$(\eta_E/L) \sim (\eta_D/L)^{1-(E-D)/4}.$$

The necessary and sufficient condition $E - D < 4$ for the existence of a non-vanishing inner scale is peculiar but not very demanding. One amply sufficient condition is $E < 4$. (However, in order that curdling continue forever, meaning that $\eta_D = 0$, the converse condition $E > 4$ is necessary, but it is *not* sufficient.) Another amply sufficient condition for $\eta_D > 0$ is $E - D < 1$, which we know expresses that, with probability > 0, the linear cross-sections are *not* empty. These various conditions make it clear that a vanishing inner scale cannot be observed for turbulent dissipation, or even for the Leray-Scheffer conjectural singularities of the Navier-Stokes equations. It can at most be observed for phenomena that are very much sparser than either of the above.

Nevertheless, odd as the result may be, our criterion does indicate the following: When $E - D > 4$, *a curdling cascade will continue forever, without any physical cutoff, even when the viscosity is positive.* I do not know what this result means, and what implications it has concerning dissipation. It seems to be trying to tell us something about the singularities in the ultimate solution of the equations of motion of some physical system, but I cannot guess which one.

5.4. Inner scale of curdling when intermittency is generated by weighted curdling: first role of $\Psi'''(1)$

In this case the inner scale is best studied by two approximations. The first one yields a single typical value. The interesting fact is that this value turns out to be much smaller than the quantity η'_D obtained through the truncation of $E(k)$. The second approximation shows that this typical value is not very significant and must be replaced by a random variable, that is, described by a statistical distribution. Strictly speaking, the same situation had already prevailed in all-or-nothing curdling leading to fractally homogeneous intermittency, but in that case $\eta^{(m)}$ was simply binomial, equal to either $\eta_3 b^{-m(3-D)/4}$ or infinity, and the latter value could be neglected. In the case of weighted curdling, nothing can be neglected.

Recall that if the curdling cascade could continue forever, the dissipation density $\varepsilon(\mathbf{x})$ at the point $\mathbf{x}$ would be a product of weights W, one per cascade stage. This product takes the form $W(i_1) \, W(i_1, i_2) \, W(i_1, i_2, i_3) \ldots$, where the real number $0.i_1 i_2 \ldots$ denotes $\mathbf{x}$ in the counting base b, and the Ws are an infinite sequence of independent random weights. Similarly, η^m will simply be written in the form $\eta_3 [\Pi_{1 \le n \le m} W_n]^{-1/4}$. Curdling will stop when this random $\eta^{(m)}$ first overtakes the nonrandom $b^{-m}L$. Taking logarithms, we find that m is the first integer where

$$\Sigma_{1 \le n \le m} [(-1/4) \log W_n + \log b] = \log[L/\eta_3].$$

The left-hand side of the above expression defines a random walk with nonrandom drift equal to $\langle \log b - \log W/4 \rangle$ and an absorbing barrier. The above-defined value of m is when absorption (or ruin) takes place. Absorption will almost surely occur because the drift is positive (digression: this hold true only as long as $E - D < 4$).

First approximation. When $L/\eta_3 \gg 1$, the drift tends to overwhelm the random component of the walk. Therefore, it is permissible to approximate m by the value $\tilde{m}$ which is obtained by replacing the random walk by its expectation. The proper choice of weights in the above expectation is not obvious, but there is room here only for the results, and not for a full justification. In order to avoid irrelevant notational complication, we add the assumption that the values w_g of W are discrete with probabilities p_g. Then the proper intrinsic probability of w_g is not p_g itself, but $p_g w_g$. Since $\langle W \rangle = 1$, $\Sigma p_g w_g = 1$, therefore the $p_g w_g$ are acceptable as probabilities. Continuing to use the symbols $\langle \ \rangle$ to designate expected values under the probabilities p_g, our criterion yields

$$\eta_3/L = [b(\exp \langle W \log W \rangle)^{-1/4}]^{-\tilde{m}}.$$

The result stated in the last form turns out to apply also to nondiscrete W s. Since $-\langle W \log_b W \rangle = D - 3$, the definition of η_D reduces formally to that applicable in the fractionally homogeneous case.

Summary of the first approximation. In weighted curdling the order of magnitude of m is the same as in the all-or-nothing curdling having the same value of D, hence of $\Psi'(1)$. In particular, the order of magnitude of $1/\eta_D$ is much *greater* than the $1/\eta'_D$ obtained in Section 1.1.

Second approximation. The actual values of m scatter around $\tilde{m}$. For fixed W, the scatter increases with L/η_3. For fixed L/η_3, it is useful to define a standard scatter, to be denoted by σm. It is approximately the ratio of two factors. The first is the standard deviation of the sum of $\tilde{m}$ factors of the form $-\log W/4$. The variance of W is $\langle W \log^2 W \rangle - \langle W \log W \rangle^2$, which happens to equal $\log C \Psi''(1) = 3 \log b \Psi''(1)$. Hence, the first factor is

$$\frac{1}{4}[3\tilde{m} \log b \Psi''(1)]^{1/2} \frac{1}{4}\sqrt{[3 \log(L/\eta_3)/(1-(3-D))/4]}.$$

The second factor is the expected value of $\log b - \log W/4$, that is $\log b - \Psi'(1) \log C/4 = \log b[1 - (3-D)/4]$. Combining the two factors, we obtain the two alternative forms

$$\sigma m = (2/\log b)(1+D)^{-3/2}\sqrt{3 \log(L/\eta_3)\Psi''(1)}$$

$$= (3/\log b)^{1/2}(1+D)^{-1}\sqrt{\tilde{m}\Psi''(1)}.$$

This is the first time that the value of $\Psi''(1)$ enters in the present discussion. The value of $\Psi'(1)$ enters through D, but σm is not very sensitive to $\Psi'(1)$.

5.5. The dissipative range

The methods that Chapters 3 and 4 used to evaluate exponents and exponent changes apply only to scales for which curdling has a small probability of having stopped, that is, roughly, the scales from $(1/L)$ to $k \sim (1/L)b^{\tilde{m}-\sigma m}$. As higher wave numbers are approached, one encounters the range from $k \sim (1/L)b^{\tilde{m}-\sigma m}$ to $k \sim (1/L)b^{\tilde{m}+\sigma m}$. Here, some dissipation is likely to occur in a substantial region of the fluid.

Let us make a few more comments on this topic. By the last result of Section 5.4, $\log_b k$ is proportional to $[\Psi'''(1)]^{1/2}$. When log W is lognormal, $\Psi(q)$ is parabolic and $\Psi'''(1)$ is proportional to μ. More generally, unless the distribution of W is very bizarre, one has approximately $\Psi(2/3) \sim \Psi(1) - \Psi'(1)(1/3) + \Psi'''(1)(1/3)^2/2$. That is, $(3-D)/3 - B \sim \Psi'''(1)/6$. This relation holds even if B and $(3-D)/3$ are not linked by a numerical relationship of the kind that holds when W is lognormal and $B = (3-D)/4.5$. In other words, the width of the dissipative range – measured on the $\log_b k$ scale – is typically the square root of the defect of B with respect to the fractally homogeneous approximation.

It was to be expected that each of these quantities is a monotone increasing function of the other. Indeed, the inequality $1/\eta_D \gg 1/\eta'_D$ expresses that the spectrum $k^{-5/3-B}$ relative to the inertial range cannot be extrapolated consistently. The corresponding distribution of energy among the wave numbers decreases much too slowly as k increases, which implies that the whole energy would be completely exhausted well before reaching $k \sim 1/\eta'_D$. The greater the difference $1/\eta - 1/\eta'_D$, the sooner must this inertial range law $k^{-5/3-B}$ cease to apply.

The expressions that apply in the dissipative range and replace coefficients such as B, will be described elsewhere. {P.S.1998: the text describing those expressions was not published in the 1970s and my files for that period cannot be located.}

ACKNOWLEDGMENT

Before the present final version, I had the benefit of penetrating comments by Uriel Frisch: by listing a few of the things he did not understand, he motivated me to substantial further development.

&&&&&&&&&& **ANNOTATIONS** &&&&&&&&&&

Editorial changes. The notation was edited to conform to present usage. Thus, q was h, $\Psi(q)$ was $f(h)$, b was Γ, E was Δ, m was i, λ was η, ν or ε. The sections in this reprint were originally called chapters.

How this paper came to be written. Having been invited to Roger Temam's 1975 Symposium on turbulence, I suggested two other speakers, whose presence would lead to a nice session devoted to fractals.

The first was my old friend and Temam's Orsay colleague Jean-Pierre Kahane. I felt he should be asked to present the facts about fractals that were known before fractal geometry started becoming organized.

The second was Vladimir Scheffer, now at Rutgers. He was not yet known in Paris, but I claimed no travel expenses so he could come. We had met him a year before, when I lectured at Princeton in the seminar of Frederick J. Almgren Jr. (1933 - 1997). There I first heard of geometric measure theory and met several experts on Hausdorff dimension.

As I recall, that Princeton lecture was the first I gave to a major pure mathematics department in the USA, hence it marked a turning point towards warmer relations with that community.

On this occasion, I promoted the Hausdorff dimension as a possible tool in the study of the equations of fluid motion, and presented broad new conjectures about the fractal nature of the singularities and their possible relation to phenomena of turbulence. Scheffer, a Princeton Ph. D. student of Almgren who had stayed on as an instructor, set out to work on these conjectures, and soon reported that the Hausdorff dimension was indeed the right tool. A classic paper in the field, Leray 1934, ends rather abruptly after the proof of a certain formal inequality that no one had picked. Scheffer noticed immediately that it simply means that the instants where singularities occur have a Hausdorff dimension at most equal to 0.5. (This explains the choice of Leray as Chairman of the fractals session at Temam's Symposium in 1975.) Scheffer then went on to prove a number of different bounds on various other Hausdorff dimensions.

This line of work has, since then, become practiced by many mathematicians but the last reference I know is Lax 1984.

Fractal dimension, dispersion, and singularities of fluid motion

✦ **Abstract.** It is conjectured that turbulent dispersion in a closed vessel involves surfaces whose fractal dimension exceeds 2. The different singularities and quasi-singularities of the motion are carried by a hierarchy of sets whose dimensions are fractions. The quasi-singularities are viewed as being singularities of the Euler equations, after they have been smoothed by viscosity. ✦

1. Introduction

M 1975o{N18} showed that "fractal" sets, whose main feature is a fractional Hausdorff dimension, play a role in numerous branches of science, in particular the study of turbulence. For example, taking the Gaussian approximation to homogeneous turbulence and the Kolmogorov and Burgers velocity spectra, M 1975o shows that the iso-surfaces of passive scalars have dimensions $3 - 1/3$ and $3 - 1/2$.

On the basis of intuitive considerations and experimental measurements, the present Note states two kinds of conjectures, both related to more fundamental but less developed aspects of turbulence. The first conjecture concerns the geometry of turbulent dispersion. Next, generalizing from the models of intermittency described in M 1974f{N15}, I conjecture that each of the multiple aspects of the notion of "turbulence" corresponds to either a singularity or a "quasi- singularity" of the equations of motion, and that this singularity is carried by a space-time set whose dimension is in general a fraction.

Scheffer 1976 shows how he has succeeded in evaluating two of these dimensions, starting from the Navier-Stokes equations.

2. Dispersion.

It is widely accepted that, when turbulence acts on a material line, the length of this line increases exponentially in time. On the other hand, the radius of the smallest sphere containing one of these lines grows only slowly, or even remains bounded when the vessel is closed. Hence, the line must increasingly curl up on itself. The same is true of material surfaces. Let us first consider the effect of a Richardsonian eddy as it cascades in self-similar fashion towards higher and higher frequencies, before it dissipates. Within a critical zone of eddy intensity, one observes that a regular blob of passive contaminant transforms into a kind of octopus, then each arm subdivides into branches, then (repeatedly) into subbranches, down to the threshold of dissipation. Experimental diagrams (see Corrsin 1959b) remind us of the first steps of Peano's construction of a plane-filling curve (a section of a space-filling surface). Viscosity and molecular diffusion have a regulating character, but in their absence the process in question would imply the following: that if the contaminant and the solvent asymptotically mix in a uniform fashion, the separating surface tends towards a fractal surface with dimension 3.

I conjecture that this picture is indeed applicable if the initial eddy is very strong. To the contrary, if the eddy is weak, the mixture will not be perfect; I conjecture that if the threshold results from an admixture of eddies with Kolmogorov velocity spectrum (like the iso-surfaces in M 1975f), the fractal dimension would be $3 - 1/3$. This last result also holds in the case of an infinite vessel. No conjecture could yet be formulated for the transition from $D = 3 - 1/3$ to $D = 3$ as the force of the initial eddy is increased.

3. Singularities and quasi-singularities

Inspired by Oseen's view of turbulence, Leray 1934 has investigated the singularities of the Navier-Stokes equations, defined as points in $R^3 \times R^+$ where the local dissipation rate is infinite. Another view of turbulence, distinct historically and (no doubt) also logically, leads to the homogeneous turbulence of G. I. Taylor, which implies that the distribution of dissipation is statistically uniform. However, real dissipation is "intermittent," as noted by Batchelor and Townsend and investigated by Kolmogorov 1962 and Obukhov 1962. It may be defined on linear sections of the flow...along the X-axis. The values of the velocity $u(x)$ oscillate moderately around their mean value. But the measurements of $(\partial u/\partial x)^2$, which is viewed as an approximation of the turbulent dissipation, can take (at irregular intervals) values that are far from the norm, while finite (if only because a theoretically infinite peak would be smoothed out by the process of measurement).

Having analyzed diverse stochastic *ad-hoc* models of this phenomenon (see M 1974f{N15}), I have made a conjecture which was immediately proven in J. Peyrière 1974: that dissipation concentrates on a random set, a variant of either the Cantor set, or the sets I have named after Besicovitch. (An example is the set of points whose decimal representation contains the integers from 0 to 9 with positive and un-equal frequencies.) More precisely, the set in question is assumed to be a finite approximation of a fractal set, smoothed by viscosity. While dimension is an asymptotic concept, yet it remains useful to measure the degree of irregularity in the self-similar zone that is present in those models. An effect of viscosity is that there are no real singularities involved, only what may be called quasi-singularities. Other empirical data lead to suspect that higher derivatives of u have different, thinner, sets of intermittency.

Extrapolating from the models in M 1974f{N15}, I am led to believe that intermittency and the accompanying role of fractal sets constitute the most distinct characteristic of turbulence, and should be placed in the center of its study.

More precisely, I make two groups of conjectures.

• Turbulence in a fluid is the result of a range of different phenomena, each of which is concentrated, *either* on a set with Hausdorff dimension less than 3 (in space) and less than 4 (in space-time), *or* on such a set after it has been subjected to an inner cutoff related to viscosity.

• The most apparent of those phenomena is related to the presence of quasi-singularities, which are smoothed out singularities of the Euler equations.

&&&&&&&&&& **ANNOTATIONS** &&&&&&&&&&

The French original was presented to the Académie des Sciences on June 23, 1975 by Jean Leray.

NB

Cumulative Bibliography, including copyright credits

Foreword to the Bibliography. *Contents.* *This list puts together all the references of the reprinted and new chapters in this book. The sources being very diverse and some being known to few readers, no abbreviation is used and available variants are included.*

In this list, the Selecta *volumes are flagged by a mention of the form* *N, *which refers to Volume N. Publications reprinted in this* Selecta *volume are flagged by being preceded by a mention of the form* *N16, *which refers to Volume N, Chapter 16. Those items are followed by the name of the original copyright holder. In the case of publications scheduled for the* Selecta *Volumes H and L, the flags are tentative. Finally, the papers and unpublished reports that are not reprinted as such but quoted at length or paraphrased in a chapter of this book are also marked by that chapter's number preceded by an asterisk* *.

Style. A reviewer chided M1997E for not providing a balanced view of finance. I may have invited this criticism by expanding beyond a simple collection of reprints. However, both M1997E and this book hold to the spirit of Selecta, *to the extent that their goal is not to provide a balanced exposition of the field, or even of the publications that challenge or expand my work. I can only beg those denied adequate credit for understanding and forgiveness, and ask them to educate me when they have a chance.*

ARMITRANO, C., CONIGLIO, A. & DI LIBERTO, F. 1986. Growth probability distribution in kinetic aggregation processes. *Physical Review Letters*: **57**, 1016-1019.

ATHREYA, K. B., & NEY, P. E. 1972. *Branching Processes*. New York: Springer.

BANDT, C. 1997. Review of a book by P. Mattila. *Bulletin of the American Mathematical Society*: **34**, 323-327.

BARNSLEY, M. 1988. *Fractals Everywhere*. Orlando, FL: Academic Press.

BELL, D. A. 1980. A survey of $1/f$ noise in electrical conductors. *Journal of Physics C: Solid State Physics*: **13**, 4425.

BEN NASR, F. 1986. *Étude de mesures aléatoires et calculs de dimensions de Hausdorff.* Thèse soutenue le 18 Juin 1986 à l'Université Paris-Sud (numéro d'ordre 43032).

BEN NASR, F., 1987. Mesures aléatoires de Mandelbrot associées à des substitutions. *Comptes Rendus* (Paris): **304-I**, 255-258

*N6 BERGER, J. M. & MANDELBROT, B. B. 1963. A new model for the clustering of errors on telephone circuits. *IBM Journal of Research and Development*: **7**, 224-236. © IBM.

BERNAMONT 1934-7. Fluctuations dans un conducteur métallique de faíble volume (two letters and one paper with similar titles and devoted to one topic.) *Comptes Rendus* (Paris): **198**, 1755-1758 & 2144-2146; *Annales de Physique*: **X1-7**, 71-140.

BILLINGSLEY, P. 1967. *Ergodic Theory and Information*. New York: Wiley.

BLUMENFELD, R. & AHARONY, A. 1989. Breakdown of multifractal behavior in diffusion limited aggregates. *Physical Review Letters*: **62**, 2977-2980.

BLUMENTHAL, R. M. & GETOOR, R. K. 1960. Some theorems on stable processes. *Traditions of the American Mathematical Society*: **95**, 263-273.

BOAS, R. P. 1987. *Invitation to Complex Analysis*. New York: Random House

BOCHNER, S. 1959. *Lectures on Fourier Integrals, with and Author's Supplement* (translated by M. Tenenbaum and H. Pollard). Princeton: Princeton University Press.

BOLZANO, B.

BOREL, E. 1914. *Le hasard*. Paris: Alcan.

BREIMAN, L. 1965. On some limit theorems similar to the arc sin law. *Teoria Verojatnostii i. Primenenia*: **10**, 323-331.

BRODÉN, T. 1897. Beiträge zur Theorie der stetigen Funktionen einer reellen Veränderlichen. *Journal für reine und angewandte Mathematik*: **118**, 1-60.

BUCKINGHAM, M. J., 1983. *Noise in Electronic Devices and Systems*. New York: Wiley-Halsted.

CALOYANNIDES, M. A. 1974. Microcycle spectral estimates of $1/f$ noise in semiconductors. *Journal of Applied Physics*: **45**, 307-316.

CAMPBELL, M. J., JONES, B. W., 1972. Cyclic changes in insulin needs of an unstable diabetic. *Science*: **177**, 889

CAUCHY, A. 1853. Sur les résultats les plus probables. *Comptes Rendus* (Paris): **37**, 198-206.

CHERNOFF, H. 1952. A measure of asymptotic efficiency for tests of a hypothesis based on the sum of observations. *Annals of Mathematical Statistics*: **23**, 493.

CHRISTENSON, J. & PEARSON, G. L. 1936. *Bell System Technical Journal*: **15**, 197-223.

CIOCZEK-GEORGES, R., MANDELBROT, B. B. SAMORODNITSKY, G. & TAQQU, M. S. 1995. Stable fractal sums of pulses: the cylindrical case. *Bernoulli*: **1**, 201-216.

CIOCZEK-GEORGES, R. & MANDELBROT, B. B. 1995. A class of micropulses and antipersistent fractional Brownian motion. *Stochastic Processes and their Applications*: **60**, 1-18.

CIOCZEK-GEORGES, R. & MANDELBROT, B. B. 1996. Alternative micropulses and fractional Brownian motion (with Renata Cioczek-Georges). *Stochastic Processes and their Applications*: **64**, 143-192.

CIOCZEK-GEORGES, R. & MANDELBROT, B. B. 1996. Stable fractal sums of pulses: the general case.

CLARKE, J. & HAWKINS, G. 1976. Flicker $(1/f)$ noise in Josephson tunnel junctions. *Physical Review*: **B14**, 2826-2831.

CLARKE, J. & HSIANG, T. Y. 1976. Low-frequency noise in tin and lead films at the superconducting transition. *Physical Review*: **B13**, 4790-4800.

CORRSIN, S. 1962. Turbulent dissipation fluctuations. *Physics of Fluids*: **5**, 1301-1302.

COURNOT, A. 1843. *Exposition de la Théorie des Chances et des Probabilités*. Paris: Hachette.

COX, D.R. 1962. *Renewal Theory*. New York: Wiley.

DE WIJS, H. J. 1951 & 1953. Statistics of ore distribution. *Geologie en Mijnbouw* (Amsterdam): **13**, 365-375 & **15**, 12-24.

DEUTSCHEL, J. D., & STROOK, D. W., 1989. *Large Deviations*. Boston: Academic Press.

DOMB, C. 1989. Of men and ideas (after Mandelbrot). *Fractals in physics (Essays in honor of Benoit B. Mandelbrot)* A. Aharony & J. Feder. (Eds.) *Physica*: **D38**, 64-70.

DOMB, C. 1996. *The Critical Point: a Historical Introduction to the Modern Theory of Critical Phenomena*. London: Taylor & Francis.

DOOB, J. L. 1953. *Stochastic Processes*. New York: Wiley.

DOOB, J.L. 1948. Renewal theory from the viewpoint of the theory of probability. *Transactions of the American Mathematical Society*: **63**, 422-438.

DURRETT, R. & LIGGETT, T. M. 1983. Fixed points of the smoothing transformation. *Zeitschrift fü Wahrscheinlichkeitstheorie un Verwandte Gebiete*: **64**, 275-301.

DUTTA, P. & HORN, P. M. 1981. Low-frequency fluctuations in solids: $1/f$ noise. *Reviews of Modern Physics*: **53**, 497-516.

EGGLESTON, H. G. 1949. The fractional dimension of a set defined by decimal properties. *Quarterly Journal of Mathematics, Oxford Series*: **20**, 31-36.

FAMILY, F. & VICSEK, T. (Eds) 1991. *Dynamics of Fractal Surfaces*, Singapore: World Scientific.

FARMER, D., OTT, E. & YORKE, J. 1983. The dimension of chaotic attractors. *Physica*: **7D**, 153.

FELLER, W. 1949. Fluctuation theory of recurrent events. *Transactions of the American Mathematical Society*: **67**, 98-119.

FELLER, W. 1950. *An Introduction to the Theory of Probability and Its Applications*. New York: Wiley.

FELLER, W. 1951. The asymptotic distribution of the range of sums of independent random variables. *Annals of Mathematical Statistics*, **22**, 427.

FELLER, W. 1967. On regular variation and local limit theorem. *Proceedings of the Fifth Berkeley Symposium on Mathematical Statistics and Probability*: **2**, Part II, 373-388. Berkeley and Los Angeles: University of California Press.

FISHER, M.E. 1967. The theory of equilibrium critical phenomena. *Reports on Progress in Physics*: **30**, 615-

FISHER, M.E. 1998. Renormalization group theory: its basis and formulation in statistical physics. *Reviews of Modern Physics*: **70**, 653-681.

FONTAINE, A. B. 1961. Applicability of coding to radio teletype channels. *Report 25 G-.3* Lexington, MA: M.I.T. Lincoln Laboratories.

FONTAINE, A. B. & GALLAGHER, R. G. 1961. Error statistics and coding for binary transmission over telephone circuits. *Proceedings of the IRE*: **49** (June), 1059-1065.

FRIEDLANDER, S. K. & TOPPER, L. (Eds.) 1961. *Turbulence: Classic Papers on Statistical Theory*. New York: Interscience.

FRISCH, U. 1983. In *Les Houches, Session XXXVI, Chaotic Behavior in Deterministic Systems*. Eds. G. Ioos, H. G. Helleman, & R. Stora. Amsterdam: North-Holland.

*N2 FRISCH, U. & PARISI, G. 1985. Fully developed turbulence and intermittency, in *Turbulence and Predictability in Geophysical Fluid Dynamics and Climate Dynamics*. International

School of Physics "Enrico Fermi." Course 88, Eds. M. Ghil et. al. Amsterdam: North-Holland, 84-88. © North-Holland.

FRISCH, U. SULEM, P. L. & NELKIN, M. 1978. A simple dynamical model of intermittent fully developed turbulence. *Journal of Fluid Mechanics*: **87**, 719-736.

GARDNER, M. 1978. Mathematical games -- white and brown music, fractal curves and $1/f$. *Scientific American*: **238 (4)**, 16-32.

GELFAND, I.M. & VILENKIN, N. YA. 1964. *Generalized Functions*, Vol. 4. New York: Academic Press.

GILBERT, E. N. 1961. Capacity of a burst-noise channel. *Bell Systems Technical Journal*: **39**, 1253-1265.

GNEDENKO, B. V. & KOLMOGOROV, A. N. 1954. *Limit distributions for sums of independent random variables*. Translated by K. L. Chung. Reading, MA: Addison Wesley.

GOOD, I. J., **BUG**

GRASSBERGER, P. 1983. Generalized dimensions of strange attractors. *Physics Letters*: **97A**, 227-230.

GRIFFITHS, D. J. 1995. *Introduction to Quantum Mechanics,* Prentice Hall: Englewood Cliffs, NJ.

GRIMMETT, G. *Percolation*. New York: Springer.

GUIVARC'H, Y. 1987. Remarques sur les solutions d'une équation fonctionnelle non linéaire de Benoît Mandelbrot. *Comptes Rendus* (Paris): **305I**, 139-141.

GUIVARC'H, Y. 1990. Sur une extension de la notion de loi semi-stable. *Ann. Inst. Henri Poicaré*: **26 (2)**, 261-285.

GURVICH, A. S. & YAGLOM, A. M. 1967. Breakdown of eddies and probability distribution for small scale turbulence. *Boundary Layers and Turbulence* (Kyoto International Symposium, 1966), a supplement to *Physics of Fluids*: **10**, S59-S65.

GURVICH, A. S. & ZUBKOVSKII, S. L. 1963. On the experimental evaluation of the fluctuation of dissipation of turbulent energy. *Izvestia Akademii Nauk SSSR (Geofizicheskaya Seriia)*: **12**, 1856-___.

HALSEY, T.C., JENSEN, M.H., KADANOFF, L.P., PROCACCIA, I. & SHRAIMAN, B.I. 1986. Fractal measures and their singularities: the characterization of strange sets. *Physical Review*: **A 33**, 1141-1151. IMPORTANT ERRATA: *Physical Review*: **A34**, 1986, 1601.

HARDY, G.H. 1940. *A Mathematician's Apology*. Cambridge: The University Press.

HARRIS, T. E. 1963. *The Theory of Branching Processes*. New York: Springer.

HENTSCHEL, H. G. E. & PROCACCIA, I. 1983. The infinite number of generalized dimensions of fractals and strange attractors. *Physica (Utrecht)*: **8D**, 435-444.

HEYDE, C. C. 1963. On a property of the lognormal distribution. *Journal of the Royal Statistical Society* : **25**, 392-393.

HEYDE, C. C. & SENETA, E. 1977. *I. J. Bienaymé: Statistical Theory Anticipated*. New York: Springer.

HOOGE, N. & HOPPENBROUWERS, A. M. H. 1969. *Physica*: **45**, 386-392.

HOOGE, N. 1970. $1/f$ noise in the conductance of aqueous solutions. *Physics Letters*: **33A**, 169-170.

HOOGE, F. N., 1976. $1/f$ noise. *Physica*: **83B**, 14-23.

HOOGE, F. N., KLEINPENNING, T. G. M. & VANDAMME, L. K. J. 1981. Experimental studies on $1/f$ noise. *Reports of Progress of Physics*: **44**, 481-532.

HURST, H. E. 1951. Long-term storage capacity of reservoirs. *Traditions of the American Society of Civil Engineers*: **116**, 770-808.

HURST, H. E. 1955. Methods of using long-term storage in reservoirs. *Proceedings of the Institution of Civil Engineers*: Part I, 519-577.

HUTCHINSON, J. E. 1981. Fractals and self-similarity, *Indiana University Mathematics Journal* **30**, 713-747.

JOHNSON, J. B. 1925. The Schottky effect in low frequency circuits. *Physical Review*: **26**, 71-85.

JONA-LASINIO, G. 1975. The renormalization group: a probabilistic view. *Nuovo Cimento*: **26B**, 99-119.

KADANOFF, L.P. 1966. Scaling laws for Ising models near T_c. *Physics*: **2**, 263-272. Reprinted in Kadanoff 1993, 165-274.

KADANOFF, L.P. 1993. *From Order to Chaos; Essays: Critical, Chaotic and Otherwise*. Singapore: World Scientific.

KAGAN, S. 1997. *Electronic Noise and Fluctuations in Solids*. Cambridge: The University Press.

KAHANE, J. P. 1974. Sur le modèle de turbulence de Benoit Mandelbrot. *Comptes Rendus* (Paris): **278A**, 621-623.

KAHANE, J. P. 1976. Measures et dimensions. *Turbulence and Navier-Stokes Equations* Ed. R. Temam. New York: Springer. Lecture Notes in Mathematics: **565**, 94-103.

KAHANE, J.P. 1987a. Multiplications aléatoires et dimensions de Hausdorff. *Annales de l'Institut Henri Poincaré*: **B23**, 289-296.

KAHANE, J. P. 1987b. Positive martingales and random measures. *Chinese Annals of Mathematics*: **8B**, 1-12.

KAHANE, J.P. 1989. Random multiplications, random coverings, and multiplicative chaos. *Proceedings of the Special Year in Modern Analysis*. Eds. E. Berkson et al. (London Mathematical Society Lecture Note Series: **137**). Cambridge University Press, 196-255.

KAHANE, J.P. 1991a. Produits de poids aléatoires indépendants et applications. *Fractal Geometry and Analysis (Montreal, PQ, 1989)*. Eds. J. Bélair & S. Dubuc. Dordrecht-Boston : Kluwer, 277-324..

KAHANE J.P. 1991b. in *La France mathématique, 1870-1914*. Ed. H. Gispert. *Cahiers d'histoire et de philosophie des sciences*: **34**, 277-297.

*N11 KAHANE, J. P. & MANDELBROT, B. B. 1965. Ensembles de multiplicité aléatoires. *Comptes Rendus* (Paris): **261**, 3931-3933. © *Académie des Sciences*.

*N17 KAHANE, J. P. & PEYRIÈRE, J. 1976. Sur certaines martingales de B. Mandelbrot. *Advances in Mathematics*: **22**, 131-145. © Academic Press.

KAHANE, J. P. & SALEM, R. 1963. *Ensembles parfaits et séries trigonométriques*. Paris: Hermann.

KESHNER, M. S. 1982. $1/f$ noise. *Proceedings of the IEEE*: **70**, 212-218.

KNOPP, K. 1918. Ein einfaches Verfahren zur Bildung stetiger nirgends differenzierbarer Funktionen. *Mathematische Zeitschrift*: **2**, 1-26.

KOGAN, Sh. 1996. *Electronic noise and fluctuations in solids*. Cambridge University Press.

KOLMOGOROV, A. N. 1941. The local structure of turbulence in incompressible viscous fluid for very large Reynolds numbers. *Comptes Rendus (Doklady) Académie des Sciences de l'URSS* (N.S.): **30**, 301-305. Reprinted in Friedlander & Topper 1961, 159-161.

KOLMOGOROV, A. N. 1962. A refinement of previous hypotheses concerning the local structure of turbulence in a viscous incompressible fluid at high Reynolds number. *Journal of Fluid Mechanics*: **13**, 82-85. Original Russian text and French translation in *Mécanique de la Turbulence* (Marseille, 1961), Paris: CNRS, 447-458.

KOLMOGOROV, A. N. 1985-1991. *Selected Works* (3 volumes). Dordrecht & Boston: Kluwer.

KOOSIE, P. 1988 - 92. *The Logarithmic Integral.* Cambridge University Press.

KUO, A. Y. S. & CORRSIN, S. 1972. Experiments on the geometry of the fine structure regions in fully turbulent fluid. *Journal of Fluid Mechanics*: **56**, 477-479.

LAMPERTI, J. 1958m. Some limit theorems for stochastic processes. *Journal of Mathematics and Mechanics*: **7**, 433-448.

LAMPERTI, J. 1958t. An occupation time theorem for a class of stochastic processes. *Transactions of the American Mathematical Society*: **88**, 380-387.

LANDAU, L. D. & LIFSHITZ, E. M. 1953-1959. *Fluid Mechanics.* London: Pergamon. Reading, MA: Addison Wesley.

LAX, P. 1984. *Proceedings of the International Congress of Mathematicians* (Warsaw, 1983). Ed. Z. Cieselski. Amsterdam: North Holland.

LERAY, J. 1934. Sur le mouvement d'un liquide visqueux emplissant l'espace. *Acta Mathematica*: **63**, 193-248.

LÉVY, P. 1925. *Calcul des probabilités.* Paris: Gauthier Villars.

LÉVY, P. 1937 & 1954. *Théorie de l'addition des variables aléatoires.* Paris: Gauthier Villars.

LEWIS, P. A. W. & COX, D. R. 1966. A statistical analysis of telephone circuit error data. *IEEE Transactions on Communication Technology*: **COM-14**, 382-389.

LI, W. & KANEKO, K 1992. Long-range correlation and partial $1/f$ spectrum in noncoding DNA sequence. *Europhysics Letters*: **17** (7), 655-660.

LOÈVE, M. 1955. *Probability Theory.* New York: Van Nostrand.

LORENZ, E. 1980. Noisy periodicity and reverse bifurcation. *Annals of the N.Y. Academy of Science*: **357**, 282-291.

*L LOVEJOY, S. & MANDELBROT, B. B. 1985. Fractal properties of rain, and a fractal model. *Tellus*: **A 37**, 209-232.

LOVEJOY, S. & SCHERTZER, D. 1983b. Buoyancy, shear, scaling and fractals. *Sixth Symposium on Atmospheric and Oceanic Waves and Stability* (Boston).

MANDELBROT, B.B. 1955b. On recurrent noise-limiting coding. *Information Networks, the Brooklyn Polytechnic Institute Symposium*: 205-221. E. Weber Ed. New York: Interscience. Translation into Russian.

MANDELBROT, B.B. 1955e. Théorie de la précorrection des erreurs de transmission. *Annales des Télécommunications*: **10**, 122-134.

MANDELBROT, B.B. 1956c. La distribution de Willis-Yule, relative au nombre d'espèces dans les genres taxonomiques. *Compte Rendus* (Paris): **242**, 2223-2225.

MANDELBROT, B.B. 1956w. On the language of taxonomy: an outline of a thermo-statistical theory of systems of categories, with Willis (natural) structure. *Information Theory,* the *Third London Symposium*. Ed. C. Cherry. London: Butterworth; New York: Academic, 1956, 135-145.

MANDELBROT, B.B. 1957p. Linguistique statistique macroscopique. *Logique, language et théorie de l'information* (avec Leo Apostel & Albert Morf). Paris: Presses Universitaires de France, 1-80.

*FE MANDELBROT, B. B. 1959p. Variables et processus stochastiques de Pareto-Lévy et la répartition des revenus, I & II. *Comptes Rendus* (Paris): **249**, 613-615 & 2153-2155.

*E10 MANDELBROT, B. B. 1960i. The Pareto-Lévy law and the distribution of income. *International Economic Review*: **1**, 79-106.

MANDELBROT, B. B. 1961b. On the theory of word frequencies and on related Markovian models of discourse. *Structures of language and its mathematical aspects*. Ed. R. Jakobson. New York: American Mathematical Society, 120-219.

*E11 MANDELBROT, B. B. 1961e. Stable Paretian random functions and the multiplicative variation of income. *Econometrica*: **29**, 517-543.

MANDELBROT, B. 1962c. Sur certains prix spéculatifs: faits empiriques et modéle basé sur les processus stables additifs de Paul Lévy. *Comptes Rendus* (Paris): **254**, 3968-3970.

*E12 MANDELBROT, B. B. 1962e. Paretian distributions and income maximization. *Quarterly Journal of Economics*: **76**, 57-85.

*E14,15 MANDELBROT, B.B. 1962i. *The Variation of Certain Speculative Prices*. IBM Research Report **NC-87**, March, 1962.

MANDELBROT, B. B. 1963. *Towards a Revival of the Statistical Law of Pareto*. IBM Research Report **NC-227**, March, 1963.

*E14 MANDELBROT, B. B. 1963b. The variation of certain speculative prices. *Journal of Business* (Chicago): **36**, 394-419. Reprint followed by discussions by Eugene F. Fama and Paul H. Cootner: *The Random Character of Stock Market Prices*. Ed. P.H. Cootner. Cambridge, MA: MIT Press, 1964: 297-337.

*E3 MANDELBROT, B. B. 1963e. New methods in statistical economics. *Journal of Political Economy* **71**, 421-440. Reprint in *Bulletin of the International Statistical Institute*, Ottawa Session: **40** (2), 669-720.

MANDELBROT, B. B. 1963o. *Oligopoly, Mergers, and the Paretian Size Distribution of Firms*. IBM Research Note: **NC-246**, March 1963.

*N5 MANDELBROT, B. B. 1964w. *Self-similar Turbulence and Non-Wienerian Conditioned Spectra*. IBM Research Report: **RC-134**.

*N8 MANDELBROT, B. B. 1965b. Time-varying channels, $1/f$ noises and the infrared catastrophe, or: why does the low frequency energy sometimes seem infinite. *IEEE Communication Convention*. Boulder, CO. © *Institute of Electrical and Electronics Engineers*.

*N7 MANDELBROT, B. B. 1965c. Self-similar error clusters in communications systems and the concept of conditional stationarity. *IEEE Transactions on Communications Technology*: **13**, 71-90. © *Institute of Electrical and Electronics Engineers*.

*H MANDELBROT, B. B. 1965h. Une classe de processus stochastiques homothétiques à soi; application à la loi climatologique de H. E. Hurst. *Comptes Rendus* (Paris): **260**, 3274-3277.

MANDELBROT, B. B. 1965m. Very long-tailed probability distributions and the empirical distribution of city sizes. *Mathematical Explorations in Behavioral Science* (Cambria Pines CA, 1964). Eds. F. Massarik & P. Ratoosh. Homewood, Ill.: R. D. Irwin, 322-332.

*N10 MANDELBROT, B. B. 1967b. Sporadic random functions and conditional spectral analysis; self-similar examples and limits. *Proceedings of the Fifth Berkeley Symposium on Math-*

ematical Statistics and Probability **3**, 155-179. Eds. L. LeCam & J. Neyman. Berkeley: University of California Press. © The University of California Press.

*N9 MANDELBROT, B. B. 1967i. Some noises with $1/f$ spectrum, a bridge between direct current and white noise. *IEEE Transactions on Information Theory*: **13**, 289-298. © IEEE.

*N12 MANDELBROT, B. B. 1967k. Sporadic turbulence. *Proc. International Symposium on Boundary Layers and Turbulence, including Geophysical Applications (Kyoto, 1966)*. Supplement to *The Physics of Fluids*: **10**, Sept. 1967, S302-303. © American Institute of Physics.

MANDELBROT, B. B. 1967s. How long is the coast of Britain? Statistical self-similarity and fractional dimension. *Science*: **155**, 636-638.

*N13 MANDELBROT, B. B. 1969b. On intermittent free turbulence: Abstract, followed by unpublished draft. *Proceedings of the Symposium on Turbulence of Fluids and Plasmas*. (Polytechnic Institute of Brooklyn). New York: Interscience.

MANDELBROT, B. B. 1969e. Long-run linearity, locally Gaussian process, H-spectra and infinite variances. *International Economic Review*: **10**, 82-111.

*H MANDELBROT, B. B. 1971f. A fast fractional Gaussian noise generator. *Water Resources Research*: **7**, 543-553. IMPORTANT ERRATA: in the first fraction on p. 545, 1 must be erased in the numerator and added to the fraction.

*E14 MANDELBROT, B. B. 1972b. Correction of an error in "The variation of certain speculative prices (1963)". *Journal of Business*: **40**, 542-543.

*N5 MANDELBROT, B.B. 1972f. Draft of M 1974f {N15}.

*N14 MANDELBROT, B. B. 1972j. Possible refinement of the lognormal hypothesis concerning the distribution of energy dissipation in intermittent turbulence. *Statistical Models and Turbulence*. Eds. M. Rosenblatt & C. Van Atta. Lecture Notes in Physics. New York: Springer, **12**, 333-351. © Springer-Verlag.

*H MANDELBROT, B.B. 1972w. Broken line process derived as an approximation to fractional noise. *Water Resources Research*: **8**, 1354-1356.

*N16 MANDELBROT, B. B. 1974c. Multiplications aléatoires itérées et distributions invariantes par moyenne pondérée. *Comptes Rendus* (Paris): **278A**, 289-292 & 355-358. © Académie des Sciences.

*N15 MANDELBROT, B. B. 1974f. Intermittent turbulence in self-similar cascades: divergence of high moments and dimension of the carrier. *Journal of Fluid Mechanics*: **62**, 331-358. © Cambridge University Press.

*H MANDELBROT, B. B. 1975f. On the geometry of homogeneous turbulence, with stress on the fractal dimension of the iso-surfaces of scalars. *Journal of Fluid Mechanics*: **72**, 401-416.

*N5 MANDELBROT, B. B. 1975O-1984O-1989O-1995O (*OF*). *Les objets fractals: forme, hasard et dimension*. Paris: Flammarion.

*N18 MANDELBROT, B. B. 1976o. Intermittent turbulence and fractal dimension: kurtosis and the spectral exponent $5/3 + B$. In *Turbulence and Navier Stokes Equations*. Ed. R. Temam. New York: Springer. © Springer-Verlag.

*N19 MANDELBROT, B. B. 1976c. Géométrie fractale de la turbulence. Dimension de Hausdorff, dispersion et nature des singularités du mouvement des fluides. *Comptes Rendus* (Paris): **282A**, 119-120. © Académie des Sciences.

MANDELBROT, B. B. 1977F. *Fractals: form, chance, and dimension*. San Francisco: W. H. Freeman & Co.

MANDELBROT, B. B. 1977l. Physical objects with fractional dimension: seacoasts, galaxy clusters, turbulence and soap. *The Institute of Mathematics and its Applications* (Great Britain)

Bulletin: **13**, 189-196. Also in *Fluid Dynamics – les Houches*. 1973. Eds. R. Balian & J. L. Peube. New York: Gordon & Breach, 557-578.

*N5 MANDELBROT, B. B. 1978h. Geometric facets of statistical physics: scaling and fractals. *Annals of the Israel Physical Society*. **2** (1), 225-233. © The Israel Physical Society..

*N5 MANDELBROT, B. B. 1982F *(FGN)*. *The Fractal Geometry of Nature*. New York: W. H. Freeman.

MANDELBROT, B.B. 1983i. Fractal curves osculated by sigma-discs, and construction of self-inverse limit sets. *Mathematical Intelligencer:* **5 (2)**, Front and back covers and pp. 9-17.

MANDELBROT, B. B. 1984e. Fractals and physics: squig clusters, diffusions, fractal measures and the unicity of fractal dimensionality. *Statistical Physics 15, International IUPAP Conference* (Edinburgh, 1983). Eds. D. Wallace & A. Bruce. *Journal of Statistical Physics:* **34**, 1984, 895-910.

*N5 MANDELBROT, B. B. 1984w. On fractal geometry and a few of the mathematical questions it had raised. *Proceedings of the Twelfth International Congress of Mathematicians* (Warsaw, 1983). Ed. Z. Ciesielski. Warsaw: PWN & Amsterdam: North-Holland, 1984, 1661-1675. © Polish Scientific Publishers.

MANDELBROT, B.B. 1985l. Self-affine fractals and fractal dimension. *Physica Scripta:* **32**, 257-260.

MANDELBROT, B.B. 1986k. Multifractals and fractals (letter to the Editor). *Physics Today* (September issue): 11-12.

MANDELBROT, B.B. 1986t. Self-affine fractal sets, I: The basic fractal dimensions, II: Length and area measurements, III: Hausdorff dimension anomalies and their implications. *Fractals in Physics (Trieste, 1985)*. Ed. L. Pietronero and E. Tosatti. North-Holland, 3-28.

*L MANDELBROT, B. B. 1988c. An introduction to multifractal distribution functions. *Fluctuations and Pattern Formation* (Cargèse, 1988). Eds. H. E. Stanley & N. Ostrowsky. Dordrecht-Boston: Kluwer, 345-360.

*L MANDELBROT, B. B. 1989e. A class of multifractal measures with negative (latent) values for the "dimension" $f(\alpha)$. *Fractals' Physical Origin and Properties* (Erice, 1988). (Ed.) L. Pietronero. New York: Plenum, 3-29.

*L MANDELBROT, B. B. 1989g. Multifractal measures, especially for the geophysicist. *Pure and Applied Geophysics*: **131**, 5-42 Also *Fractals in Geophysics*. Eds. C. H. Scholz & B. B. Mandelbrot. Boston: Birkhauser.

*L MANDELBROT, B. B. 1990d. New "anomalous" multiplicative multifractals: left-sided $f(\alpha)$ and the modeling of DLA. *Condensed Matter Physics, in Honor of Cyril Domb*. *Physica*: **A168**, 95-111.

*L MANDELBROT, B. B. 1990r. Negative fractal dimensions and multifractals. *Statistical Physics 17, International IUPAP Conference*. Ed. C. Tsallis. *Physica A:* **163**, 306-315

*L MANDELBROT, B. B. 1990t. Limit lognormal multifractal measures. *Frontiers of Physics: Landau Memorial Conference* (Tel Aviv, 1988). E. A. Gotsman, Y. Ne'eman & A. Voronel (Eds). New York: Pergamon, 309-340.

*L MANDELBROT, B. B. 1991k. Random multifractals: negative dimensions and the resulting limitations of the thermodynamic formalism. *Proceedings of the Royal Society* (London): **A434**, 79-88. Also in *Turbulence and Stochastic Processes: Kolmogorov's ideas 50 years on*. Eds. J. C. R. Hunt, O. M. Phillips & D. Williams. London: The Royal Society, 1991.

MANDELBROT, B. B. 1992h. Plane DLA is not self-similar; is it a fractal that becomes increasingly compact as it grows? Physica: **A 191**, 95-107.

*L MANDELBROT, B. B. 1993s. The Minkowski measure and multifractal anomalies in invariant measures of parabolic dynamic systems. *Chaos in Australia* (Sydney, 1990). Eds. G. Brown & A. Opie. Singapore: World Publishing, 1993, 83-94. Slightly edited reprint: *Fractals and Disordered Systems*. Second edition. Eds. A. Bunde & S. Havlin. New York: Springer: 1995.

*L MANDELBROT, B. B. 1995f. Measures of fractal lacunarity: Minkowski content and alternatives. *Fractal Geometry and Statistics*. Eds. C. Bandt, S. Graf & M. Zähle. Basel & Boston: Birkhauser. 12-38.

*H MANDELBROT, B. B. 1995h. Introduction to fractal sums and pulses. *Lévy Flights and Related Phenomena in Physics*. (Nice, 1994). Eds. G. Zaslawsky, M. F. Shlesinger & U. Frisch. New York: Springer, 110-123.

*L MANDELBROT, B. B. 1995k. Negative dimensions and Hölders, multifractals and their Hölder spectra, and the role of lateral preasymptotics in science. *J. P. Kahane's meeting* (Paris, 1993). Eds. A. Bonami & J. Peyrière. *The Journal of Fourier Analysis and Applications*: special issue, 409-432.

MANDELBROT, B. B. 1997E. *Fractals and Scaling in Finance: Discontinuity, Concentration, Risk*. New York: Springer-Verlag.

MANDELBROT, B.B. 1997FE. *Fractales, hasard et finance*. Paris: Flammarion.

MANDELBROT, B.B. 1998e. Fractality, lacunarity and the near-isotopic distribution of galaxies. *Current Topics in Astrofundamental Physics*. Eds. N. Sanchez & A. Zichichi. Dordrecht: Kluwer, 585-603.

H MANDELBROT, B. B. 1998H. *Gaussian Fractals & Kin: R/S, 1/f, Globality Reliefs & Rivers*. New York: Springer-Verlag.

L MANDELBROT, B.B. 1998L. *Multifractals and Lacunarity*. New York: Springer-Verlag.

MANDELBROT, B. B., CALVET, L & FISHER, A.1997. *The Multifractal Model of Asset Returns; Large Deviations and the Distribution of Price Changes; The Multifractality of the Deutschmark/US Dollar Exchange Rate*. New Haven CT: Cowles Foundation Discussion Papers 1164, 1165 and 1166. These reports are available on the Internet under the following three addresses: First paper: http://papers.ssrn.com/sol3/paper.taf?ABSTRACT_ID = 78588. Second paper: the same except that ID = 78608. Third paper: the same except that ID = 78628.

MANDELBROT, B. B. & EVERTSZ, C. J. G. 1990. The potential distribution around growing fractal clusters. Nature: **378** (6296), front cover & pp. 143-145.

L MANDELBROT, B. B. & EVERTSZ, C. J. G. 1991n. Exactly self-similar multifractals with left-sided $f(\alpha)$. *Fractals and Disordered Systems*. Eds. A. Bunde & S. Havlin. 323-346.

MANDELBROT, B.B. & RIEDI, R.H. 1995. Multifractal formalism for infinite multinomial measures. *Advances in Applied Mathematics*: **16**, 132-150.

MANDELBROT, B.B. & RIEDI, R.H. 1997. Inverse measures, the inversion formula, and discontinuous multifractals. *Advances in Applied Mathematics*: **18**, 50-58.

*E21 MANDELBROT, B.B. & TAYLOR, H. 1967. On the distribution of stock price differences. *Operations Research*: **15**, 1057-1062.

*H MANDELBROT, B. B. & VAN NESS, J. W. 1968. Fractional Brownian motions, fractional noises and applications. *SIAM Review*: **10**, 422-437.

*H MANDELBROT, B. B. & WALLIS, J. R. 1969b. Some long-run properties of geophysical records. *Water Resources Research*: **5**, 321-340.

*H MANDELBROT, B. B. & WALLIS, J. R. 1969c. Robustness of the rescaled range R/S in the measurement of noncyclic long-run statistical dependence. *Water Resources Research*: **5**, 967-988.

● Additional papers co-authored by Mandelbrot found under Berger, Cioczek-Georges, Kahane, Lovejoy and Riedi.

MATHERON, G. 1962. *Traité de Géostatistique Appliquée*. Tome 1, Paris: Technip.

MENGER, K. 1932. *Curventheorie*. Berlin: Springer.

MERTZ, P. 1961. Statistics of hyperbolic error distributions in data transmission. *IRE Trans. on Communications Systems*: **CS-9**, 377-382.

MEYER, E. & THIEDE, H. 1935. Widerstandsschwankungen duenner Kohleschichten. *Elektrische Nachrichten-Technik*: **12**, 237-42.

MEYER, P. A. 1966. *Probabilités et potentiels*. Paris: Hermann.

MONIN, A. S. & YAGLOM, A. M. 1963. On the laws of small scale turbulent flow of liquids and gases. *Russian Mathematical Surveys* (translated from the Russian): **18**, 89-109.

MONIN, A. S. & YAGLOM, A. M. 1971 & 1975. *Statistical fluid mechanics, Volumes 1 and 2* (translated from the Russian). Cambridge, MA: M.I.T. Press.

MORIARTY, B. J. 1963. Blocks in error and Pareto's law. *Lincoln Laboratory of M.I.T.*: Report 255 or 25 G-15.

NELKIN, M. 1984. Review of "The Fractal Geometry of Nature" by B.B. Mandelbrot. *American Meteorological Society Bulletin*: **63**.

NOVIKOV, E. A. 1969. Scale similarity for random fields. *Doklady Akademii Nauk SSSR*: **184**, 1072-1075. (English trans. *Soviet Physics Doklady*: **14**, 104-107.)

NOVIKOV, E. A. 1971. Intermittence and scale similarity in the structure of a turbulent flow. *Prikladnaia Matematika i Mekhanika*: **35**, 266-277. English in *P.M.M. Applied Mathematics and Mechanics*

NOVIKOV, E. A. & STEWART, R.W. 1964. Intermittence of turbulence and the spectrum of fluctuations of energy dissipation (in Russian). *Isvestia Akademii Nauk SSR* Seria Geofizicheskaia: **3**, 408-413.

OBUKHOV, A. M. 1941. On the distribution of energy in the spectrum of turbulent flow. *Comptes Rendus (Doklady) Académie des Sciences de l'URSS* (N.S.): **32**, 22-24.

OBUKHOV, A. M. 1962. Some specific features of atmospheric turbulence. *Journal of Fluid Mechanics*: **13**, 77-81. Also in *Journal of Geophysical Research*: **67**, 3011-3014.

OLSEN 1996. This global reference calls for a large collection of preprints and hard-to-find articles from Olsen & Associates in Zürich. The contributing authors include the following, listed alphabetically: B. Chopard, M. M. Dacarogna, R. D. Davé, C. G. de Vries, C. L. Gauvreau, D. M. Guillaume, C. Jost, C. Morgenegg, U. A. Müller, R. J. Nagler, M. Oudsaidene, O. V. Pictet, R. Schirru, M. Schwarz, M. Tomassini, J. E. von Weizsäcker, J. R. Ward. Their work can be printed from "the web".

ORSZAG, S. A. 1970. Indeterminacy of the moment problem for intermittent turbulence. *Physics of Fluids*: **13**, 2211-2212.

PEYRIÈRE, J. 1974. Turbulence et dimension de Hausdorff. *Comptes Rendus* (Paris): **278A**, 567-569.

PIRANIAN, G. 1989. A review of "Invitation to Complex Analysis", by R. P. Boas. *American Mathematical Monthly*: **96**, 376-378.

POND, S. & STEWART, R. W. 1965. Measurements of the statistical characteristics of small scale turbulent motion. *Izvestia Akademii Nauk SSSR, Fisika* Atmosfery i Okeana: **12**, 914-919.

PRESS, W. 1978. Flicker noise in astronomy and elsewhere. *Comments on Astronomy*: **7**, 103-119.

RÉNYI, A. 1955. On a new axiomatic theory of probability. *Acta Mathematica Hungarica*: **6**, 285-335.

RÉNYI, A. 1959. On the dimension and entropy of probability distributions. *Acta Mathematica Hungarica*: **10**, 193-215.

RÉNYI, A. 1976. *Selected Papers*. Ed. P. Turan. Budapest: Akademiai Kiado.

RICHTERS, J. S. 1965. Applications of Pareto error statistics to Hagelberger Codes, *IEEE Transactions of Information Theory*: **IT-11**, 571-596.

RIEDI, R. & MANDELBROT, B.B., 1998. Exceptions to the multifractal formalism for discontinuous measures. *Mathematical Proceedings of the Cambridge Philosophical Society*: **123**, 133-157.

ROSENBLATT, M. & VAN ATTA, C. (Eds.) 1972. *Statistical models and turbulence*. Lecture Notes in Physics: **12**. New York: Springer.

RUDIN, W. 1960. Fourier-Stieltjes transforms of measures of independent sets. *Bulletin of the American Mathematical Society*: **66**, 199-204.

SALEM, R. 1951. Lacunary power series and Peano curves. *Arkiv för Matematik*: **1**, 353-365.

SANDBORN, V. A. 1959. Measurements of intermittency of turbulent motion in a boundary layer. *Journal of Fluid Mechanics*: **6**, 221-240.

SHOHAT, J.A.O. TAMARKIN, J. D. 1943. *The Problem of Moments*. Providence, RI: American Mathematical Society.

SIGGIA 1977.

SMITH, W. L. 1960. Infinitesimal renewal processes. *Contributions to Probability and Statistics, in honor of Harold Hotelling*. Stanford: Stanford University Press, 396-413.

STAUFFER, D. & AHARONY, A. 1992. *Introduction to Percolation Theory*. Second edition. London: Taylor & Francis.

STENT, G. 1978. *Paradoxes of Progress*. New York: W.H. Freeman.

STIELTJES, T. J. 1914. *Oeuvres complètes de Thomas Jan Stieltjes*; 2 vols. Groningen: Noordhoff. Reprinted as *Collected Papers*. Ed. G. Van Dijk. New York: Springer, 1993.

SUSSMAN, S. M. 1963. Analysis of the Pareto model for error statistics on telephone circuits. *IEEE Transactions on Communication Systems*: **C5-11**, 213-221.

TAQQU, M. S. 1975. Weak convergence to fractional Brownian motion and to the Rosenblatt process. *Z. für Wahrscheinlichkeitstheorie*: **31**, 287-302.

TAQQU, M. S. 1979. Convergence of integrated processes of arbitrary Hermite rank. *Z. für Wahrscheinlichkeitstheorie*: **50**, 53-83.

VAN DER ZIEL, A. 1970. *Noise: sources, characterization, measurement*. Englewood Cliffs, NJ: Prentice-Hall.

VAN DER ZIEL, A. 1987. Noise in electronic circuits. *Encyclopedia of Physical Science and Technology*: New York: Academic Press, 47-60.

VERVEEN, A. A. & DERKSEN, H. E. 1968. Fluctuation phenomena in nerve membranes. *Proceedings of the IEEE Institute of the Electronics and Electrical Engineers*. **56**, 906-916.

VON NEUMANN, J. 1949. Recent theories of turbulence. First printed in *Collected Works*. Ed. A. H. Traub. New York: Pergamon, 1963: **6**, 437-472.

VOSS, R. F. & CLARKE, J. 1975. $1/f$ noise in music and speech. *Nature*: **258**, 317-318.

VOSS, R. G., CLARKE, J. 1976. Flicker $(1/f)$ noise: equilibrium temperature and resistance fluctuations. *Physics Review B*: **13**, 566.

VOSS, R. F. & CLARKE, J. 1978. $1/f$ noise in music: music from $1/f$ noise. *Journal of the Acoustical Society of America*: **63**, 258-263.

VOSS, R. F. 1978a. Linearity of $1/f$ noise mechanism. *Physical Review Letters*: **40**, 913-916.

VOSS, R. F. 1978b. $1/f$ Noise and percolation in impurity bands in inversion layers. *Journal of Physics C: Solid State Physics*: **11**, L923.

VOSS, R. F. 1992. Evolution of long-range fractal correlations and $1/f$ noise in DNA base sequences. *Physical Review Letters*: **68**, 3805-3808.

WIENER, N. 1930. Generalized harmonic analysis. *Acta Mathematica*: **55**, 117-258.

WILLIAMS, L. & BURDETT, R. K. 1969. *Journal of Physical Chemistry*: **2**, 298-307.

WILSON, K.G. 1972. Feynman-graph expansions for critical exponents. *Physical Review Letters*: **28**, 548-551.

WILSON, K.G. & FISHER, M.E. 1972. Critical exponents in 3.99 dimensions. *Physical Review Letters*: **28**, 240-243.

WITTEN, T. A. & SANDER, L. M. 1981. Diffusion-limited aggregation, a kinetic critical phenomenon. *Physical Review Letters*: **47**, 1400-1403.

YAGLOM, A. M. 1966. The influence of fluctuations in energy dissipation on the shape of turbulence characteristics in the inertial interval. *Doklady Akademii Nauk SSSR*: **16**, 49-52. (English trans. *Soviet Physics Doklady*: **2**, 26-29.)

ZYGMUND, A. 1959. *Trigonometric Series*. 2nd edition, Cambridge University Press.

ZOLOTAREV, V. 1986. *One-Dimensional Stable Distributions*. Providence, RI: American Mathematical Society.

Index

α, 89
α_1, 319, 323, 325
α_3, 326
$\mathcal{A}$, 251
absolute curdling, 81–83, 124–125, 390–391, 392(figure), 397, 405
astronomy
 —clustering of galaxies, 123, 346
 —sunspots, 76
Aulin, T., 162

β, 362, 369
"β model," 65, 82–83, 125
b (base), 35–37, 41, 81–83, 358
$\mathcal{B}$, 252
B (correction to 5/3 and 2/3 laws), 402–408
B (scaling exponent), 20, 27(figure), 54, 74, 76–79. See also scaling exponent
$B_H(t)$, 35, 38–39, 42
"baby theorem," 19, 34(figure), 47, 52, 53(figure), 54(figure)
Barnsley, M., 130
Barron, M. E., 160–161, 161(figure)
base, 35–37, 41, 81–83, 358
Berger and M 1963 model, 81–82, 107–108, 132–165, 251
Berger and M 1965 model, 166–204
Berger, J. M., 69
Bernoulli measure. See binomial measure
Bernoulli random variables, 298, 360
Besicovitch, A. S., 83, 89–90
Besicovitch measure, 363
bifurcation theory, 106–107
binary symmetric channels without memory, 132–133, 169

binomial measure, 18, 21, 40, 83–84, 89–91
binomial random variables, 169, 298, 339, 350, 397
 —and distribution of inter-error intervals, 136, 137(figure)
biology
 —chronic illness, 75
 —nerve membranes, 75, 216
 —taxonomy, 108
Birkhoff's ergodic theorem, 236, 238, 272
birth-and-death process, 325–326, 340, 359–360, 374
blocks. See boxes
Bochner, S., 32
Bolzano, B., 40, 91
Borel, E., 134–135
Borel field, 251
Borel sets, 374, 375, 387
box dimension, 364–366
boxes, 35–37, 39, 42–46, 179
 —boxes in error and inter-boxes, 167
 —and conditioned laws, 173–174
 —k-boxes, 167
 —and renormalization group theory, 167–168
 —and self-similarity, 173
 —unibox vs. multibox constructions, 44
 —See also generators
Brodén, T., 90–91
Brownian motion. See fractional Brownian motion; Wiener Brownian motion
Burgers turbulence, 407

$C(\alpha)$, 371
$C_W(t, n)$, 222
Callen, H., 103

canonical cascades, 326–327, 372–389
 —defined, 332–334
 —and limit lognormal processes, 328
 —moments of eddy averages, 335–337
canonical multifractal measure, 67
Cantor devil staircase, 40, 77, 92
Cantor dust, 28, 40, 81–83, 92–93, 284
Cantor, G., 93, 284–285
Cantor set, 70
carbon resistors, 75
cartoons, 21, 22–28
 —additive vs. multiplicative operations, 32, 58–59
 —"baby theorem," 19, 34(figure), 47, 52, 53(figure), 54(figure)
 —cartoon embedding, 49–51
 —compound representations, 47–52
 —diagonal axial self-affinity, 22, 32, 33(figure), 34(figure), 36(figure)
 —dimension-generating equation, 48–49
 —graph dimension of, 54
 —intrinsic time of, 51–52
 —of Lévy-stable motion, 39
 —mesofractal, 44–45, 51–52
 —multifractal, 45
  —intrinsic time of, 51–52
 —oscillating, 49
 —oscillating constructions, 47, 49
 —unifractal, 23, 43–44, 48, 51–52, 53(figure)
  —fractional Brownian motion, 23, 37–38, 47, 53(figure)
  —intrinsic time of, 51–52

INDEX   433

—visual perception of, 29, 30(figure)
—of Wiener Brownian motion, 23, 25(figure), 26(figure), 27(figure), 33(figure), 35–38, 36(figure), 47
—*See also* generators; grid-bound recursive constructions
cascades, 67, 81–83, 120, 317–357
—"β model" term, 65, 82–83, 125
—Cantor dust, 81–82, 92–93, 284
—cascade base, 81–83, 358
—conservative. *See* microcanonical cascades
—curdling. *See* curdling
—de Wijs–Yaglom cascade arguments, 300–302, 320–332
—degenerate classes, 344
—and determining functions, 337–343
—direct (intepolative), 369–371
—intrinsic dimension of the carrier, 338, 345–347
—inverse (extrapolative), 369–371
—irregular (nondegenerate) class, 344
—macrocanonical. *See* macrocanonical cascades
—microcanonical. *See* microcanonical cascades
—mixed classes, 344–345
—moments of eddy averages, 335–337, 343–344
—multiplicative factors, 329–331
—regular classes, 343–344
Cauchy, A., 104–105
Cauchy distribution, 105
Cauchy functional equation, 104–105
Cauchy-Lévy stability, 110
Cesaro triangle sweep, 39
Chernoff inequality, 367
city size, 108–109
clustering of errors. *See* error clustering
clustering of galaxies, 123, 346
co-indicator functions, 60–61, 115–116, 215–245, 266–267
—average noise energy and average dc component, 236–238

—conditioned, 223, 230, 234–236, 238
—covariance properties of, 223, 225, 233, 234, 266–267
—deltavariance of, 274–276
—"derivative" of, 216, 227
—ergodicity of, 224–225, 236
—for $E(U) < \infty$, 233–234
—examples of variants, 242–244
—for f^{-2} functions, 222–226
—for general renewal processes, 226–229
—and infrared catastrophe, 215–216
—"locally Gaussian" sums of, 242–244
—multifractals as products of, 244–245
—orthogonal, 267
—and self-similarity, 267–269
—Wiener-Kinchin spectra of, 230
—and wild randomness, 238–240
coarse-graining, 66, 78–79
coastlines, 21, 28, 71, 345, 396
codimension, 83, 371, 392–393, 399
coding theory, 146–147, 195
coin function. *See* co-indicator functions
coin tossing, 134–135, 135(figure), 142, 144, 174–175
communication circuits, 132–133, 156, 169–170, 177. *See also* error clustering; telephone circuits
compound cartoon representations, 47–52
condensed matter physics, 38
conditional covariance, 220, 223, 225, 265
conditional covariance stationarity, 220–221, 223, 265
conditional deltavariance, 263–264, 267–269
conditional restricted stationarity, 255
conditional self-similarity, 267–271
conditional spectra, 7, 223, 247–284
—of co-indicator function, 230
—of conditionally covariance stationary random function, 220–221

—and f^{-1} functions, 212–213
—and f^{-2} functions, 222–226
—and f^{D-2} functions, 217–221
—of indicator function, 230
—for intermittent turbulence, 291
—and low frequency divergence, 78
—of sporadic generalized random functions, 221, 277–280
conditional stationarity, 220, 254–259
—defined, 176–177
—examples of, 256–257
—and measure preservation, 254–259
conditional Taylor's scale, 269
conditioned sample moments, 272
conditioning
—of co-indicator function, 223, 234–236, 238
—conditioning events, 254
—conditioning length, 78
—for error clustering, 173–174
—of indicator function, 234–235
—and infrared catastrophe, 208–214
conservative cascades. *See* microcanonical cascades
core function. *See* co-indicator function
correlation dimension, 116, 339, 389
correlation exponent, 56
correlation function, 3
Corrsin, S., 401
covariance
—of co-indicator function, 266–267
—conditional, 220, 223, 225, 265
—for limit lognormal random processes, 301
—of stationary ordinary random function, 263
—unweighted, 265
covariance-stationary random functions, 219–221
Cox, D. R., 157–160
Cramèr, H., 64
Cramèr theory of large deviations, 86, 371
"creasing" of grid-bound constructions, 31
curdling, 390–397, 398(figure), 403–414

—absolute, 81–83, 124–125, 390–391, 392(figure), 397, 405
—and determining function, 397, 412–413
—inner scale of, 409–414
—in spaces of Euclidean dimension $E > 3$, 411–412
—weighted, 394, 397, 403, 406–407, 412–413
current, 218, 222–226
Cvitanovic-Feigenbaum equation, 106–107

D, 173–174, 291, 292. See also scaling exponent
$\mathcal{D}$, 317
D_2, 116
D_E, 338, 345–347
D_{elna}, 72
D_G, 54, 55
D_{HB}, 72
D_{heur}, 72
D_T, 27(figure), 48–51, 52, 56
$D(q)$, 87, 98–99
Darwin-Fowler method of steepest ascents, 86
data transmission, 160–161, 161(figure), 162(figure), 165, 165(figure). See also communication circuits; error clustering; telephone circuits
de Gennes, P.-G., 103
de Wijs, H. J., 88, 297, 318
de Wijs–Yaglom cascade arguments, 300–302
decomposable dynamical systems, 125–126, 127(figure), 130
deltavariance, 263–264, 271–278
derivative, meaning in physics vs. meaning in mathematics, 79
determining functions, 337–343, 397–408, 398(figure)
—asymptotically rectilinear, 340
—and dissipative range, 399
—and inertial range, 399
—linear, 339–340, 397, 400
—modifications to 5/3 and 2/3 laws, 402–408
—parabolic, 340, 342, 347–348, 397
—and second-order (spectral) properties, 399
—sensitivity to W, 397–399
—$-\Psi(2)$, 397, 400–402
—$-\Psi(2/3)$, 397, 399, 402–408
—$-\Psi'(1)$, 397, 399

—$-\Psi''(1)$, 397, 399, 412–413
—See also $\tau(q)$
devil staircase, 40, 77, 92
diagonal self-affinity, 22, 32, 33(figure), 34(figure), 36(figure), 80
diffusion
—Fickian, 5–6, 23, 35, 36, 76
—non-Fickian, 20, 35, 76
dimension, 39, 87, 98–99
—codimension, 83, 371, 392–393, 399
—correlation dimension, 76, 116, 389
—dimension-generating equation, 48–49, 51
—entropy-like, non-averaged, 72
—Fourier dimension, 285
—fractal. See fractal dimension
—generalized from sets to measures, 98–99
—graph dimension, 54, 55
—Hausdorff dimension, 72, 90, 285, 365, 375, 391, 415, 418
—heuristic dimension, 72
—information dimension, 88, 90, 389
—mass dimension, 82
—negative, 71–72
—pointwise, 89
—similarity dimension, 82, 124–125, 393(figure)
—trail dimension, 27(figure), 48–51, 54, 57
Dimock, P. V., 165, 165(figure)
discontinuous mesofractals, 45
discontinuous multifractals, 45, 54(figure)
dispersion, turbulent, 416–418
dissipation. See turbulent dissipation
dissipative range, 403, 409–414
domain of attraction, 105–106, 271, 371
Domb, C., 90
Doob, J. L., 302
Durrett, R., 364
dustborne noise, 39–40, 79–80, 115–116. See also Lévy dust; white noise

ε, 167
ε_r, 291, 403
$\varepsilon(x, r, \eta, L)$, 297
η, 297, 317, 394, 409–414
economics. See finance
eddies, 317–323
—concentration of dissipation in, 350–351
—dissipation as energy transfer between eddy sizes, 326
—and fractional intrinsic dimenstion, 345
—and Kolmogorov's "second hypothesis of similarity," 327–328
—and lognormality, 322–323
—moments of eddy averages, 335–337
—shape of domain, 320, 322–323, 326–327, 350–352, 403–406, 407–408
—and turbulent dispersion, 417
embedding (cartoons), 49–51
entropy-information, 90, 352
Ephremides, A., 161, 162(figure)
ergodicity, 224–225, 236, 238, 272
error clustering, 107–108, 132–168
—and coin tossing, 134–135, 135(figure), 174–175
—competing models of, 169
—and conditioning, 173
—distribution of inter-boxes, 179–181
—distribution of sums of successive inter-boxes, 201–202
—early contributors, 157–165
—empirical verifications, 190–191, 202–203
—error generators, 171
—excess noise, 182–183
—generation of pathological distributions through mixtures of nonpathological random variables, 149–150, 183
—illustrations of, 138–140, 140(figure), 141(figure), 142(figure), 168(figure), 202–203
—infinite message length, 146–147
—and infrared catastrophe, 209
—inter-error interval, 133–134, 136, 137(figure)
—interdependence of inter-boxes, 192–203, 198–201
—joint probability distribution, 195–198
—marginal distribution of inter-boxes, 184–191
—mean duration of a burst, 181

INDEX

—mechanism of cluster
   generation, 181–182
—population moments of
   inter-error interval,
   142–144
—and self-similarity, 166–
   204, 173–174
—and Shannon's information
   theory, 194–195
—tail behavior of, 156–158,
   167
Euclidean space, 83
Euler equations, 395–396
Euler, L., 22
exponential time, 114–115

F (prefactor), 112
$f(\alpha)$, 84
—graph of, 86–87
—historical sources, 95–98
—Legendre transform
   construction, 86–87
$F(\omega)$, 358
$P(x, \lambda, L)$, 301
Feller, W., 144, 192, 231, 239,
   274
Fick, A. E., 76
Fickian diffusion, 5–6, 23, 35,
   36, 76. See also white
   noise
Fickian property, 23, 37
filtered functions, 110–117
finance, 3, 23–24, 29, 30(figure),
   58, 68, 109–110
—lognormality and
   multiplicative effects,
   314
—Noah Effect, Joseph
   Effect, 35, 244
—Pareto's law, 314
—privileged statistical
   models, 108
firm size, 108
fixed points
—and diagonal self-affinity,
   80
—and Gaussian processes,
   106
—as "privileged models,"
   107–110
—of smoothing
   transformations of
   probability
   distributions, 128
fluid equations, 395–396, 417–
   418
Fourier dimension, 285
Fourier transform, 104–105, 402
fractal codimension, 28–29
fractal dimension, 19, 28
—and Hurst-Hölder
   exponent, 38–39
—and absolute curdling,
   391, 405

—and behavior of cross-
   sections of a fractal,
   394–395
—in Corrsin's model, 401
—$D = 0$ limit, 405
—and determining function,
   397
—and embedding, 51
—experimental values, 400–
   401
—forms of, 39
—geometric interpretation of,
   345–347
—and intermittent
   turbulence, 123–125,
   338, 345–347
—as metric, not topological,
   characteristic, 393
—and singularities of fluid
   motion, 417–418
—and telephone circuits,
   143
—in Tennekes' model, 401
fractal homogeneity, 390–396
fractal sums of pulses, 59–60
fractal time, 40, 48, 51–52
fractality vs. randomness, 31–32
fractals
—additive operations, 32
—behavior of cross-sections,
   394–395
—and curdling, 390–396,
   392(figure)
—elusive meaning of term,
   14–15
—Gaussian, 38
—and Navier-Stokes and
   Euler equations, 395–
   396
—origin of term, 8–9
fractional Brownian motion, 23,
   32, 35–38, 52, 54, 57–
   58, 69
—and cartoon embedding,
   49–50
—cartoons of, 37–38, 47,
   53(figure)
—and fractal sums of pulses,
   60
—specification of, 42
Fréchet functional equation, 106
Fréchet, M., 106
Frenkiel, F. N., 293
Frisch, U., 71, 72, 82, 96–97,
   125
functional equation for
   martingales, 373, 388

$G(x)$, 113
galaxies, clustering of, 123, 346
gamma random variables, 349,
   388
Gauss-Markov process, 74
Gaussian limit theorem, 5
Gaussian processes, 38

—fixed points, 106
—fractional Brownian
   motion. See fractional
   Brownian motion
—fractional Gaussian
   noises, 35
—and generalized derivative
   of Wiener Brownian
   motion, 79
—Hermite polynomials as
   nonlinear transforms,
   114
—linear transformation
   ratios, 74
—"locally Gaussian" sums of
   co-indicator functions,
   242–244
—low frequency divergence
   as indicator of non-
   stationarity, 78
—non-stationary processes
   with stationary
   increments, 74
—scaling exponent for, 76
—stationary stochastic
   processes, 74
—visual perception of, 79–
   80
—volatility of, 80
—white noise, 25(figure)
Gelfand, I. M., 250–251
generalized random functions
—defined, 252
—degrees of intermittency,
   259
—and Schwartz
   distributions, 251
—See also sporadic
   generalized random
   functions
generators, 22
—for cartoons of fractional
   Brownian motion,
   53(figure)
—for cartoons of Lévy-stable
   motion, 39
—for cartoons of Wiener
   Brownian motion, 22,
   24(figure), 24–25,
   25(figure), 35–38,
   36(figure)
—defined, 41
—for discontinuous
   multifractals, 45,
   54(figure)
—initiator, 22, 35, 41, 43
—for intermittent
   multifractals, 45,
   54(figure)
—for intermittent $1/f$ noise,
   34(figure), 39–40
—iterated generator
   systems, 46
—and lacunarity, 46

–for mesofractal cartoons, 44–45
–miscellaneous examples, 40
–for multifractal cartoons, 45, 48, 53(figure), 54(figure)
–and multifractal intrinsic time, 40
–nondecreasing, 45, 48
–oscillating, 45
–randomized, 46
–unibox vs. multibox constructions, 44
–for unifractal cartoons, 43–44, 48, 53(figure)
geometric distribution, 132, 136, 137(figure), 144(figure), 169
geometric measure theory, 91–92
geometry, 9–10
–and fractional intrinsic dimension, 345–347
geomorphology, 296, 318
Gibert, E. N., 169
Gibrat, R., 314
global statistical dependence, 23
globality of wild randomness, 3–4
graph dimension, 54, 55
grid-bound recursive constructions, 18–61
–construction method, 41–42
–and "creasing," 31
–with diagonal generators, 35–38
–and "dustborne" intermittent $1/f$ noises, 34(figure), 39–40
–and intrinsic time, 40
–Lévy-stable motion, 39
–and randomness, 31–32
–See also cartoons
grid-free recursive constructions, 18–19, 29, 31. See also limit lognormal random processes
Guivarc'h, Y., 364
Gurvich, A. S., 296–297
Gurvich-Zubkovskii correlation, 297, 299

H. See Hurst-Hölder exponent
H_G, 52, 54, 55
H_h, 42–46, 49
H_T, 19, 48–49
Hadamard, J., 81
Halsey, T. C., 73, 84
Hardy, G. H., 90
Hausdorff dimension, 72, 90, 96–97, 285, 365, 375, 391, 415, 418

Hausdorff, F., 90, 345
Hermite polynomials, 97, 114
Hermite rank of nonlinear transform, 97
heuristic dimension, 72
high frequency divergence, 77–79, 208, 221
historical development of methods, 64–73, 101–110
Hölder, L. O., 29, 89
Hurst, H. E., 29, 76
Hurst-Hölder exponent, 19, 28–29, 42–46, 50, 71–72, 76, 407
–and fractal dimension, 38–39
–for Wiener Brownian motion, 35
Hurst puzzle, 69, 76, 244

IFS. *See* iterated function systems
income, personal, 108
indicator function, 215–216, 225–227, 230–240
–average noise energy and average dc component, 236–238
–conditioned, 230, 234–235
–covariance of, 231, 234
–ergodic theorem for, 236
–for $E(U) < \infty$, 231–233
–for general renewal processes, 226–227
–as rectified form of the "derivative" of the co-indicator function, 216, 227
–spectral density of, 230–231
–and wild randomness, 238–240
information dimension, 88, 389
–*See also* error clustering; telephone circuits
information theory
–infinite message length, 146–147
–noiseless transmission distinguished from theory with vanishingly small noise, 195
–Shannon's, 194–195
–word length, 195
infrared catastrophe, 77–79, 208–216, 220, 278–279
–and co-indicator function, 215–216
–conditioning for, 78, 208–214, 216–217
–and intermittent turbulence, 249

–and Wiener-Kinchin theory, 210–213, 217
initiator, 22, 35, 41, 43
inner scale, 297, 317, 394, 409–414
inter-boxes, 179–181, 184–203
–distribution laws, 179–180, 189, 191
–distribution of sums of, 201–202
–doubly logarithmic plots, 168(figure)
–interdependence of, 192–203
–marginal distribution of, 184–191
–renormalization of, 167
–tail distribution of, 167
inter-error interval, 170, 174–178
–average information required for specifying, 146
–Barron's figure, 161(figure)
–and conditional stationarity, 174–175
–Dimock's figure, 165(figure)
–distribution of, 180–182
–binomial distribution, 136, 137(figure), 181
–exponential distribution, 182
–Gaussian distribution, 182
–geometric distribution, 136, 137(figure), 169
–joint probability distribution, 140–141
–scaling distribution, 136, 137(figure)
–self-similar distribution, 182
–truncated self-similar law, 181
–doubly logarithmic plots, 138–140, 140(figure), 140–141, 141(figure), 142(figure), 162(figure), 164(figure), 190–191
–Ephremides and Snyder's figure, 162(figure)
–infinite average duration of, 174–175
–information required for specifying, 194
–Mertz's figure, 164(figure)
–Moriary's figure, 158(figure)
–population moments, 142–144

INDEX

437

—Stuck and Kleiner's figure, 163(figure)
—successive error-free intervals, 158–160
—Sussman's figures, 159(figure)
intermittent mesofractals, 45
intermittent multifractals, 34(figure), 39–40, 45, 54(figure). *See also* turbulent dissipation
intermittent turbulence. *See* turbulent dissipation
intrinsic time. *See under* time
inverse Legendre transform, 86–87, 96
irregularity, fractals as measurements of, 28–29
isochrone time, 19, 50, 56
iterated function systems (IFS), 66, 126–127, 130
iterated generator systems, 46
iterated random multiplications, 358–372
—and Besicovich measure, 363
—birth-and-death process, 359–360
—and box dimension, 364–366
—Chernoff inequality, 367
—conservation relations, 363–364
—convergence conditions for, 361–362
—degeneracy conditions, 364
—divergent moments of, 363
—fixed point property of the limit measure, 360–361
—hyperbolically distributed weights, 367
—and Legendre transform, 371–372
—limit laws, 364–366
—martingale property, 361–362
—multiplicative measures, 358–359
—normalized, 360, 366–367
—q_{crit} and convergence of moments, 362
—recursion relations, 360
—renormalized, 352, 370
—symmetric binomial case, 359–360
—tail behavior of, 369, 371

Jensen, M. H., 73
joint probability distribution
—of first and last errors in a box, 195–198
—and renewal processes, 282–283

—of successive inter-error intervals, 140–141, 145(figure)
Jona-Lasinio, G., 101
Joseph Effect, 35, 244
Josephson junctions, 75

Kadanoff, L. P., 73
Kahane, J.-P., 70, 72, 284–288, 315–316, 372–389
Kampé de Fériet, J., 292
Karamata's class, 241–242
Kleiner, B., 161–162, 163(figure)
Knopp, K., 59
Kolmogorov −5/3 spectrum, 76, 118–119
—corrections to, 124, 402–408
—and Yaglom's microcanonical cascades, 320
Kolmogorov, A. M., 67, 69–70, 88, 93–94, 118–119
—and inner scale for intermittent turbulence, 410
—lognormal (third) hypothesis, 294–316, 297, 353–354
—similarity (second) hypothesis, 302–303, 327–328
kurtosis, 399–402

lacunarity, 28, 46
Lagrange multipliers, 86
Landau, L. D., 297
Laplace transform, 192
lateral parameter, 243
laws of large numbers, 5, 272–273
Lebesgue measure, 36, 166–167, 266, 285
Legendre transform, 66, 86, 96, 371–372
Leray, J., 417
Lévy dust, 81–82, 115, 166–167
—and "broken-line" process, 245
—and co-indicator function, 215, 242–243
—coarse-grained, 167–168
—as sets of multiplicity, 93, 284–288
Lévy, P., 105
Lévy semi-stable distributions, 369–371
Lévy-stable motion, 39, 60, 68
Lévy-stable random processes, 68, 109–110, 269, 361
Lewis, P. A. W., 157–160
Lewitan, H., 304
Lifshitz, E. M., 297
Liggett, T. M., 364
limit lognormal random

processes, 18–19, 301–303, 304–307(figures), 352–353, 359
—and base-free measures, 315–316
—and canonical cascade, 328
—degenerate limit, 304–310
—and determining function, 397
—nondegenerate limit, 310–312
limit theorems, and renewal processes, 272–276, 280–283
linear transformations, 74, 80
Lipschitz, R. O., 89
log-gamma distribution, 349
logarithmic transform, 116
lognormality, 94–95, 125, 294–316, 300(figure), 304–307(figures), 330–331, 345–354
—approximate vs. strict, 348–349
—critiques of lognormal hypothesis, 298–301
—and de Wijs–Yaglom cascade arguments, 300–302
—and finance, 314
—limit lognormal random processes as alternative to, 301–303, 304–307(figures), 304–312
—lognormal weights, 345, 347–348
—moments of lognormal, 299
—Obukhov's approximate hypothesis, 297–298
—and shape of domain, 322–323
—and $\tau(q)$, 88
—and turbulent dissipation, 294–316, 398(figure)
—undetermined by moments, 342, 349–350
Lovejoy, S., 71
low frequency divergence. *See* infrared catastrophe

M 1956c model, 104, 108
M 1963b model, 68, 109–110
M 1963e model, 108
M 1964w model, 118–120
M 1965b model, 110
M 1965c model, 156
M 1967k model, 70
M 1969b model, 88
M & Wallis 1969 model, 76

INDEX

M 1972f model, 120
M 1972j model, 31, 66, 72, 328
M 1974c model, 66, 110
M 1974f model, 31, 71, 72, 82, 110, 120, 317–357
M 1975O model, 120–121
M 1976c model, 82
M 1977F model, 63
M 1978h model, 123–125, 129
M 1982f model, 7, 20, 66, 75, 122, 126–127
M 1984w model, 128
Markov-Gauss processes, 78
martingales, 302, 330–331, 361–362, 372–389
 –conditions for finite moments, 374–383
 –conditions of nondegeneracy, 374–379
 –convergence condition for, 361–362
 –functional equation for, 373, 388
 –and measure μ, 375, 384–387
measures, 83–84
 –base-bound vs. base-free, 315–316
 –Besicovitch measure, 363
 –binomial measure, 21, 83–84, 89
 –geometric measure theory, 91–92
 –Lebesgue measure, 36, 166–167, 266, 285
 –and martingales, 372–375, 384–387
 –measure preservation and conditional stationarity, 254–259
 –measure space, 251
 –microcanonical, 364
 –multifractal, 45, 48
 –multinomial measure, 18, 21, 89, 363
 –narrow multifractals, 84
 –self-affinity and self-similarity, 83–84
 –for sets of multiplicity, 285
 –shift-invariant, 254–258
 –sigma-finite, 252
 –for sporadic generalized random functions, 251–272
medicine: chronic illness, 75
Mertz, P., 164, 164(figure)
mesofractal cartoons, 44–45, 51–52
mesofractal time, 40, 48, 51–52
mesofractal trading time, 25
mesofractals, 5
metallic films, 216
meteorology, non-Fickian character of, 76

microcanonical cascades, 320–332
 –critiques of, 326–327
 –defined, 331–332
 –and lognormal hypothesis, 321–322, 326, 353–354
 –moments of eddy averages, 335–337
 –postulate of independence, 321–322, 326
 –regularity of, 343
microcanonical measures, 18, 364
mild randomness, 3–4, 6
mineral distribution, 123, 296, 318
Minkowski sausage, 167
Mittag-Leffler distribution, 273–274
mixed multifractals, 45
modern statistical physics, 65, 101–110
Moffatt, K., 356–357
Moriarty, B. J., 157, 158(figure)
multibox vs. unibox constructions, 44
multifractal cartoons, 45, 49, 51–52, 53(figure)
multifractal intrinsic time, 23–24, 40, 47, 48, 51–52, 57–58
multifractal measures, 45, 48
multifractals, 4–5
 –canonical, 67
 –coarse-grained measures, 66
 –discontinuous, 45
 –early notions of (excerpts in French), 120–122, 129
 –formal analysis vs. constructive geometric approach, 63–64
 –generated by products of co-indicator functions, 244–245
 –grid-bound, 18–61
 –grid-free, 18–19
 –limit lognormal, 18–19
 –and logarithmic transform, 116
 –microcanonical, 67
 –multiplicative operations, 32
 –narrow, 84
 –origin of term, 72, 97
 –oscillating, 45
 –and power law transform, 116
 –proto-multifractals, 87, 89–93
 –and relative curdling. See weighted curdling

 –successive points in a discrete orbit, 66, 127(figure)
 –visual perception of, 79–80
multinomial measure, 18, 21, 89, 363
multiplications, iterated random, 358–368
multiplicative perturbations, 294–316
music, 75

$N(t)$, 238
Navier-Stokes equations, 395–396, 417–418
Nelkin, M., 82
nerve membranes, 75, 216
Neyman, J., 283–284
Nile River, 69, 76, 110
Noah Effect, 244
non-Fickian diffusion, 20, 35, 76
nonlinear transforms, 113–117
nonstationary processes
 –Gaussian, 78
 –resolution by conditioning, 208–214
 –sums of oscillatory term and a slowly varying trend, 208–209
Novikov, E. A., 69–70, 81–82, 88, 296, 350, 354, 355, 356

Ω, 251
$(\Omega, \mathcal{A}, \mu)$, 250
$(\Omega, \mathcal{A}, \mathcal{B}, \mu)$, 251–252
Obukhov, A. M., 69, 88, 118, 296, 298, 402–408
Obukhov sphere, 403–404
$1/f$ noises, 20, 35, 56, 75–77, 80, 111
 –cartoons of, for different exponents, 27(figure)
 –co-indicator functions. See co-indicator functions
 –defined, 73–74
 –f^{-0} functions. See white noise
 –f^{-1} functions, 208–214
 –f^{-2} functions, 222–226
 –f^{D-2} functions, 216–242
 –formal analysis vs. constructive geometric approach, 63–64
 –high frequency divergence. See ultraviolet catastrophe
 –low frequency divergence. See infrared catastrophe
 –obtained as fractal sums of pulses, 59–60
 –and self-affinity, 55–56
 –sound of, 79

INDEX

-sufficient criteria for, 80
-and trail dimension, 48
-underspecified meaning of, 80–81
-visual perception of, 79–76
Ornstein-Uhlenbeck processes, 78
Orszag, S. A., 299, 349–350
oscillating cartoon constructions, 47, 49

$\phi(q)$, 373
$P^*(u)$, 179, 189, 191
Pareto, N., 108–109
Pareto's law, 136, 138, 314
Parisi, G., 71, 72, 96–97
partition function, 84
Peano motion, 39
personal income, 108
perturbations, power-law decay of, 75
Peyrière, J., 70, 72, 372–389
Pharoah's Breastplate, 126, 127(figure)
Pietronero, L., 67
Poincaré, H., 81
pointwise dimension, 89
Poisson distribution, 60
Poisson noise contamination, 160
Poisson random variables, 298
Poisson set, 245
Polyà, G., 105
power law transform for multifractals, 116
pre-asymptotics, 243
pre-Gaussian noises, 116–117
prefactor
-examples of, 325
-for f^{-1} noises, 209–210
-redefined, 113
-for renormalized tail distribution of inter-box length, 168
-and turbulent dissipation, 113, 325
-in Wiener-Kinchin theory, 112–113
price variation, 23–24, 29, 30(figure), 58, 68, 103, 109–110
probability theory, 104–107
-asymptotic distribution of sums of nearly independent random variables, 325
-birth-and-death process, 325–326, 340, 374
-Cramèr theory of large deviations, 86, 371
-domain of attraction, 105–106
-fixed points, 104–110
-sums of random variables, 104–106
Procaccia, I., 73
proto-multifractals, 87, 89–93
pseudomarginal distributions, 257

q, 84–86
q_{crit}, 362
Q (correlation dimension), 76, 339
$Q(T, \omega)$, 220
quasi-singularities of fluid motion, 417–418

R/S plots, 76
random functions, defined, 251. See also generalized random functions; iterated random multiplications; limit lognormal random processes; sporadic generalized random functions; stationary random functions
random variables, 128
-Bernoulli, 298, 360
-binomial, 169, 298
-and co-indicator function, 227
-coin tossing, 134–135, 135(figure)
-conditioned, 173
-gamma random, 349
-Gaussian, 282
-higher moments of different types, 298
-and indicator function, 227
-maxima of, 106
-maximally skew Lévy-stable random variable, 239
-Mittag-Leffler distribution, 273–274
-Poisson, 298
-Schuster periodogram, 280
-Schwartz distributions, 221
-and white function, 227
randomized generators, 46
randomness
-vs. fractality, 31–32
-mild, 3–4, 6
-slow, 3–4, 241–242
-wild, 3–4, 238–240
ratio self-similarity, 83–84
recursive constructions
-grid-bound vs. grid-free, 29, 31
-types of, 29–32
-See also cartoons; grid-bound recursive constructions; grid-free recursive constructions

Redner, S., 67
relative curdling. See weighted curdling
relative intermittence, 122
renewal process
-co-indicator function for, 234
-generalization of, 260–262
-indicator function for, 226–227
-infinitely divisible process technique, 261
-limit theorems, 271–276, 280–283
-and self-similarity, 269–276
-and sporadic generalized random functions, 259–262
-stationary, 228–229, 260
-synchronized classical renewal set, 260
-white function for, 227–228
renormalization, 74–75
-base-free, 105
-Cauchy-Polyà-Lévy renormalization, 105
-Cvitanovic-Feigenbaum equation, 106–107
-and error clustering, 167–168
Rényi, A., 99
Reynolds number, 400
river discharges, 35, 69, 76, 110
roughness, fractals as measurements of, 28–29

Salem, R., 288
sample deltavariance of co-indicator function, 274–275
sample estimators, 280–283
sample length, 112
sample moments, 84–86, 85, 142–143, 272
sample spectrum, 218, 220, 224
San Marco dragon, 126, 127(figure)
scale invariance. See self-affinity
scaling distribution, 133–134, 136–137, 137(figure), 325. See also error clustering
scaling exponent, 42–46, 54, 56–59
-coin tossing, 142
-intermittent turbulence, 291, 292
-Pareto's law, 136
-self-similar law, 173–174
-telephone circuits, 134, 137

Schuster periodogram, 277–278, 280, 283
Schwartz distributions, 251
self-affinity, 5, 18–61, 80
 —and additive operations, 58–59
 —defined, 22
 —diagonal self-affinity, 3, 32, 33(figure), 34(figure), 36(figure), 80
 —distinguished from self-similarity, 4, 8, 20–22
 —for measures, 83–84
 —multifractality, 21
 —and 1/f noise, 21, 55–56
 —origin of term, 63
 —of turbulent dissipation, 76
 —underspecified meaning of, 80–81
 —unifractality, 21
 —See also cartoons; recursive constructions
self-similarity, 80, 173–174, 304–307(figures)
 —asymptotic, 267–271, 270–271
 —and co-indicator function, 267–269
 —conditional, 267–271
 —and conditional stationarity, 166–204
 —defined, 21
 —distinct aspects of consequences, 402
 —distinguished from self-affinity, 4, 8, 20–22
 —distinguished from statistical stationarity, 178
 —as expression of invariance of the generating mechanism with respect to multiplication of time by a constant, 178
 —for inter-error intervals, 189, 191
 —for limit case $D = 0$, 174
 —for measures, 83–84
 —mechanism of cluster generation, 181–182
 —ratio self-similarity, 83–84
 —and recurrence relations, 334
 —and renewal processes, 269, 272–276
 —and self-similar point processes, 171–183
 —self-similar recursive constructions, 20
 —truncated, 190
 —of turbulent dissipation, 321
 —uniform, 272–276

 —See also error clustering
semiconducting devices, 75, 216
sets
 —co-indicator function for, 215, 266
 —indicator function for, 215, 216
 —of multiplicity, 93, 284–288
 —self-similarity, 80
 —shift-invariant measures, 254–258
 —and sporadic generalized random functions, 251–272
 —of unicity, 93, 284–285
 —white function of, 216
 —See also measures
shift-invariant measures, 254–258
shot noise, 74, 78
Shraiman, B. I., 73
sigma-finite measures, 252
similarity dimension, 82, 124–125, 346, 393(figure)
slow randomness, 3–4, 241–242
Snyder, R. O., 161, 162(figure)
sounds, 79
speculative prices, 109–110
sporadic generalized random functions, 218, 247–284
 —conditional spectra of, 221, 277–280
 —conditionally covariance stationarity, 265
 —conditioned deltavariance of, 263–264
 —defined, 250, 252–253
 —examples of, 259–262
 —as infinitely intermittent random functions, 259
 —intermissions of, 253
 —limit theorems, 271–276, 280–283
 —measure preservation and conditional stationarity of, 254–259
 —notation for, 251–252
 —population infrared catastrophe, 278–279
 —and renewal processes, 259–276
 —Schuster periodogram, 277–278
 —set of variation of, 254
stationarity, 7, 115
 —conditional, 220
 —conditional covariance stationarity, 220–221, 223, 265
 —conditional deltavariance stationarity, 263–264
 —as expression of the invariance of the generating mechanism

with respect to the addition of a constant to time, 178
 —statistical stationarity distinguished from self-similarity, 178
stationary random functions, 32, 219, 221
stationary renewal process, 228, 229
statistical dependence, uniformaly global, 20
statistical thermodynamics, Darwin-Fowler approach, 86
statistics
 —blind spot of spectral analysis, 57–58
 —inadequacy of, 28
Stewart, R. W., 69–70, 81–82, 292, 296
stopping rule, 114
string generator, 41
Stuck, B. W., 161–162, 163(figure)
Sulem, P. L., 82
sunspots, 76
Sussman, S. M., 157, 159(figure), 191

$\tau(q)$, 55, 70, 82, 88, 99, 389
 —and correlation dimension, 116
 —defined through population or sample moments, 84–86
 —See also determining functions
$\tau'(1)$, 88, 389
$\tau(2)$, 56, 389
$\theta(t)$, 57–58
T (conditioning length), 78
T (sample length), 112
Tauberian theorem, 211
taxonomy, 108
Taylor, G. I., 269, 322
Taylor homogeneous turbulence, 391, 403, 409
telephone circuits, 69, 132–165, 177
 —amendments to model, 147–149
 —competing models, 132–133
 —consequences of model, 142–147
 —data mixing, 149–150
 —doubly logarithmic plots, 143(figure), 144(figure), 150(figure), 151(figure), 152(figure), 153(figure)

INDEX 441

—experimental verification, 138–142, 147–149
—fractal dimension, 143
—infinite message length, 146–147
—model described, 133–138
—phase modulation tests, 153–154
—Poisson noise contamination, 160
—signal to noise ratio, 156
—successive error-free intervals, 158–160
—tail behavior of, 156, 157–158
—two-parameter laws, 148–149
—*See also* communication circuits; error clustering
television transmission, 165, 165(figure)
texture, fractals as measurements of, 28–29
thin films, 75, 216
time
—"baby theorem." *See* "baby theorem"
—intrinsic, 19, 23–24, 27(figure), 47, 48
—and clock time, 53(figure)
—defined, 51–52
—fractal, 40, 48, 51–52
—linear, 51–52
—multifractal, 40, 47, 48, 51–52
—isochrone, 19, 50, 56
—trading time, 24, 25
—and trail exponent H_T, 19
time quantization, 208, 221, 226–227
topology, 407–408
—inadequacy of, 28
trading time, 24, 25
trail dimension, 27(figure), 48–51, 52, 54, 56, 57
trail exponent, 19, 48–49
trigonometric series
—and f^{D-2} functions, 221
—sets of multiplicity for, 284–288
Tufts, D., 69
turbulent dispersion, 416–418
turbulent dissipation, 69–70, 76, 80, 118–119, 124–125, 129, 291–294, 304–307(figures), 317–357, 341(figure)
—and absolute curdling, 124–125
—Burgers threshold, 407
—conditional probability for, 291
—conservation of, 301, 326

—correlation between averages, 339
—correlation dimension, 339
—correlation of, 76
—covariance of, 400
—and cross-sections of a fractal, 394–395
—degenerate class, 319–320, 323, 324(figure)
—degrees of intermittency, 259, 291, 292, 400
—determined/un-determined by moments, 349–350
—determining functions, 337–343, 347–348, 397–399
—dissipation as energy transfer between eddy sizes, 326
—dissipative range, 403, 409–414
—exponents for, 123–125, 395
—fractally homogeneous, 390–396, 398(figure), 400–412
—fractional intrinsic dimension, 320, 338, 345–347
—homogeneous (Taylor's concept of), 391, 403, 409
—inner scale, 297, 317, 394, 409–414
—interdependence of Yaglom's ratios, 326
—irregular (nondegenerate) class, 319–320, 324(figure), 325
—kurtosis of, 400–402
—laminar solution, 292
—and limit lognormal processes, 352–353
—lognormal approximation to W, 347–348
—lognormal hypothesis, 294–316, 398(figure)
—M 1964w model, 118–119
—M 1972f model, 120
—moments of, 124, 319, 322, 335–337
—infinite, 249, 293, 399
—multiplicative factors, 329–331
—and multiplicative perturbations, 294–316
—non-Fickian character of, 76
—non-normalized data sets, 113
—prefactor for, 113
—properties of intermittency, 292–293
—recurrence relations, 334–337

—regular class, 319–320, 323, 324(figure)
—and self-similar cascades, 317–357
—shape of domain, 320, 322–323, 326–327, 403–406, 407–408
—similarity dimension, 124–125
—and topological connectedness, 407–408
—volatility of, 80
—and weighted curdling, 394, 401–402, 406–407, 412–413
—Yaglom's microcanonical cascade argument, 300–302, 320–332
—*See also* eddies
turbulent velocity, 124, 404, 409

U, 167
U_c, 173
U_{max}, 172
UHF data channels, 162(figure)
Uhlenbech-Ornstein process, 74
ultraviolet catastrophe, 77–79, 208, 221
unibox vs. multibox constructions, 44
uniformly global statistical dependence, 20
unifractal generators, 48
unifractals, 5, 65, 81–83
—cartoons of. *See under* cartoons
—fractional Brownian motion. *See* fractional Brownian motion
—H for, 43–44
—parameters needed for specifying, 28
—*See also* absolute curdling
universality, class of, 105–106
universe, mass distribution of, 123, 346

V (indicator function), 115, 175, 176, 216
vacuum tubes, 75
Van Ness, J. W., 69
variance
—conditioned, and infrared catastrophe, 208–214
—and determining function, 399
—of $V_\Sigma(t)$, $W_\Sigma(t)$, $X_\Sigma(t)$, 236–237
VHF data channels, 162(figure)
viscosity, 409, 412, 417, 418
visual perception of $1/f$ noise, 79–76
visual perception of cartoons, 29, 30(figure)

volatility, 53(figure), 80

W. *See* weights
Wallis, J. R., 69, 76
Weierstrass function, 59, 92
Weierstrass, K., 59
weighted curdling, 97, 394, 397, 401–403, 406–407, 412–413
weights, 58–59, 358
 –binomial, 350, 359–360
 –and determining function, 397–399
 –hyperbolic, 367
 –for Lévy-stable processes, 361
 –lognormal, 350–352
 –and Yaglom's ratio, 329, 332
white function, 40, 216, 227–228
 –average noise energy and average dc component, 236–238
 –conditional spectra of, 231
 –ergodic theorem for, 236
 –Wiener-Kinchin spectra of, 231
white noise, 25(figure), 26(figure), 27(figure), 36–37, 60, 74–75, 79
 –blind spot, 57–58, 115
 –with continuous marginal distribution, 226
 –as $D = 2$ limit of f^{D-2} noises, 217
 –financial data, 58
 –noise energy and dc component, 237
 –nonlinear transforms distinguish between unstable and stable occurrences, 115
 –time-modulating white noise, 226
 –trail dimension for, 57
 –*See also* Fickian diffusion
Wiener Brownian motion, 23, 79, 114–115
 –and cartoon embedding, 49–50
 –cartoons of, 24(figure), 25(figure), 26(figure), 27(figure), 33(figure), 35–38, 36(figure), 47
 –H for, 35
Wiener-Kinchin theory, 7, 77–78, 218–220, 272–276
 –and conditional spectra, 217
 –and infrared catastrophe, 210–213
 –prefactor, 113–114
 –second-order stationary random processes defined, 272–273
 –spectra of co-indicator function, 230
 –spectra of indicator function, 230
 –spectra of white function, 231
 –Wiener-Kinchin deltavariance, 278
Wiener, N., 77
wild randomness
 –and co-indicator function, 238–240
 –globality of, 3–4
 –and indicator function, 238–240
wild self-affine variability, 18, 20
$W(t)$. *See* co-indicator function
$\psi(q)$. *See* determining functions

$X(t)$. *See* white function

Y. *See* Yaglom ratios
Y_n, 373
Y_s, 321
Yaglom, A. M., 94–95, 124, 296–297
 –microcanonical cascades, 320, 321–322, 326–327
Yaglom ratios, 323, 329–332

zero-crossing function, 114, 115

QA 614.86 .M27 1999
Mandelbrot, Benoit B.
Multifractals and 1/f noise